Photonic Crystal Fibres

Photonic Crystal Fibres

By
Anders Bjarklev
Jes Broeng
Araceli Sanchez Bjarklev

KLUWER ACADEMIC PUBLISHERS
Boston / Dordrecht / London

Distributors for North, Central and South America:
Kluwer Academic Publishers
101 Philip Drive
Assinippi Park
Norwell, Massachusetts 02061 USA
Telephone (781) 871-6600
Fax (781) 871-6528
E-Mail <kluwer@wkap.com>

Distributors for all other countries:
Kluwer Academic Publishers Group
Post Office Box 322
3300 AH Dordrecht, THE NETHERLANDS
Telephone 31 78 6576 000
Fax 31 78 6576 474
E-Mail <orderdept@wkap.nl>

Electronic Services <http://www.wkap.nl>

Library of Congress Cataloging-in-Publication

CIP info or:

Title: Photonic Crystal Fibres
Author (s): Anders Bjarklev, Jes Broeng and Araceli Sanchez Bjarklev
ISBN: 1-4020-7610-X

Printed on acid-free paper.

Printed in the United States of America

CONTENTS

Preface ix

Acknowledgements xi

1. Introduction 1
   1.1 From classical optics to photonic bandgaps 1
   1.2 Microstructures in nature 3
   1.3 Photonic crystals and bandgaps in fibres 5
   1.4 Development of the research field 6
   1.5 Different classes of microstructured optical fibres 9
   1.6 Organisation of the book 11
   References 13

2. Fundamentals of photonic crystal waveguides 19
   2.1 Introduction 19
   2.2 The development from 1D over 2D to 3D photonic crystals 20
      2.2.1 One-dimensional photonic crystals 20
      2.2.2 Two- and three-dimensional photonic crystals 23
      2.2.3 Fabrication of photonic crystals for the optical domain 25
      2.2.4 Spatial defects in photonic crystals 26
   2.3 Terms involved in photonic crystal fibre technology 29
   2.4 Efficient structures for creating photonic bandgaps in fibres 30
      2.4.1 In-plane photonic crystals 31
      2.4.2 From in-plane to out-of-plane photonic crystals 34
      2.4.3 The supercell approximation and efficient plane-wave method 37
   2.5 Summary 42
   References 42

3. Theory and modelling of microstructured fibres 53
3.1 Introduction 53
3.2 The effective-index approach 55
3.3 The method of localized basis functions 61
3.4 Full-vectorial plane-wave expansion method 67
3.4.1 Solving Maxwell's equation using a plane-wave expansion method 68
3.4.2 Two-dimensional photonic crystal with hexagonal symmetry 71
3.5 The biorthonormal-basis method 75
3.5.1 The basic equations of the biorthonormal-basis method 77
3.5.2 Biorthonormal-basis method with periodical boundary conditions 80
3.6 The multipole method 81
3.7 The Fourier decomposition method 85
3.8 The Finite-Difference method 90
3.8.1 The Finite-Difference Time-Domain method 90
3.8.2 The Finite-Difference Frequency-Domain method 94
3.9 The Finite-Element Method 97
3.10 The Beam-propagation method 99
3.11 The equivalent averaged index method 101
3.12 Summary 104
References 106

4. Fabrication of photonic crystal fibres 115
4.1 Introduction 115
4.2 Production of photonic crystal fibre preforms 116
4.3 Drawing of photonic crystal fibre 119
4.4 Photonic crystal fibres in new materials or material combinations 123
4.4.1 Fabrication of hole-assisted lightguide fibres 124
4.4.2 Chalcogenide fibres with microscopic air-hole structures 124
4.4.3 Microstructured polymer optical fibres 125
4.4.4 Extruded non-silica glass fibres 126
4.5 Summary 128
References 128

5. Properties of high-index core fibres 131
5.1 Introduction 131
5.2 Background-"Single-material" fibres 132

5.3 Fibres with periodic cladding structures 134
5.3.1 Basic properties of high-index core photonic crystal fibres 135
5.3.2 Cut-off properties of index-guiding photonic crystal fibres 138
5.3.3 Macro-bending losses of index-guiding PCFs 146
5.3.4 Dispersion properties of index-guiding PCFs 148
5.4 Fibres with non-periodic or non-circular cladding structures 151
5.5 Hole-assisted lightguiding fibres 154
5.6 Summary 155
References 156

6. Low-index core fibres – the true photonic bandgap approach 161
6.1 Introduction 161
6.2 Silica-air photonic crystals 162
6.2.1 Simple triangular structures 162
6.2.2 Effective-index considerations 164
6.2.3 Hexagonal or honeycomb structures 167
6.2.4 Modified triangular and honeycomb photonic crystals 170
6.3 Designing large-bandgap photonic crystals 173
6.4 The first experimental demonstration of waveguidance by photonic bandgap effect at optical wavelengths 177
6.4.1 Considerations on fabrication of honeycomb fibres 177
6.4.2 Basic properties of honeycomb-based fibres 178
6.4.2.1 The waveguiding principle of honeycomb fibres 179
6.4.2.2 Simple core design considerations of PBG fibres 183
6.4.3 Basic characterisation of the first honeycomb fibres 185
6.4.4 Modelling of realistic photonic bandgap fibres 187
6.5 Properties of PBG-guiding fibres 191
6.5.1 Dispersion properties of honeycomb-based fibres 191
6.5.1.1 Single-mode fibres with high anomalous dispersion and strongly shifted zero dispersion wavelength 192
6.5.1.2 Single-mode fibres with broadband, near-zero, dispersion-flattened behaviour 194
6.5.2 Polarization properties of honeycomb-based fibres 197
6.5.2.1 Polarization effects from non-uniformities 198
6.5.2.2 High-birefringent fibres and single-polarization state fibres 200
6.6 Air-guiding fibres 206
6.6.1 Cladding requirements for obtaining leakage-free waveguidance in air 206

6.6.2 Core requirements of air-guiding fibres 208
6.6.3 Advanced properties of air-guiding photonic crystal fibres 210
6.6.4 Loss properties of air-guiding fibres 213
6.7 Summary 214
References 215

7. Applications and future perspectives 219
7.1 Introduction 219
7.2 Large-mode-area photonic crystal fibres 221
7.2.1 Characteristics of large-mode-area photonic crystal fibres 222
7.2.2 Key parameters in describing large-mode-area photonic crystal fibres 225
7.2.3 New approaches and recent improvements of large-mode-area photonic crystal fibres 227
7.3 Highly nonlinear photonic crystal fibres 232
7.3.1 Design considerations of highly nonlinear PCFs 233
7.3.2 Dispersion management in highly nonlinear photonic crystal fibres 234
7.3.3 Supercontinuum generation in silica-based index-guiding photonic crystal fibres 238
7.3.4 Device demonstrations 241
7.3.5 Comparison of key parameters of highly nonlinear fibres 242
7.4 High numerical aperture fibres 245
7.5 Photonic crystal fibre amplifiers 247
7.6 Tuneable photonic crystal fibre components 250
7.7 Highly birefringent fibres 252
7.8 Dispersion managed photonic crystal fibres 255
7.9 Coupling and splicing 259
7.10 Loss properties of photonic crystal fibres 260
7.11 Summary 264
References 265

Acronyms 277

List of variables (including dimensions) 279

Index 285

PREFACE

Photonic crystal fibres represent one of the most active research areas today in the field of optics. The diversity of applications that may be addressed by these fibres and their fundamental appeal, by opening up the possibility of guiding light in a radically new way compared to conventional optical fibres, have spun an interest from almost all areas of optics and photonics. The aim of this book is to provide an understanding of the different types of photonic crystal fibres and to outline some of the many new and exciting applications that these fibres offer. The book is intended for both readers with a general interest in photonic crystals, as well as for scientists who are entering the field and desire a broad overview as well as a solid starting point for further specialized studies. The book, therefore, covers both general aspects such as the link from classical optics to photonic bandgap structures and thoughts of inspiration from microstructures in nature, as well as classification of the various photonic crystal fibres, theoretical tools for analysing the fibres and methods of their production. Finally, the book points toward some of the many future applications, where photonic crystal fibres are expected to break new grounds.

ACKNOWLEDGEMENTS

The authors would like to express a few personal feelings concerning the work leading to the present book. The work in the area of photonic crystal fibres has been extremely rewarding and we feel in many respects thankful for being able to participate in a tremendously exciting research area where pioneering work in both optical fibre technology and fundamental physics have been – and still is being – achieved.

Our work would not have been possible without the collaboration and support from a large number of persons; most importantly we would like to express our appreciation to our colleagues at COM, Technical University of Denmark, and Crystal Fibre A/S. We would further like to acknowledge the financial and academic support of several companies and institutions, including NKT A/S, DTU Innovation A/S, the Danish Research Council, University of Bath, the European Optical Society (awarding the "EOS Optics Prize 1999"), the Danish Optical Society (awarding the "DOPS Annual Prize 1999"), and Dana Lim A/S (awarding the "Dana Lim Prisen 2000").

Finally, we would like to thank our families for loving support and patience.

Chapter 1

INTRODUCTION

1.1 FROM CLASSICAL OPTICS TO PHOTONIC BANDGAPS

In the world of scientific investigations, we are often facing the situation that a complex problem or the fascination of a given idea forces us to focus so much that we have a tendency to overlook the things, which should have been the true inspiration of our work. Many times, it is first at the point, when we have determined a solution to a simple sub-problem that we might generalize the finding and obtain a wider perspective. Then the situation becomes even more rewarding, when we from the more general theory become able to solve new and unforeseen problems, which at first glance had very little to do with the first sub-problem.

If we look at the subject of this book - the photonic crystal fibre - it is possible to recognize the same kind of pattern, and we will in this introduction sketch some of the elements of this.

We will start by taking offset in the widespread application of guided-wave optics. Here, it is clear that optical fibres and integrated optical waveguides today are finding extensive use, covering areas such as telecommunications, sensor technology, spectroscopy, and medicine [1.1-1.3]. Their operation usually relies on light being guided by the physical mechanism known as total internal reflection, or index guiding. In order to achieve total internal reflection in these waveguides (which are normally formed from dielectrics or semiconductors), a higher refractive index of the core compared to the surrounding media is required. Total internal reflection is, thus, a physical mechanism that has been known and exploited technologically for many years. However, within the last decade, the research in new purpose-built materials has opened up the possibilities of localizing and controlling light in cavities and waveguides by a new physical mechanism, namely the photonic bandgap effect. Such waveguides are now beginning to break new grounds in the field of photonics [1.4-1.12].

Looking at the present technological development, it is very important to realize that the photonic bandgap effect was first described in 1987 by Yablonovitch and John, who independently studied spontaneous emission control and localization of light in novel periodic materials [1.12-1.13]. The name photonic crystals (or photonic bandgap structures) was coined to describe those periodic materials exhibiting frequency intervals within which electromagnetic waves irrespectively of propagation direction could be rigorously forbidden - and accordingly these forbidden frequency intervals were named photonic bandgaps. The initial idea behind using a photonic crystal for spontaneous emission control was that for a radiative relaxation of an excited atom or a recombination of an electron-hole pair (exciton) in a semiconductor to occur, there must be a non-zero number of electromagnetic states available for the emitted photon. Therefore, optical transitions of atoms/excitons should, in principle, be inhibited within a photonic crystal, if the crystal was designed to have a photonic bandgap overlapping the frequencies of the optical transitions. This ability of photonic crystals to inhibit the propagation of photons with well-defined frequencies has a close resemblance with the electronic properties of semiconductors [1.7,1.14-1.15], and the prospects of having an optical analogue to electronic bandgap materials have caused a huge interest in photonic crystals [1.8,1.16-1.19]. The analogy between photonic crystals and semiconductors is appealing; in both cases a spatially, periodically distributed potential causes the opening of forbidden gaps in the dispersion relations for photon and electron waves. However, where the electronic bandgaps are caused by a periodically varying *electric* potential from the lattice arrangement of the atoms constituting the semiconductor, a periodic distribution of a *dielectric* potential causes the bandgaps to open in photonic crystals. Considering the tremendous impact on electronics from exploiting the electronic bandgap in silicon and other semiconductors, a major impact on photonics has been predicted by the exploitation of photonic crystals [1.20-1.23].

The initial studies of the photonic bandgap effect was, therefore, not directly related to waveguiding, but certainly related to the question of highly improved localization of light in future optical components. It is probably also correct to say that most researchers, which came across the new ideas of photonic bandgaps in the late 1980'ies or early 1990'ies, primarily saw these ideas as being rather exotic, and - although fascinating –

not very near any practical applications. To understand this, there are a number of aspects, which made photonic crystals much more difficult to explore compared to semiconductors. The main reason for this - and probably the reason why the concept of photonic crystals was realized so relatively late in science - is that there is a lack of naturally occurring materials exhibiting three-dimensional photonic bandgaps (i.e., for arbitrary propagation direction of the light).

For a three-dimensional photonic bandgap to occur, there are severe requirements, which the periodic structure must fulfill. These concern realization of high refractive-index contrasts, properly designed crystal structures and wavelength-scale lattice dimensions. In contrast to semiconductors, where the atoms are naturally arranged in a lattice with dimensions corresponding to the electron wavelength, photonic crystals must, for operation at visible or near-infrared wavelengths, have artificially manufactured "optical atoms" (dielectric scatterers) arranged on a scale of approximately one micron. Even though this is around thousand times larger than the electronic wavelength scale, this makes the realization of photonic crystals for use in the optical domain a severe technological challenge [1.24-1.26]. Regarding the appealing analogy between semiconductors and photonic crystals, it is, therefore, important to bear in mind that adding to the difficulties in manufacturing micro-scale photonic crystals, there is a fundamental difference between electrons and photons. Hence, although photonic crystals may provide both novel, unique photonics components as well as add extra complexity to existing components, ideas of a direct transfer of electronic components and technologies to photonic counterparts using photonic crystals must be considered to be over simplified. Photonic crystal technology should, therefore, be seen in its own respect as a research field, which still has a large fundamental interest, and where the application-driven research is just beginning to demonstrate the first fruitful aspects. One of these application-oriented aspects is, as previously mentioned, new types of optical waveguides.

1.2 MICROSTRUCTURES IN NATURE

Although it is difficult to find the previously mentioned examples of three-dimensional photonic bandgap structures in nature, we should, however, not forget the fact that most of us actually are familiar with - at least two-dimensional – photonic crystals, from various well known examples. One of the most classical examples is the colourful spots on certain butterfly wings (see the example in Figure 1-1). However, as mentioned in the first paragraph of this chapter, these quite obvious expressions of the effects that we were actually looking after, was to some degree overlooked by many engineers and scientists in the early phase of the work.

If we look at the highly periodic structure (with a period close to the wavelength of visible light), it seems like an obvious source of inspiration for the scientist. The well-known wavelength selectivity, which most of us have experienced by looking at a butterfly moving its wings, and letting the spots change colour, e.g., from blue to green, is of clear interest in modern optical research. However, despite its broad application in nature, it is first recently that the principles have been brought into use in optical components considered to operate by the photonic bandgap effect. The principle of one-dimensional (1D) periodic structures have, however, been exploited quite extensively within optics for many years, because Bragg stacks [1.27] and optical gratings [1.28] have been used for optical coatings, wavelength and diffraction selective reflectors etc. for many years.

Photonic bandgaps have actually been found in many other places in nature than "just" in butterfly wings, and recent publications on structural colours through photonic crystals [1.29] describe interesting examples such as the seamouse and colourful beetles.

Figure 1-1. An example of photonic bandgaps in nature. A photograph of a butterfly of the species Madagascar Boisdiural. The butterfly has been kindly provided by Professor Niels Peder Kristensen from Zoological Museum, University of Copenhagen, and the picture has been taken by Lasse Rusborg, COM, DTU.

1.3 PHOTONIC CRYSTALS AND BANDGAPS IN FIBRES

Photonic crystal waveguides are today explored for both planar and fibre applications [1.5,1.10,1.20,1.30]. Whereas planar photonic crystal waveguides have the attractive potential of providing loss-less transmission of light around 90° sharp bents, and may be the key to large-scale integrated photonics [1.9], photonic crystal fibres may, through a lifting of the usual core-cladding index requirements and restrictions, exhibit radically new properties and potentially push the limits for e.g., high-power laser deliverance, spectral location of transmission windows and data transmission speeds. As an important milestone for planar photonic crystal technology, sharply-bent waveguides were recently demonstrated by Lin *et al.* [1.11]. However, this result was obtained at millimetre wavelengths and although the properties of photonic crystals may be scaled to different wavelength regimes by scaling the dimensions of the crystals, a severe problem for operation at optical wavelengths concerns the reduction of high scattering losses resulting from the finite vertical extend of the planar photonic crystal structures. The problem of vertical scattering losses was circumvented in the experiment by Lin *et al.* by sandwiching the planar photonic crystal between two metallic plates. While this procedure is feasible at millimeter wavelengths, it is, as mentioned, still a partly unsolved problem to provide a strong vertical confinement in planar photonic crystals operating at optical wavelength.

In contrast to the planar structures, photonic crystal waveguides of practically infinite vertical extent may readily be fabricated in fibre forms using traditional fibre manufacturing techniques with only modest modifications [1.4]. This makes optical fibres appear the most mature technology today for exploring photonic crystals operating at optical wavelengths.

In Figure 1-2, an example of a highly periodic cladding structure surrounding a solid glass core of a photonic crystal fibre is shown. We note that the air-holes, which are equally spaced in the cladding, have almost identical magnitudes.

As we will describe in much more detail later on in this book, the fibre structure shown in Figure 1-2, does not operate by the photonic bandgap effect, but rather by the principle of modified total internal reflection. Note that the diameters of the air-holes are around 2 microns for the example shown in Figure 1-2.

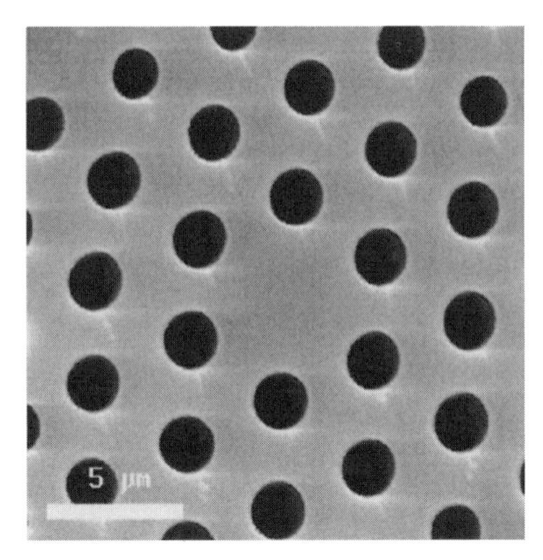

Figure 1-2. A photonic crystal fibre consisting of a solid glass core and a cladding, which contains numerous periodically placed air-holes or voids in a glass base material. The photograph is kindly provided by Crystal Fibre A/S.

1.4 DEVELOPMENT OF THE RESEARCH FIELD

Optical fibres in which air holes are introduced in the cladding region and extended in the axial direction of the fibre has been known, since the early days of silica waveguide research through the work of Kaiser *et al.* [1.31]. This early work demonstrated the first low-loss single-material fibres made entirely from silica. In the demonstrated structures, very small silica cores were held by thin bridges of silica in air.

Due to the technological development of doped-silica fibres – primarily within the area of optical communications – the so-called “single-material” (or more accurately “air-silica”) fibres were not developed much further over a number of years. However, in 1996, Russell and co-workers demonstrated fibres having a so-called photonic crystal cladding [1.32].

By the work of Russell and co-workers [1.32], the field of photonic crystal fibres was founded, and the realization of microstructured silica fibres with a periodic array of several hundred air holes running down their length [1.4,1.33] was demonstrated. The initial aim of Russell *et al.* was to realize fibres that would be able to guide light using photonic bandgap effects [1.34]. The first fabricated fibres used a triangular arrangement of air holes in the cladding structure and a single missing air hole to form the fibre core. The fibres showed robust, and they were relatively easy to couple light into as well as they were guiding light efficiently [1.32]. It was, however, early realized that the operation of the fibres was not based on the photonic bandgap effect, but a type of modified index-guidance through a higher refractive index of the core region compared to the effective index of the cladding structure [1.35]. Despite the resemblance with conventional optical fibres, the index-guiding photonic crystal fibres were, however, found to display important differences and are now the subject of significant research activities [1.5,1.35-1.37]. While progress were made rapidly for index-guiding photonic crystal fibres, the realization of a truly photonic bandgap-guiding fibre was initially hindered by the difficulties in realizing triangular photonic crystal cladding structures with sufficiently large air holes at the required dimensions. A second limiting factor was the lack of accurate numerical tools, which could be used to analyse and design other types of fibre structures. The complex nature of the cladding structure of the photonic crystal fibres does not allow for the direct use of methods from traditional fibre theory as we will discuss in detail in Chapter 3 of this book.

An important turning-point for the realization of photonic bandgap-guiding fibres was the implementation of an accurate, full-vectorial numerical method. Using this method, novel silica-air photonic crystal structures, which would exhibit the photonic bandgap effect at realistic dimensions, were discovered [1.38] and allowed design and later experimental demonstration of the first truly bandgap-guiding photonic crystal fibres [1.30,1.39].

Since the realisation of the first photonic crystal fibres in the mid 1990'ies, an increasingly growing interest has been seen – not only from the academic community, but also from industrial side. In order to get an idea of the increasingly growing interest in the research area of photonic crystal fibres, we may look at the illustration in Figure 1-3. This illustration contains a diagram showing the number of published documents registered in international databases per year from 1987 and on to 2001. Although these data due to the limited number of search terms might not cover every publication (i.e., conference or journal paper or published patent application), it may provide a rather good indicator for the international research activity. We also note that the two databases do not show the same number of documents - and not always having more registered documents in one rather than in the other. Figure 1-3 shows that from a relatively limited activity from 1987 and onto the mid 1990'ies, an annual increase may be seen since 1995 leading to about 140 published documents in 2001, and pointing towards even more papers in this and the following years to come.

The rapidly increasing number of new papers and patents within the research field of photonic crystal fibres does naturally indicate that a tremendous development takes place, and with all the resources directed into these activities, every month brings forward new and exiting developments. For the same reason, it is a very challenging task to try to cover the latest developments within a text book at this time, since new results are moving the frontiers of our understanding almost every day, and it will, therefore, be very difficult to include all the latest news. We have, consequently, chosen to aim this presentation towards a relatively broad and fundamental description of the field of photonic crystal fibres. We, further, hope that all our colleagues, which every day are developing this exiting field of research with their contributions, will have the understanding that it is certainly not our intention to overlook their work - if it is not mentioned - but rather that we have focussed on the basic properties and chosen some of the representative examples from the most recent literature.

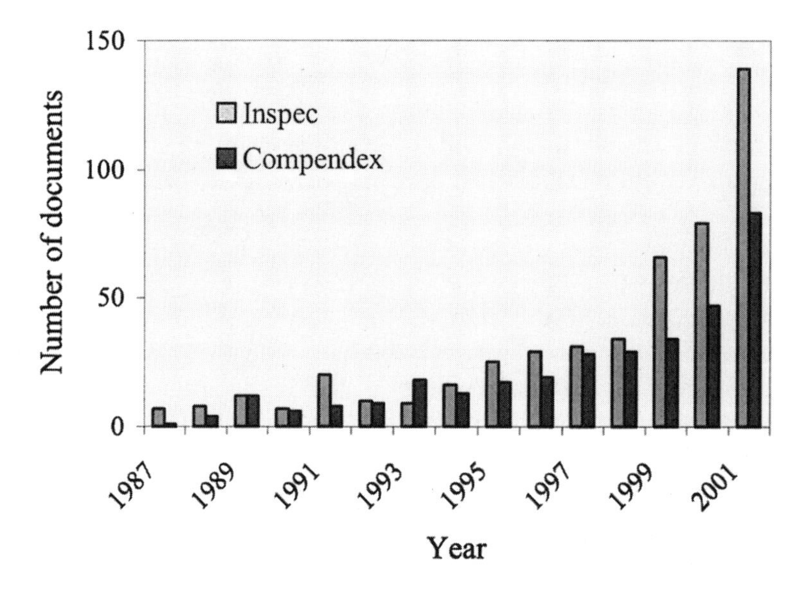

Figure 1-3. Number of published documents (conference or journal papers, patents etc.) per year registered in the respective databases "Inspec" or "Compendex". The data has been kindly supplied by librarian Ilse Linde. DTV, Kgs.Lyngby, Denmark, based on search on the following terms: "Crystal fibres/fibers, Microstructured fibres/fibers, Bandgap fibres/fibers, or Holey fibres/fibers".

1.5 DIFFERENT CLASSES OF MICROSTRUCTURED OPTICAL FIBRES

In this section, we will introduce some of the names and terms, which have become commonly used in the field of research described in this book. It should be noted that, since the field is still relatively young, and also because different national "schools" have developed their own preferences, there may be some disagreement on the correct use of words and terms. We have – of course – tried to mention the most commonly applied names, but at the same time, we will use this section to specify our own choice, which will be used thoughout the book.

If we look at Figure 1-4, the overall terms used for the research field in general is shown at the top. The most generally used terms are Photonic Crystal Fibres (PCF), MicroStructured Fibres (MSF) [1.40], or Microstructured Optical Fibres (MOF) [1.41]. In this connection, it is relevant to note that some researchers in the field prefer to reserve the name photonic crystal fibre to fibres with highly periodic distribution of air-holes in the cladding. However, throughout this book we will use the term "photonic crystal fibre" as our choice of general term describing a fibre with a number of high-index contrast structures in the cladding (and sometimes even in the core), which are used to form the waveguiding properties of the fibre.

Any PCF or microstructured fibre may be placed in one of two main classes, namely either fibres operating by a principle comparable to that of standard optical fibres, i.e., total internal reflection (TIR), or those guiding light by the photonic bandgap (PBG) effect. The fibres belonging to the first main class may – as indicated in Figure 1-4 – often be referred to as either High-Index Core (HIC) fibres, Index-Guiding (IG) fibres [1.42], or Holey Fibres (HF) [1.43]. In the second main class, the fibres are generally referred to as PBG fibres or Bandgap-Guiding (BG) fibres.

We may now divide each of the two main classes into a number of sub-classes, which primarily are determined by the dimensions of the fibre structures, and their specific properties. For the index-guiding fibres, three subclasses have today immerged, and they are as follows: High-numerical-aperture (HNA) fibres having a central part surrounded by a ring of relatively large air-holes, Large-Mode-Area (LMA) fibres [1.44] using relatively large dimensions and small effective refractive index contrasts to spread out the transverse optical field, and, finally, Highly-Non-Linear (HNL) fibres [1.47] that applies very small core dimensions to provide tight mode confinement.

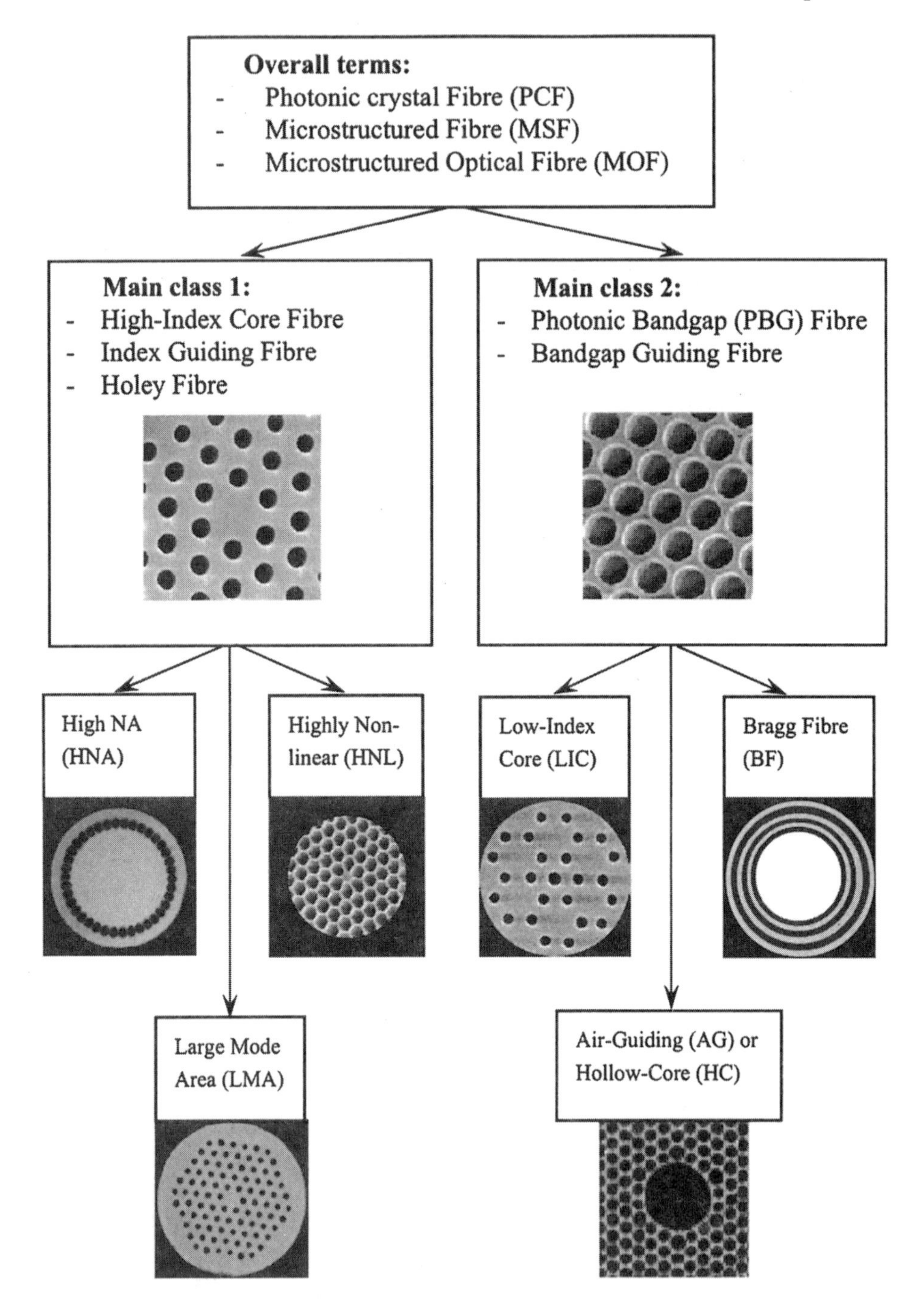

Figure 1-4. Diagram showing the most commonly used terms and typical structures for the major classes and sub-classes of photonic crystal fibres. The included photos are kindly provided by Crystal fibre A/S.

The photonic bandgap fibres may be separated into the sub-classes of Low-Index Core (LIC) fibres, and Air-Guiding (AG) [1.45] or Hollow-Core (HC) [1.46] fibres, and in addition to this, the rotation symmetrical Bragg fibres may be added, although the latter with its ring structure is somewhat different from the rest of the fibres having individual holes/voids distributed in the cladding. The LIC fibres are waveguides, which guide light by the photonic bandgap effect, and, therefore, confine light to the centre of the fibre (the so-called defect referring to the fact that the fibre core appears, where the periodicity is broken). Although the effective refractive index of the core region is lower than that of the cladding region, the major part of the optical power is propagating in the high index material (most commonly the silica). In contrast to this, the air-guiding fibres provide a bandgap, which allow the majority of the optical power to propagate in the central hole of the fibre structure – and hereby the name air-guiding naturally appears.

1.6 ORGANISATION OF THE BOOK

We have chosen to organise the book based on our wish to provide a fundamental overview of the field of photonic crystal fibres, their properties and unique features. The following chapters are accordingly divided into the following elements:

Chapter 2 describes fundamentals of photonic crystal waveguides. In this connection, it is important to note that a significant part of Chapter 2 is devoted to the new properties of the photonic bandgap effect, and this is done in spite of the fact that many crystal fibres today do not operate by this effect but by what may be called Modified Total Internal Reflection (MTIR), which is much closer to the operation principle of standard optical fibres. However, due to the relative novelty of the photonic bandgap effect, we have chosen to keep the focus on this in Chapter 2, which describes the development from one-dimensional to three-dimensional photonic crystal optical waveguides. We also address the issue of the design af efficient structures for the creation of photonic bandgaps and silica-air photonic crystals are introduced as a first step towards the silica-based PCF technology.

Chapter 3 contains a description of eleven different theoretical and numerical methods applied in the analysis of photonic crystal fibres. Point of reference is taken in the simple (and relatively early) Effective-index approach, whereafter the Localised Basis Function Method is introduced. The major effort is placed on the full-vectorial Plane-Wave Expansion Method, because we find the general properties of this method very useful in providing the fundamental understanding of the key issues in PCF modelling. Another vectorial method is represented in the short description of the Biorthonormal-Basis Method. Thereafter, two recently applied methods, which are aimed at the prediction of fibre loss, are described, namely the multipole method and the Fourier-Decomposition Method. After this, a number of more "classical" numerical approaches are presented, including the Finite-Difference Method, the Finite-Element Method, and the Beam-Propagation Method. Finally, a very recent qualitative method – the Equivalent Averaged Index Method – is presented, and the chapter is concluded by an overview of advantages and disadvantages of the different methods.

In Chapter 4, the fundamental issues of the fabrication of photonic crystal fibres are described. This include elements such as preform realisation, fibre drawing, as well as a description of microstructured fibres in new materials, i.e., chalcogenides, polymers, and low-melting point glasses.

Chapter 5 describes the basic issues of the presently most widely used class of photonic crystal fibres, namely the high-index-core fibres. The chapter includes a description of fundamental waveguiding properties, e.g., cutoff properties, dispersion, mode field confinement etc., and it addresses the need for periodic or non-periodic cladding structures.

In Chapter 6, the focus is on the photonic bandgap class of PCFs. We will here refer to the steps leading to the first demonstration of photonic bandgaps at optical wavelengths. Furthermore, the concept of air-guiding fibres will be discussed and reviewed, and we will present the possibility of providing low-index contrast PBG fibres.

Chapter 7 contains descriptions of some of the most significant application of photonic crystal fibres known at present time. Note here that especially the area of applications is developing at a very high speed, because the photonic crystal fibres very often provide completely new and alternative functionalities compared to standard optical fibres. It is, consequently, a very hard task to make a fully up-to-date description of the numerous applications that are published every week. Our choice has, therefore, been to select some of the most significant areas of applications, and present these at a level, which hopefully may serve as an inspiration to the reader. The selected applications are as follows: Large-mode-area PCFs (representing large-core designs), Highly nonlinear PCFs (representing very-small-core designs), Highly birefringent PCFs (showing enhanced polarisation properties), and High NA PCFs indicating exiting possibilities within areas such as high-power fibre lasers and amplifiers. Also the issue of coupling and splicing to photonic crystal fibres is addressed in this chapter.

REFERENCES

[1.1] A. Bjarklev,
"Optical Fiber Amplifiers: Design and System Applications",
Artech House, Boston-London, August 1993.

[1.2] J. Dakin and B. Culshaw,
"Optical fiber sensors. Vol.4: Applications, analysis, and .future trends",
Artech House, Boston-London, 1997.

[1.3] S. Takahashi, M. Futamata, and I. Kojima,
"Spectroscopy with scanning near-field optical microscopy using photon tunnelling mode,"
Journal of Microscopy, Vol. 194, No.2-3, pp. 519-522, 1999.

[1.4] J. Knight, T. Birks, D. Atkin, and P. Russell,
"Pure silica single-mode fibre with hexagonal photonic crystal cladding,",
Optical Fiber Communication Conference, Vol. 2, p. CH35901, 1996.

[1.5] J. Broeng, D. Mogilevtsev, S. Barkou, and A. Bjarklev,
"Photonic crystal fibres: a new class of optical waveguides",
Optical Fiber Technology, Vol. 5, pp. 305-330, July 1999.

[1.6] P. Rigby,
"A photonic crystal fibre",
Nature, Vol. 396, pp. 415-416, 3 Dec. 1998.

[1.7] J. Joannopoulos, J. Winn, and R. Meade,
"Photonic Crystals: Molding the Flow of Light",
Princeton University Press, 1995.

[1.8] J. Joannopoulos, P. Villeneuve, and S. Fan,
"Photonic crystals: putting a new twist on light,"
Nature, Vol. 386, pp. 143-149, March 1997.

[1.9] A. Mekis, I. Chen, I. Kurland, S. Fan, P. R. Villeneuve, and J. D. Joannopoulos,
"High transmission through sharp bends in photonic crystal waveguides,"
Physical Review Letters, vol. 77, pp. 3787-3790, *act.* 1996.

[1.10] H. Benisty,
"Modal analysis of optical guides with two-dimensional photonic band-gap boundaries",
Journal of Applied Physics, Vol. 79, pp. 7483-7492, May 1996.

[1.11] S.-Y. Lin, E. Chow, V. Hietala, P. Villeneuve, and J. Joannopoulos,
"Experimental demonstration of guiding and bending of electromagnetic waves in a photonic crystal",
Science, Vol. 282, pp. 274-276, *Oct.* 1998.

[1.12] E. Yablonovitch,
"Inhibited spontaneous emission in solid-state physics and electronics",
Physical Review Letters, Vol. 58, pp. 2059-2062, May 1987.

[1.13] S. John,
"Strong localization of photons in certain disordered dielectric superlattices",
Physical Review Letters, Vol. 58, No.23, pp. 2486-2489, 1987.

[1.14] E. Yablonovitch,
"Photonic band-gap structures",
Journal of the Optical Society of America B, Vol. 10, pp. 283-295, Feb. 1993.

[1.15] S. John,
"Localization of light: Theory of photonic band gap materials",
Photonic band gap materials (C. Soukoulis, ed.), Vol. 315 of *NATO ASI series. Series E, Applied sciences,* pp. 563-666, Dordrecht : Kluwer, 1996.

[1.16] C. Soukoulis, ed.,
"Photonic band gaps and localization"
Proceedings of the NATO advanced research workshop, Heraklion 1992, Vol. 308 of *NATO ASI series. Series B, Physics,* Dor- drecht: Kluwer, 1993.

[1.17] E. Burstein and C. Weisbuch, eds.,
"Confined electrons and photons New physics and applications"
Proceedings of a NATO Advanced Study Institute, Erice 1993, vol. 340 of *NATO ASI series. Series B, Physics,* New York,N .Y. : Plenum Press, 1995.

[1.18] C. Soukoulis, ed.,
"Photonic band gap materials"
Proceedings of the NATO advanced study institute, Elounda 1995, vol. 315 of *NATO ASI series. Series E, Applied sciences,* Dordrecht : Kluwer, 1996.

[1.19] J. Rarity and C. Weisbuch, eds.,
"Microcavities and photonic band gaps"
Physics and applications, vol. 324 of *NATO ASI series. Series E, Applied sciences,* Dordrecht : Kluwer, 1996.

[1.20] T. Krauss, R. De La Rue, and S. Brand,
"Two-dimensional photonic-bandgap structures operating at near-infrared wavelengths",
Nature, Vol. 383, pp. 699-702,24 Oct. 1996.

[1.21] T. Baba and T. Matsuzaki,
"Polarisation changes in spontaneous emission from GaInAsP/InP two-dimensional photonic crystals,"
IEE Electronics Letters, Vol. 31, pp. 1776-1778, Sept. 1995.

[1.22] P. Evans, J. Wierer, and N. Holonyak,
"Photopumped laser operation of an oxide post GaAs-AlAs superlattice photonic lattice,"
Applied Physics Letters, Vol. 70, pp. 1119-1121, March. 1997.

[1.23] O. Painter, R. Lee, A. Scherer, A. Yariv, P. O'Brien, J. D. Dapkus, and I. Kim,
"Two-dimensional photonic band-gap defect mode laser",
Science, Vol. 284, pp. 1819-1821, June 1999.

[1.24] S. Lin, J. Fleming, D. Hetherington, B. Smith, R. Biswas, K. Ho, M. Sigalas, W. Zubrzycki, S. Kurtz, and J. Bur,
"A three-dimensional photonic crystal operating at infrared wavelengths",
Nature, Vol. 394, pp. 251-253, 16. *July* 1998.

[1.25] A. VanBlaaderen,
"Materials science - Opals in a new light",
Science, Vol. 282, pp. 887-888, Oct. 1998.

[1.26] B. Goss-Levi,
"Search & discovery: Visible progress made in 3d photonic crystals",
Physics Today, vol. 52, pp. 17-19, Ian. 1999.

[1.27] S. Ramo, J. Whinnery, and T.Van Duzer,
"Fields and waves in communications electronics",
Wiley, 1994

[1.28] R.Kashyap
"Fiber Bragg gratings"
Academic Press, San Diego, London, 1999, ISBN: 0-12-400560-8

[1.29] R. C. McPhedran, N. A. Nicorovici, D. R. McKenzie,
"The sea mouse and the photonic crystal",
Aust. J Chem., Vol. 54, No. 4, pp. 241-244, 2001.

[1.30] J. Knight, J. Broeng, T. Birks, and P. Russell,
"Photonic band gap guidance in optical fibers",
Science, Vol. 282, pp. 1476-1478, *Nov.* 20, 1998.

[1.31] P.V.Kaiser, and H.W. Astle,
"Low-loss single-material fibers made from pure fused silica",
The Bell System Technical Journal, Vol.53, pp.1021-1039, 1974.

[1.32] J. C. Knight, T. A. Birks, P. St. J. Russell, and D. M. Atkin,
"All-silica single-mode optical fiber with photonic crystal cladding",
Optics Letters, Vol.21, No.19, pp. 1547-1549, October 1.,1996.

[1.33] T. Birks, D. Atkin, G. Wylangowski, P. Russell, and P. Roberts,
"2D photonic band gap structures in fibre form,"
Photonic Band Gap Materials (C. Soukoulis, ed.) , Kluwer, 1996.

[1.34] T. Birks, P. Roberts, P. Russell, D. Atkin, and T. Shepherd,
"Full 2-d photonic bandgaps in silica/air structures",
IEE Electronics Letters, Vol. 31, pp. 1941-1943, Oct. 1995.

[1.35] J. Knight, T. Birks, P. Russell, and J. Sandro,
"Properties of photonic crystal fiber and the effectice index model",
Journal of the Optical Society of America A, Vol. 15, pp. 748-752, March 1998.

[1.36] T. Monro, D. Richardson, and N. Broderick,
"Efficient modelling of holey fibers",
Optical Fiber Communication Conference, San Diego, FG3, pp. 111-113, Feb.1999.

[1.37] R. Windeler, J. Wagener, and D. DiGiovanni,
"Silica-air micro structured fibers: properties and applications",
Optical Fiber Communication Conference, San Diego, FG1, pp. 106-107 , Feb. 1999.

[1.38] J. Broeng, S. Barkou, A. Bjarklev, J. Knight, T. Birks, and P. Russell,
"Highly increased photonic band gaps in silica/air structures",
Optics Communications, vol. 156, pp. 240-244, Nov. 1998.

[1.39] J. Broeng, S. Barkou, and A. Bjarklev,
"Waveguiding by the photonic band gap effect,"
Topical meeting on Electromagnetic Optics, (Hyeres, France), pp. 67-68, EOS, September 1998.

[1.40] D. Ouzounov, D. Homoelle, W. Zipfel,
"Dispersion measurements of microstructured fibers using femtosecond laser pulses",
Optics Communications, Vol. 192: (3-6), pp. 219-223, 2001.

[1.41] B. J. Eggleton, P. S. Westbrook, R. S. Windeler, S. Spalter, T. A. Strasser,
"Grating resonances in air-silica microstructured optical fibers",
Optics Letters, Vol. 24, No. 21, pp. 1460-1462, Nov. 1999.

[1.42] T. P. Hansen, J. Broeng, S. E. B. Libori, A. Bjarklev,
"Highly birefringent index-guiding photonic crystal fibers",
IEEE Photonics Technology Letters, Vol. 13, No. 6, pp. 588-590, Jun. 2001.

[1.43] R. Ghosh, A. Kumar, J. P. Meunier,
"Waveguiding properties of holey fibres and effective-V model",
IEE Electronics Letters, Vol. 35, No. 21, pp. 1873-1875, Oct. 1999.

[1.44] J. C. Knight, T. A. Birks, R. F. Cregan, P. S. Russell, J. P. de Sandro,
"Large mode area photonic crystal fibre",
IEE Electronics Letters, Vol. 34, No. 13, pp. 1347-1348, Jun. 1998.

[1.45] J Broeng, S. E. Barkou, T. Sondergaard, A. Bjarklev,
"Analysis of air-guiding photonic bandgap fibers",
Optics Letters, Vol. 25, No. 2, pp. 96-98, Jan. 2000.

[1.46] A. N. Naumov, A. M. Zheltikov,
"Optical harmonic generation in hollow-core photonic-crystal fibres: analysis of optical losses and phase-matching conditions",
Quantum Electronics, Vol. 32, No.2, pp. 129-134, Feb. 2002.

[1.47] K.P. Hansen, J.R. Jensen, C. Jacobsen, H.R. Simonsen, J. Broeng, P.M.W. Skovgaard and A. Bjarklev,
"Highly nonlinear photonic crystal fiber with zerodispersion at 1.55 um"
OFC'2002, postdeadline paper, paper FA9, pp. 1-3, March, 2002.

Chapter 2

FUNDAMENTALS OF PHOTONIC CRYSTAL WAVEGUIDES

2.1 INTRODUCTION

The aim of this chapter is to introduce the general concept of photonic crystals. The chapter will start by guiding the reader to an understanding of photonic crystals and explain their basic characteristics, and indicate methods of fabrication and potential applications. The purpose of the chapter is not to give an exhaustive presentation of the wide field of photonic crystals, but to establish a fundamental understanding of the concepts, terminology and theoretical tools that is required for conducting research in photonic crystal waveguides. Readers interested in general aspects of photonic crystals are pointed to the introductory book by Joannopoulos *et al.* [2.1], or the more recent book by Johnson *et al.* [2.2]. We may also recommend one of the very good web-sites [2.12], where relevant papers such as e.g., [2.90-2.91] may be found.

A further detailed overview of the field may be obtained by studying various review articles, special journal issues, and collections of papers [2.3 - 2.10]. Note also that at present more than 2500 international publications exist on photonic crystals (Ultimo 2002 [2.11]). The most extensive and up-to-date bibliographies on photonic crystal research including journal articles, patents and research groups may be found on the world-wide-web pages of J. Dowling *et al.* [2.11] and Vlasov [2.12].

Generally speaking, the research area of photonic crystals is a field of a (extraordinarily) wide interest, and it is characterized by research groups with at least three different scientific backgrounds, namely classical optics, solid-state physics and quantum optics. This book will not address quantum electro-dynamical aspects of photonic crystals, but focus on the passive properties of photonic crystals and mainly employ techniques adapted from solid-state physics.

2.2 THE DEVELOPMENT FROM 1D OVER 2D TO 3D PHOTONIC CRYSTALS

Photonic crystals are today used as a general term describing periodic structures both in one, two, and three dimensions. While structures with a periodicity in one dimension (1D) have been known and exploited for decades, e.g., finding use in high-reflection mirrors and fibre Bragg gratings [2.13-2.14], their two- and three-dimensional (2D and 3D) counterparts have only been explored since the publication of the original ideas of Yablonovitch and John in 1987 [2.15, 2.16].

2.2.1 One-dimensional photonic crystals

A general example of a 1D periodic structure is the multilayer stack illustrated in Figure 2-1. The stack consists of two alternating layers of different materials, with refractive indices n_1 and n_2, and has the period Λ. As known from optics textbooks [2.13], such a structure is able to transmit electromagnetic waves at certain frequencies, f_T, while reflecting others, f_R. Frequently studied cases of lD periodic structures are Bragg stacks, where maximum reflection is obtained, when the optical thickness of the layers is a quarter of a wavelength.

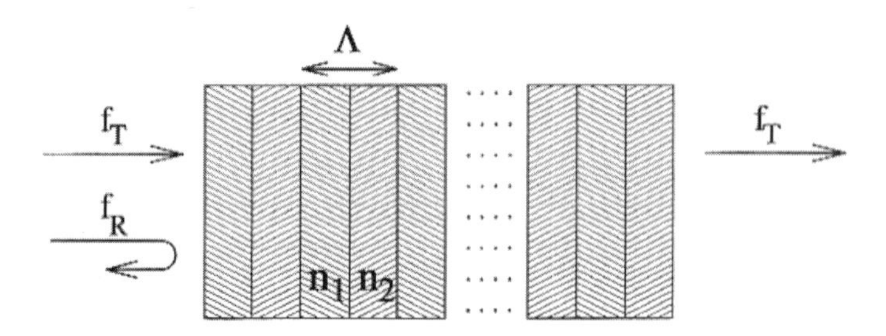

Figure 2-1. The Bragg stack: a one-dimensional photonic crystal.

The transmission spectrum for a 1D periodic structure consisting of 30% air (n_1 = 1.0) and 70% high-index dielectric (n_2 = 3.6) is presented in Figure 2-2(a). The simulated structure is of finite size and consists of ten alternating layers. The figure reveals several frequency intervals in which light may not penetrate the structure. The physical explanation of these low-transmission intervals is to be found in multiple scattering and wave interference phenomena within the periodic structure. Considering a wave propagating at normal incidence to the 1D periodic structure, each of the layers cause a fraction of the wave to be reflected as a result of the refractive-index difference. For specific frequencies (the ones falling inside the low-trans mission intervals indicated in Fig. 2-2a), the optical-path differences of the reflected wavelets, along with phase changes introduced at the layer boundaries, cause the reflecting wavelets to interfere constructively. Provided the materials are free of absorption, the 1D structure may, hence, exhibit a high reflection of light within these frequency intervals.

An important, general tool for analysis of periodic structures is the photonic band diagram, which provides the dispersion relations of infinitely extending periodic structures. The mathematical framework for calculating band diagrams will not be provided in this section, but is the subject of Chapter 3. The band diagram of the infinite version of the above-studied 1D periodic structure is presented in Figure 2-2(b), and it illustrates the relation between the frequency and the wave vector for modes, which may propagate at normal incidence to the layers. The information to be obtained from the band diagram is, however, two-fold: Apart from providing the dispersion relations, it also reveals the frequency intervals in which no mode solutions are found. Hence, no modes at such frequencies will be able to propagate throughout the structure. From a comparison of Figure 2-2(a) and Figure 2-2(b), it is seen that the frequency intervals having near-zero transmission coefficient correspond to the intervals, where no mode solutions are found in the band diagram. These frequency intervals are defined as photonic bandgaps (PBG). As, however, the transmission spectrum and the band diagram were both simulated for waves propagating at normal incidence to the layers in the 1D periodic structure, the photonic bandgaps seen from Figure 2-2 should righteously only be referred to as 1D PBGs.

The electromagnetic wave propagation is described by Maxwell's equations and as these have no fundamental length scale, the frequencies in Figure 2-2 have been normalized using the factor c/Λ, where c is the velocity of light in vacuum. As c is constant, it is easily found that the properties of the 1D structure may be scaled to any frequency interval by changing the period of the structure (provided of course that the refractive indices remain the same). To quantify this, the above-described structure could be designed to reflect light with a free-space wavelength around one micron. This could for example be done by utilizing the lowest-frequency photonic bandgap of the structure, which is centred around a normalized frequency of approximately 0.3 and, therefore, dictates a period of $\Lambda = 0.3\lambda = 0.3$ μm, where λ is the free-space wave wavelength. Since presenting normalized frequencies as Λ/λ directly provides the required structure dimensions for a specific operational wavelength, this will be preferred throughout the remainder of this book.

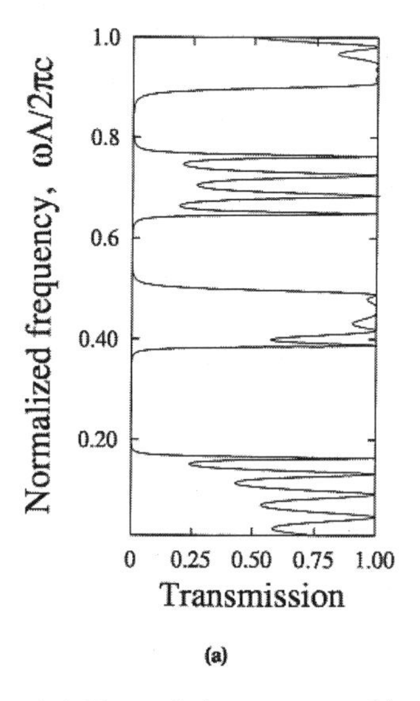

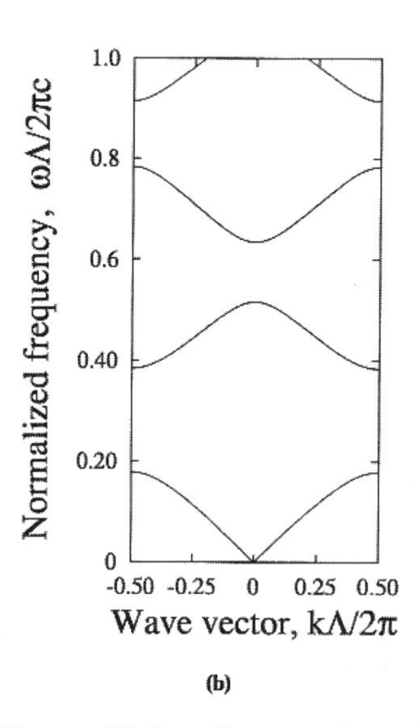

Figure 2-2. Transmission spectrum (a) and banddiagram (b) for a Bragg stack consisting of alternating layers of air (with refractive index $n = 1.0$) and a high-index dielectric material (with refractive index $n = 3.6$). The stop bands in the transmission spectrum correspond to the 1D photonic bandgaps appearing in the band diagram. The transmission spectrum was calculated for a finite-size Bragg stack having five periods.

Having discussed the case of wave propagation restricted to normal incidence in a 1D periodic structure, the discussion will be extended to more general cases of photonic crystals. Whereas, it is relatively trivial to follow the mechanisms causing the opening of PBG in the pure 1D case, it is far from trivial to realize that periodic materials exist in which waves at certain frequencies will be disallowed irrespectively of their propagation direction and even less trivial to design such materials. Since the optical path changes strongly with incident angle, 1D periodic structures are not able to display forbidden frequency intervals covering all solid angles (which is obvious in the specific case of wave propagation parallel to the interfaces in a 1D photonic crystal, where no periodic index difference will be experienced). Note that recently Chigrin *et al.* [2.17] and Fink *et al.* [2.18] discovered 1D photonic crystals, which may reflect light from any angle of incidence under the condition that light is incident from air upon the structure. The essential step is to provide structures, where the periodicity is extended to higher dimensions. Although this step may appear straight-forward, it should be emphasised that it represents an only decade-old discovery in the very classical and canonical field of wave optics. The periodicity of the structure, however, is not by itself a sufficient requirement for a bandgap covering all solid angles to occur. The requirements to fulfil are: A well-chosen lattice structure, a large refractive index contrast, and a wavelength-scale lattice dimension, as shall be discussed in more detail in the proceeding sections.

2.2.2 Two- and three-dimensional photonic crystals

Although triply periodic materials are required to obtain reflection for arbitrary angle of incidence [2.1,2.4], structures with only a 2D periodicity may still display some features of 3D photonic crystals [2.15], and it is useful to consider these first. A schematic example of a 2D photonic crystal is illustrated in Figure 2-3. Photonic crystals are characterized by discrete translational symmetry, and the 2D crystal in this example consists of a background material with cylinders (in the case of cylinders with a lower refractive index than the background material, they will be named holes, and in the case of higher refractive index they will be named rods) arranged in a hexagonal lattice with a period Λ. In the direction normal to the cross section shown in Figure 2-3(a), the 2D photonic crystal is uniform. In contrast to 1D periodic structures, where any index contrast will cause the opening of at least one 1D PBG, a high contrast is required to open 2D PBGs (the minimum required ratio between the refractive indices is believed to be around 2.6 [2.1]). The 2D PBGs are frequency intervals, where the 2D photonic crystals will reflect light with any non-zero wave vector component in the periodic plane for a given (inclusive zero) out-of-plane wave vector component. Therefore, where the 1D structure may be considered as providing reflectance within a cone of incident angles, the 2D photonic crystal may extend this to cover all angles in the plane of periodicity and provide reflectance within disk-like angular ranges, see Figure 2-3(b). Due to the invariance in the vertical direction normal to the cross section shown in Figure 2-3(a), the 2D photonic crystal may, however, not provide 3D omni-directional PBGs. The above-excepted case of wave propagation with zero in-plane wave-vector component is either the case of stationary fields or wave propagation along the invariant direction of the 2D crystals, where of course no periodic index-difference is experienced. It is important to notice that, as indicated in Figure 2-3(b), the 2D PBGs may both be displayed around $k_z = 0$, which corresponds to light propagation strictly in the plane of periodicity as well as 2D PBGs for wave propagation at non-zero angles to the plane of periodicity. In particular, 2D PBGs which do not cover the $k_z = 0$ case may be displayed (as indicated in Figure 2-3(b) between the out-of-plane angles α_2 and α_3). The fact, that these so-called out-of-plane PBGs may be displayed is the basic requirement for realization of PBG-fibres, and an in-depth analysis of the out-of-plane case will be provided in Section 2.4.

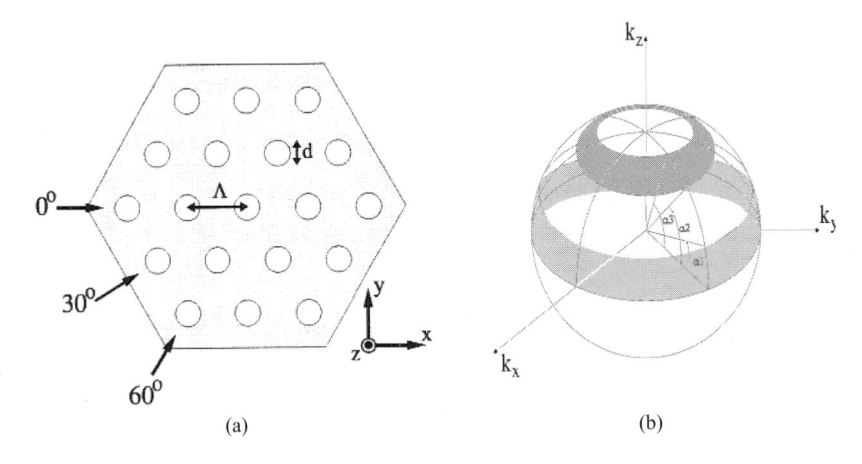

Figure 2-3. (a) Schematic example of a 2D photonic crystal. The crystal has typically circular low-index cylinders arranged periodically in a high-index background material. The 2D photonic crystal is invariant in the z-direction. The intersection of the axis of these cylinders with the *xy*-plane form a two-dimensional Bravais lattice. The lattice period is Λ and the cylinders have diameter *d*. The illustrated crystals have a hexagonal symmetry (i.e., the structures appear the same for waves incident at angles of 0° and 60°). (b) Schematic illustration of solid angles falling inside the photonic bandgap of a 2D photonic crystal operating at a fixed frequency [2.19]. k_x and k_y indicate the wave vector components in the plane, where the structure varies periodically with in-plane position. k_z is the wave vector component in the invariant direction.

Initially in-plane wave propagation in 2D photonic crystals was by far the most studied case [2.20-2.37], and only a more limited number of studies of out-of-plane propagation in 2D photonic crystals have been presented in literature (for some early references see e.g., [2.19,2.38-2.42]. There are several reasons for this. Firstly, from a theoretical point of view the in-plane case is simpler to treat than the out-of-plane case. Secondly, many investigations are focused on applications of 2D photonic crystals in integrated optics, where the wave propagation is intended to be solely in the plane of periodicity. However, this may in practice not be achieved due to the finite height of the photonic crystal in planar configurations. Therefore, out-of-plane propagation and the finite height must be incorporated for accurate analysis of e.g., wave propagation in 2D photonic crystal slabs. This was, nonetheless, neglected in most early investigations of photonic crystals intended for planar applications [2.43-2.45]. For planar applications it is, furthermore, important to point out that as a result of the invariance in the vertical direction, 2D photonic crystals are intrinsically incapable of providing vertical confinement. Indeed, fabricated 2D photonic crystals in planar technology have been found to exhibit high losses [2.20]. Sandwiching a slab of 2D photonic crystal between lower refractive index materials - in order to provide total internal reflection - is, however, not in general a sufficient requirement to obtain leakage-free bound modes. Only under specific conditions may such bound modes be obtained as addressed in a recent book of Johnson *et al.* [2.2] (or see also a few recent papers [2.47-2.49]). For results on propagation through planar optical structures of finite height the work by Søndergaard *et al.* [2.90] may be studied.

The final step to photonic crystals exhibiting full 3D PBGs seems straight-forward from the discussion of 1D and 2D photonic crystals. Naturally, a 3D periodicity is required, which must again be on the scale of the operational wavelength and have a high index contrast [2.1]. As mentioned in the introduction, the fabrication of 2D and 3D photonic crystals for the optical domain is a severe challenge. However, the scalability discussed in the 1D periodic case is valid for 2D and 3D structures as well. This makes it possible to exploit 2D and 3D photonic crystals over the whole electromagnetic spectrum and also to provide larger easy-to-fabricate test structures to be analyzed at long wavelengths [2.1]. Many experimental studies of 2D and 3D periodic structures have, therefore, been performed at microwave frequencies [2.22,2.50-2.52]. This allowed in 1991 Yablonovitch to demonstrate the first structure exhibiting a full 3D photonic bandgap [2.51]. Apart from testing crystal structures that are designed for operation at short wavelengths, the properties of photonic crystals are also increasingly being exploited in the microwave regime, e.g., for antenna applications [2.53-2.57], and also for bandgaps are considered acoustic waves (see e.g., [2.92] or [2.12] for more references).

2.2.3 Fabrication of photonic crystals for the optical domain

For operation at optical wavelengths, 1D periodic structures may readily be fabricated using techniques such as evaporation, sputtering, crystal growth, UV-writing, etc. [2.14]. To realize 2D and 3D photonic crystals, however, no manufacturing methods may yet be classified as mature, but a wide range are being investigated. For 2D crystals, these include planar realizations using lithography and etching techniques developed from the electronics industry [2.20,2.36,2.58,2.89] and fibre realizations based on traditional drawing techniques [2.59]. Among the most promising for the planar technology are realization in GaAs or InP using electron-beam lithography and reactive ion etching [2.20] and realizations of macroporous silicon using standard lithography and anisotropic etching [2.58]. Note that also the application of silicon-on-insulator (SOI) material combinations are interesting [2.89]. Both of these methods have been used to fabricate structures with 2D PBGs at near-infrared wavelengths. Whereas the GaAs/InP realizations may be of importance for incorporation of active devices (such as lasers), macroporous silicon 2D photonic crystals may, at present, be realized with a much higher height-to-period ratio (above 30) and, furthermore, with the possibility of controlling the diameter of the holes/rods along their centre axis [2.60]. This makes macro-porous silicon promising for applications, where only passive properties of 2D photonic crystals are exploited. The fabrication of 2D photonic crystals in fibre-form will be discussed in details in Chapter 4. For fabrication of micro-scale 3D structures, the methods that are being investigated are very different in nature and among the most promising are: micro-electro-mechanical fabrication of silicon layer-by-layer structures [2.61], synthetic opal-based crystals [2.62], and self-aligned, sputtered silicon/silica structures [2.63]. Whereas the fabrication of each layer in the layer-by-layer structure is very laborious and requires a multitude of processing steps, fabrication of many layers (>50) is more feasible using the two latter methods. These may, furthermore, readily be fabricated at sub-micrometer periodicities, but so far with insufficient index contrasts for full 3D PBGs to occur. On the other hand, the layer-by-layer structures do provide the index contrast required for full 3D PBGs to be exhibited and although these have not yet been achieved at sub-micrometer wavelengths, full 3D PBGs around 1.5.μm have been demonstrated experimentally [2.64]. Finally, an important issue for applications of photonic crystals is the influence of structural fluctuations resulting from fabricational errors/inaccuracies. Although becoming increasingly relevant as the realisation of components become more within reach, only limited attention has been paid to this issue so far and only a few studies have been presented in literature (see some of the earliest in [2.65-2.66]). Initial theoretical studies of 3D photonic crystals indicate a relatively large robustness of the PBGs for minor structural irregularities [2.66].

2.2.4 Spatial defects in photonic crystals

The presentation of photonic crystals has so far only concerned structures, which are fully periodic. Photonic crystals become, however, of more value by intentional introduction of spatial defects within them.

Hereby, the photonic crystals may act as reflecting boundaries and trap light at the defect region(s). The defects may be point-, line- or surface defects and allow for the creation of micro cavities, waveguides, novel types of compact, functional components or various types of substrates.

Defects in 1D periodic structures have been known and utilized for many years, e.g., for the design of distributed feed back (DFB) lasers utilizing weakly modulated 1D periodic structures [2.14,2.67]. Also strongly modulated structures have been extensively studied and in this case only a very limited number of periods are required to provide micro cavities with high-Q values [2.1,2.68]. For laser-applications, high-Q values, low rates of spontaneous emission, and small cavity volumes are among the key parameters for optimised performance [2.69]. Photonic crystals offer a unique potential of both controlling the rate of spontaneous emission and provide high-Q micro cavities with ultra-small volumes [2.70-2.72]. In the ideal situation, a 3D photonic-crystal-based micro cavity may be designed, where all spontaneous emission is suppressed except into a cavity mode. Such an ideal laser would be characterized by zero threshold power and was one of the original goals of Yablonovitch [2.15]. The experimental demonstration of suppressed spontaneous emission and high-Q micro cavities in 3D photonic crystals have been pursued ever since 1987, and especially within the past years, important experimental progress has been made using opal-based 3D photonic crystals [2.73-2.75,2.88]. Also for lower dimensional periodic structures, novel micro-cavity designs are being investigated [2.76-2.77], and, recently, the first experimental demonstration of lasing from a 2D photonic crystal-based micro cavity was presented [2.46]. Although the above-cited work on 2D photonic crystal micro-cavities does not document lifetime alterations, theoretical studies document that significant lifetime changes may, indeed, be obtained in 2D photonic crystals realized in III-V semiconductors [2.42,2.78].

With respect to waveguides in photonic crystals, 2D photonic crystals with a row of defects have been studied for both straight waveguides [2.44,2.79], sharply bent waveguides (with a bend radius comparable to the optical wavelength) [2.43,2.80] and - in combinations with points defects - novel types of compact wavelength-division multiplexed (WDM) components with a size of only a few square wavelengths [2.45]. However, the above-cited studies assume infinite extent of the 2D photonic crystals in the invariant direction as well as wave propagation strictly in the plane of periodicity. As discussed previously, pure 2D analysis may, however, not be considered sufficient for design of practical, functional components – but will (hopefully) be capable of providing qualitative information about the potential operation of these. Although the initial works by Russell, Johnson and Martin only address 2D photonic crystals without spatial defects [2.47-

2.49], they do present general design guidelines for elimination of vertical losses and theoretical approaches, which may be applied also for modelling of real 2D photonic crystals including spatial defects. See also recent publications by Johnson *et al.* [2.2], or Søndergaard *et al.* [2.90] concerning these subjects. Elaborating upon the ideas presented in the above-referenced papers, therefore, seems fruitful for design of leakage-free 2D photonic crystals waveguides. At the point in time, when low-loss, sharply bent 2D photonic crystal-based waveguides operating at optical wavelengths are demonstrated experimentally, it will undoubtedly be a break-through achievement, which opens the possibilities of realizing large-scale integrated optics.

In contrast to the line defects employed for 2D photonic crystals waveguides in planar applications, the waveguide-core in a photonic crystal fibre (PCF) may, in fact, be considered a point defect in an ideal 2D photonic crystal. Hence, both the point defect and the 2D photonic crystal are infinitely extending in the invariant direction. Light with a non-zero out-of-plane wave vector component, which falls within the 2D PBG of the photonic crystal, may, therefore, be trapped in the plane of periodicity and guided along the point defect in the direction perpendicular to the periodic plane. The out-of-plane wave vector component, hence, is identical to the propagation constant of the fibre. In this case, the fibre is operating truly by PBG-effect. The PCFs may, however, also trap light at the spatial defect in a more conventional manner, namely if the spatial defect has a higher refractive index than the effective index of the surrounding photonic crystal structure. In this case, the PCF is operating by simple index guidance. As mentioned in the introduction, the two different cases of PBG- and index-guiding PCFs shall be treated separately in Chapters 6 and 5, respectively, and applications of these in Chapter 7.

Due to the difficulties in fabricating 3D periodic structures, only few studies have addressed 3D photonic crystal waveguides for the optical domain [2.81].

The whole area of photonic bandgap investigations has also spurred a number of interesting new technologies such as the one using surface plasmons [2.87].

Finally, interesting aspects of 2D and 3D periodic structures concern their ability of displaying strong diffractive effects and so-called ultra-refractivity [2.82]. This may be utilized for providing unusually compact beam splitters or prisms as demonstrated using both 2D and 3D periodic structures [2.36,2.82-2.84]. Considering the present problems of realizing low-loss 2D photonic crystals, exploitation of the above features may be further attractive as only a very limited number of periods might be required to display strong diffractive effects.

2.3 TERMS INVOLVED IN PHOTONIC CRYSTAL TECHNOLOGY

It is important to realise that when we talk about photonic crystal materials – and especially photonic bandgap materials – the structures does not have continuous translation symmetry, but rather they have discrete translational symmetry. This means, as described by Jouannopoulos *et al.* [2.1] that the structures only are invariant under translations of distances, which are a multiple of some fixed (basic) step length in a specific direction – described by a vector.

The basic step length is the lattice constant, and the basic step vector is named the primitive lattice vector [2.1], which we here choose to name $\bar{a}$. Because of the symmetry, the permittivity of the periodic structure may be described as

$$\varepsilon(\bar{r}) = \varepsilon(\bar{r} + \bar{a}) \tag{2.1}$$

where $\bar{r}$ is a vector representing a given point in space. By repeating the translation, we have that $\varepsilon(\bar{r}) = \varepsilon(\bar{r} + \bar{R})$ for any vector $\bar{R}$ that is an integral multiple of $\bar{a}$ (i.e., $\bar{R} = l\bar{a}$, where l is an integer).

If we for simplicity choose to look at a simple structure, which only is periodic in one direction such as the Bragg stack shown in Figure 2-1, and without loss of generality chose the periodicity to be in the x-direction, the discrete periodicity leads to a x-dependence for the magnetic field $\bar{H}$ that is simply the product of a plane wave with a x-periodic function. As explained in [2.1], we can think of it as a plane wave, as it would be in free space, but modulated by a periodic function because of the periodic lattice. This means that the magnetic field may be expressed as follows:

$$\bar{H}(x,....) \propto e^{jk_x x}\bar{u}_{k_x}(x,....) \tag{2.2}$$

The relation expressed in Eq. (2.2) is commonly known as Bloch's theorem, and in solid-state physics, the form of Eq. (2.2) is known as a Bloch state [2.86], whereas it in mechanics is called a Floquet mode [2.85]. Within the field of photonic crystal research, the most common choice of terms follow the directions of Joannopoulos *et al.* [2.1], who chose to use the name Bloch states.

A central point concerning Bloch states is that the Bloch state with a wave vector k_x, and the Bloch state with wave vector $k_x + mb$ are identical, if b is chosen as $b = 2\pi/a$, where a is the distance given as $\bar{a} = \hat{x}a$. The k_x values that differ by integral multiples are not different from a physical point of view. Consequently, the mode frequencies must also be periodic in k_x, i.e., $\omega(k_x) = \omega(k_x + mb)$. This means that we only need to consider k_x that exist in the range $-\pi/a < k_x < \pi/a$. This region, which describes important, non-redundant values of k_x is called the Brillouin zone. We may recommend the book by Joannopoulos *et al.* [2.1] for a further description of the concepts of reciprocal lattices and Brillouin zones.

Before we go on, it should, however, be noted that analogous arguments may be applied, when the dielectric is periodic in three dimensions. In this case, the dielectric is invariant under translations through a multitude of lattice vectors $\overline{R}$ in three dimensions. Any one of these lattice vectors can be written as a particular combination of three primitive lattice vectors ($\overline{a}_1, \overline{a}_2, \overline{a}_3$), which are said to "span" the space of lattice vectors [2.2]. The vectors ($\overline{a}_1, \overline{a}_2, \overline{a}_3$) give rise to three primitive reciprocal lattice vectors ($\overline{b}_1, \overline{b}_2, \overline{b}_3$) defined so that $\overline{a}_i \cdot \overline{b}_j = 2\pi\delta_{ij}$. These reciprocal vectors form a lattice of their own, which is inhabited by wave vectors as described in [2.2]. The modes of a three-dimensional periodic system are Bloch states, which can be labeled by $\overline{k} = k_1\overline{b}_1 + k_2\overline{b}_2 + k_3\overline{b}_3$, where $\overline{k}$ lies in the Brillouin zone. Each value of the wave vector $\overline{k}$ inside the Brillouin zone identifies an eigenstate and an eigenvector $\overline{H}_{\overline{k}}$ of the form

$$\overline{H}_{\overline{k}}(\overline{r}) = e^{j(\overline{k}\cdot\overline{r})}\overline{u}_{\overline{k}}(\overline{r}) \tag{2.3}$$

where $\overline{u}_{\overline{k}}(\overline{r})$ is a periodic function on the lattice, i.e., $\overline{u}_{\overline{k}}(\overline{r}) = \overline{u}_{\overline{k}}(\overline{r} + \overline{R})$ for all lattice vectors $\overline{R}$.

2.4 EFFICIENT STRUCTURES FOR CREATING PHOTONIC BANDGAPS IN FIBRES

This section will address the properties of 2D photonic crystals in the case of wave propagation with a non-zero wave vector component perpendicular to the periodic plane. In other words, we will in the remainder of this chapter focus on the fundamental properties of the photonic crystal waveguides, which generally belongs to the category of photonic crystal fibres. The primay focus of this section will be on the photonic bandgap effect, and not so much on so-called index-guiding photonic crystal fibres.

The reason for this choice is the fundamentally new properties (compared to waveguiding by total internal reflection), which may be obtained in the PBG fibres.

The analysis will start out by taking basis in a high-index contrast structure, which exhibits a complete in-plane 2D PBG and illustrate the behaviour of the PBG as the out-of-plane wave vector component is varied. Thereafter, the investigations will be focused on a detailed analysis of silica-air photonic crystals. Albeit the silica-air systems have too low index contrasts for complete in-plane PBGs to be exhibited, the system may display complete PBGs in the out-of-plane case. The analysis of silica-air photonic crystals will both address simple triangular structures as well as more advanced structures with multiple silica dopants introduced.

2.4.1 In-plane photonic crystals

The general matrix equation for the $\overline{H}$-field (see Section 3.4) becomes further reduced, when wave propagation is restricted to the plane of periodicity in a 2D photonic crystal. For such wave propagation, the electromagnetic fields may be decomposed into two polarization states, where in one case the $\overline{E}$-field (electric field) has only components perpendicular to the hole/rod-axis of the 2D photonic crystal, and in the other case the $\overline{H}$-field has only perpendicular components. Hence, the two polarization cases are referred to as transverse electric (TE) and transverse magnetic (TM), respectively. Hence, the eigenvalue problems, which must be solved, concern only matrices of dimensions $N^x N$.

A calculation of the band diagram for a 2D photonic crystal with air holes arranged on a triangular lattice in a GaAs background material is illustrated in Figure 2-4. The photonic crystal has an air-filling fraction, f, of 45% and the dielectric constant of air and GaAs are 1.0 and 13.6, respectively. Figure 2-4 (a) reveals a broad 2D PBG for the TE-polarization case occurring between the first and second bands around a normalized frequency Λ/λ = 0.26, and a narrow PBG between bands seven and eight around Λ/λ = 0.63. Defining the relative size of a PBG as the frequency difference between the upper and lower PBG edges divided by the centre frequency, the two PBGs have a relative size of 40% and 2%, respectively. Hence, the lowest frequency PBG has the largest size and for practical applications, utilization of this should provide the most robust operation (with respect to tolerance towards structural inaccuracies etc.). However, utilization of the higher frequency PBG is seen to require a periodicity which is approximately 2.5 times larger than for utilization of the lower frequency PBG. Hence, for applications at short wavelengths, e.g., for components operating in the optical domain, where realization of sufficiently small structure dimensions may represent the limiting factor, exploration of the higher frequency PBG may be advantageous.

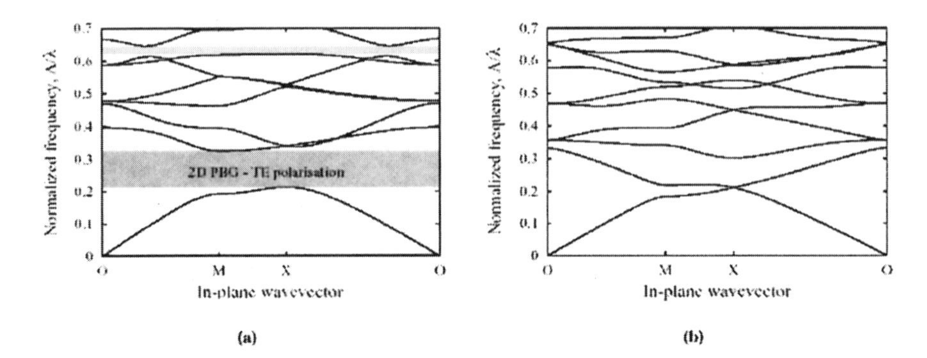

Figure 2-4. In-plane analysis of a triangular 2D photonic crystal consisting of circular air holes in a GaAs background material. The photonic band structure diagram in the case of TE-polarized wave propagation is illustrated in (a) and for TM-polarized wave propagation in (b). The air-filling fraction is 45%.

Considering the same 2D photonic crystal but in the case of TM-polarization, the band diagram is illustrated in Figure 2-4 (b). As seen, no PBGs exist in this case. Therefore, the specific 2D photonic crystal may not be utilized for reflection of electromagnetic waves of arbitrary polarization. By increasing the air-filling fraction of the photonic crystal, a 2D PBG common to both polarization may, however, be opened up. Figure 2-5 illustrates both the TE and TM bands for a 2D GaAs-air photonic crystal with $f = 70\%$. As seen from the figure, a common frequency interval exists for the large TE-polarization PBG and the more narrow TM-polarization PBG. This frequency interval, were these overlap, shall be referred to as a complete 2D PBG. A more detailed analysis of the variation of the two PBGs with respect to filling fractions and index contrast may be found in [2.1].

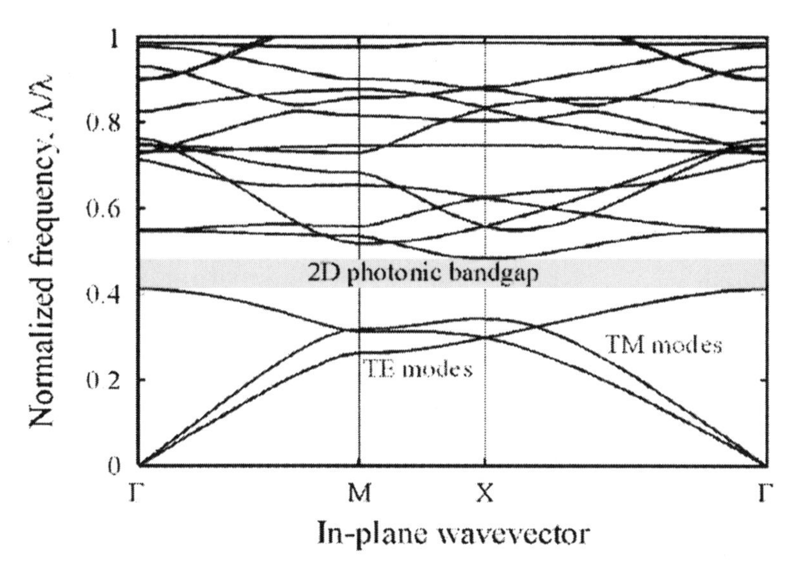

Figure 2-5. Banddiagram for a 2D photonic crystal consisting of circular air holes arranged on a triangular lattice in a GaAs background material. The photonic crystal has a relatively large air-filling fraction of 70%. Both *TE*- and *TM*-polarization band structures are illustrated, and a narrow complete PBG is present as indicated by the shaded area.

An intuitive understanding for the opening of the PBGs may be obtained by considering a series of plane waves of similar frequency - but with different wave vectors. In reciprocal space, these may be viewed as propagating in different directions, and the waves will not experience an identical periodicity of the photonic crystal (see e.g., Figure 3-6 (b) to ease the conception of this explanation). Hence, the scattering of the different waves will not be identical, but each wave may be thought of as experiencing a pseudo-lD periodic structure. Each of these pseudo-lD periodic structures may be regarded as capable of exhibiting a lD PBG and a full 2D PBG is displayed, if these have a frequency overlap. Therefore, the likelihood of a 2D periodic structure to exhibit 2D PBGs may be regarded from the degree of smoothness of the 1. Brillouin zone (i.e., a circular shape of the 1. Brillouin zone is optimum), and the likelihood will further be increased by raising the index contrast between the holes/rods and the background material. This line of argument is supported by both experimental and numerical comparisons between 2D photonic crystals with square and hexagonal symmetries, where the hexagonal structures are generally found to display wider PBGs [2.1]. Similar arguments apply to 3D photonic crystals (where a sphere-like 1. Brillouin zone is desirable), and the line of arguments has provided important design-guidelines for experimental realization of the first structures exhibiting 3D PBGs [2.4]. For the above reasons, studies of 2D periodic structures with a hexagonal symmetry will be focussed on in this book as opposed to e.g., studies of square-symmetry photonic crystals.

Finally, a few words should be added concerning the convergence criterium of the plane-wave method. This method, which forms the basis of several of the calculation results presented in this chapter, is well documented in literature [2.93-2.94] and will not be elaborate on here. However, for full periodic structures, it is noteworthy that convergence is obtained using approximately 300 shortest reciprocal lattice vectors. The band diagrams presented in this chapter were calculated using the approximately 500 shortest reciprocal lattice vectors, and the error on the frequencies was estimated to be smaller than 0.5%. These computations were readily performed on standard personal computers. However, for larger problems such as for the calculation of guided modes in PBG- fibres, a higher number of lattice vectors ($N > 10000$) are required to reach convergence. As the demands for computing resources scales as N^3 for the plane-wave method outlined in this chapter, a different plane wave implementation with a much more efficient scaling must be considered. This approach will be presented in the following section, and later employed for the treatment PBG-fibres.

2.4.2 From in-plane to out-of-plane photonic crystals

Due to the invariance in the direction perpendicular to the periodic plane of 2D photonic crystals (defined as the *z*-direction), the H-field in such a crystal may be written on the form:

$$\overline{H}(\overline{r},\omega)=\overline{H}(\overline{r}_{||},\omega)e^{-jk_z z} \tag{2.4}$$

where $\overline{H}$ is the magnetic field, $\overline{r}$ is the vector representing a point in space, and $\overline{r}_{||}$ is a vector representing a point in the periodic plane. k_z is the out-of-plane wave vector component. In the case of in-plane propagation, k_z is naturally zero. In contrast to the case of $k_z = 0$, the components of the H-field become coupled in the case of $k_z \neq 0$, and the full matrix equation presented in Chapter 3 must be solved (see Section 3.3 for details).

In Figure 2-6, the complete 2D PBG, which exists for the in-plane case of the 2D photonic crystal consisting of air holes in GaAs (the air-filling fraction $f = 70\%$), is "monitored", while moving out of the periodic plane. The figure illustrates the band diagrams for $k_z\Lambda$ values ranging from 0 to 1.5π. As seen from Figure 2-6, the complete 2D PBG around $\Lambda/\lambda = 0.4$ is narrowed and pushed towards higher frequencies as k_z is increased. It is, however, important to notice that it remains as a complete 2D PBG also in the out-of-plane case, since the PBG is extending over the whole boundary of the irreducible Brillouin zone. A further interesting property of the 2D photonic crystal is revealed in Figure 2-6 (d), namely that a complete PBG, which is not present in the in-plane case is opened up between bands 13 and 14 at a Λ/λ-value of approximately 0.85.

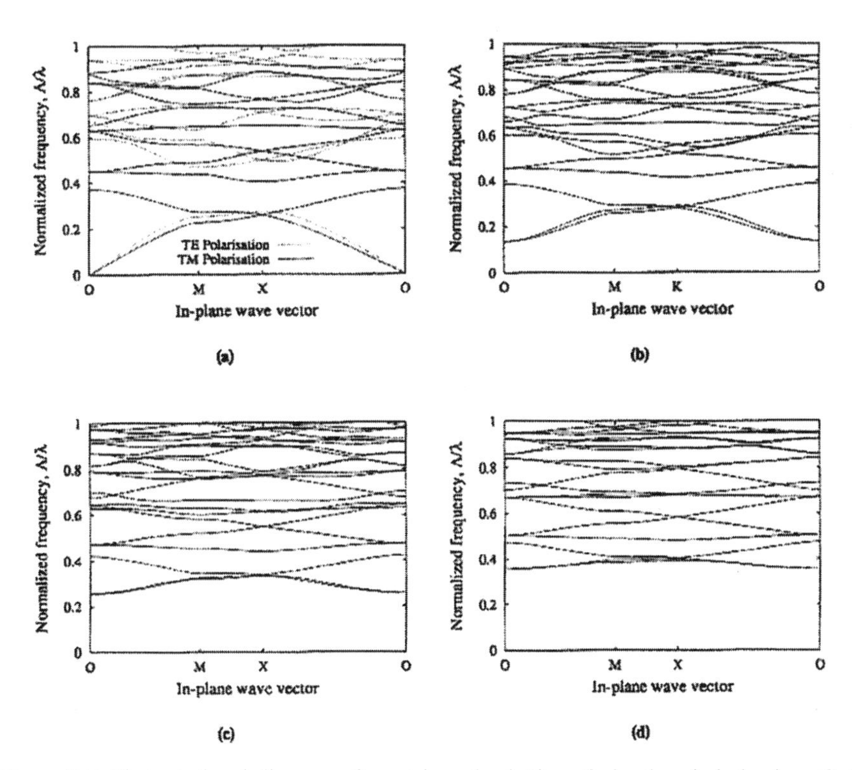

Figure 2-6. Photonic band diagrams for a triangular lattice of circular air holes in a GaAs background material. The air-filling fraction f =70%. Results are presented for: (a) $k_z\Lambda = 0$; (b) $k_z\Lambda = 0.51\pi$; (c) $k_z\Lambda = 1\ \pi$; and (d) $k_z\Lambda = 1.51\ \pi$.

The behaviour of the 2D PBGs is more clearly observed by plotting the PBG edges as function of k_z. Such a plot is presented in Figure 2-7, where the PBGs occurring within the 15 lowest-frequency bands are illustrated. As seen, the complete in-plane PBG is rapidly narrowing, when moving out of the periodic plane and closes at a $k_z\Lambda$-value of 5. In contrast to the behavior of this PBG, a range of new PBGs are seen to open up at larger k_z-values. The figure, however, reveals that no frequency interval exist, which fall within a PBG at all *k_z-values*. This means that no PBGs exist for arbitrary propagation direction and that a full 3D PBG cannot be exhibited by a 2D photonic crystal. This should not be surprising as discussed in Section 2.2.2.

Apart from the behaviour of the 2D PBGs, Figure 2-6 and 2-7 also illustrate, how the lowest frequency band is pushed upwards for increasing k_z-values and that a forbidden region is opened between the lowest band and $\Lambda/\lambda = 0$. This forbidden region and the characteristics of the lowest-order mode may be interpreted in a very simple manner using effective-index considerations (as we will see later).

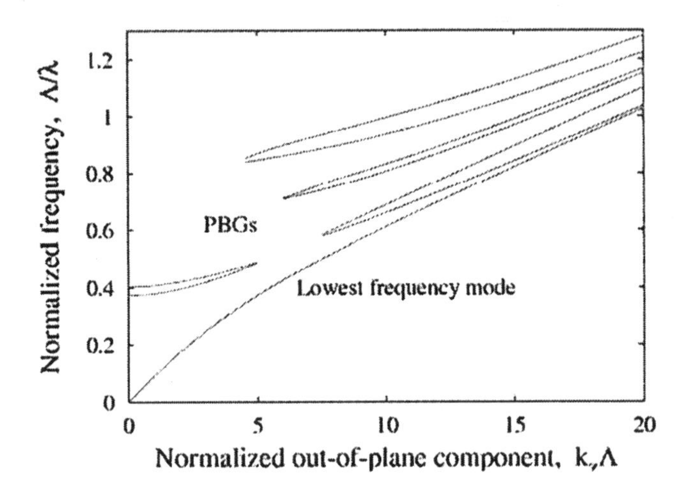

Figure 2-7. Illustration of the out-of-plane PBGs exhibited by a triangular GaAs-air photonic crystal with f = 70%. The figure also illustrates the lowest-frequency mode allowed of the photonic crystal.

This section has aimed at introducing out-of-plane wave propagation in photonic crystals and a high-index contrast system of GaAs-air was chosen for this purpose. However, exploitation of out-of-plane PBG effects in GaAs 2D photonic crystals, which have a significant extent in the invariant direction (hundreds and more optical wavelengths), is not feasible due to the technological problems in fabricating such structures. These problems are not only related to GaAs 2D photonic crystals, but generally apply to all semiconductors and other solid-state crystalline materials. In contrast to this, 2D photonic crystals of practically infinite length in the invariant direction are feasible from realization in glass materials.

2.4.3 The supercell approximation and efficient plane-wave method

As mentioned in the proceeding section, the basic plane-wave method does suffer from the fact that the computer processing time and storage requirements grow rapidly with size and complexity of the photonic crystal structure. Whereas, it is feasible with today's computers to accurately analyse full-periodic 2D photonic crystal structures using the basic plane-wave method (which typically requires less than 2000 expansion terms), the method may not be applied accurately to the case of defects in photonic crystals, where a significantly higher number of expansion terms are required [2.95]. A further limitation of the basic plane-wave method is that the discontinuous nature of the refractive index distribution causes convergences problem due to Gibbs phenomenon for the involved Fourier transformation [2.96]. In 1993 Meade *et al.* [2.94] presented an advanced method that lead to dramatic lowering in the computational demands and made accurate analysis of large photonic crystal structures feasible. This method may be applied to the modelling of out-of-plane wave propagation in 2D photonic crystals including defects using a so-called supercell approximation and the theory of this efficient method and the supercell approximation will be described in more detail in the following text.

If we for a moment borrow the results from Chapter 3 (see especially Eq. (3.18) and Eq. (3.22)), the wave equation and the Fourier-series expansion of the $\overline{H}$ -field are:

$$\nabla \times \left[\frac{1}{\varepsilon_r(\overline{r})} \nabla \times \overline{H}_{\overline{k}}(\overline{r}) \right] = \frac{\omega^2}{c^2} \overline{H}_{\overline{k}}(\overline{r}) \tag{2.5}$$

and

$$\overline{H}_{\overline{k}}(\overline{r}) = \sum_{\overline{G}} \sum_{\lambda=1,2} h_{\overline{k}+\overline{G},\lambda} e^{i(\overline{k}+\overline{G})\overline{r}} \tag{2.6}$$

where the subscript $\overline{k}$ indicate that solutions are sought for a specific wave vector. Instead of setting up the matrix equation as will be discussed in Section 3.4, the *n-th* eigensolution $\left(\frac{\omega_{\overline{k},n}^2}{c^2}, \overline{H}_{\overline{k},n} \right)$ to Eq. (3.41) may be found by minimization of the functional [2.95]:

$$E\left(\overline{H}_{\bar{k},n}\right)=\frac{\left\langle \nabla\times\left(\frac{1}{\varepsilon_r(\bar{r})}\nabla\times\overline{H}_{\bar{k},n}\right)\middle|\overline{H}_{\bar{k},n}\right\rangle}{\left\langle \overline{H}_{\bar{k},n}\middle|\overline{H}_{\bar{k},n}\right\rangle} \tag{2.7}$$

where

$$\left\langle \overline{F}\middle|\overline{G}\right\rangle=\int \overline{F}^{*}\cdot\overline{G}d^{3}R \tag{2.8}$$

When the functional $E\left(\overline{H}_{\bar{k}}\right)$ is at a minimum, $\overline{H}_{\bar{k}}$ is an eigenvector and $E\left(\overline{H}_{\bar{k}}\right)$ is the corresponding eigenvalue $\frac{\omega_{\bar{k},n}^{2}}{c^{2}}$. By inserting a trial vector on the form (2.6) in Eq. (2.7), the functional effectively becomes a function of the coefficients $h_{\bar{k}+\overline{G},\lambda}$, and the problem is reduced to varying the coefficients along a path that minimizes the functional $E\left(\overline{H}_{\bar{k}}\right)$. An efficient iterative scheme that performs this task utilizes a preconditioned conjugate gradient method for the generation of the effective change vectors [2.97]. Higher-order eigensolutions are found by restricting the trial-vectors to being orthogonal to all previously found eigenvectors and using the same minimization principle. It is vital for the minimization approach that fast evaluation of the left-hand side in Eq. (3.41) may be performed. This is achieved by performing the rotations $\nabla\times$ in reciprocal-space and the operation $\frac{1}{\varepsilon_r(\bar{r})}$ in real-space. The fast Fourier transformation (FFT) is used to go from one space to the other. The advantage of this approach is that the curl operation is diagonal in reciprocal space, whilst $\frac{1}{\varepsilon_r(\bar{r})}$ is diagonal in real-space. Hence, since all matrix operations involved in solving Eq. (3.41) are diagonal using this approach, the memory requirements scale linearly with the number of expansion terms, N_{PW}. The scaling of the computer time is dominated by the FFT that is on the order of $N_{PW}\,log(N_{PW})$.

A further important issue in order to obtain a fast convergence using the variational method is to find an appropriate representation for the refractive index distribution. The method further circumvents any problems related to Fourier transforming the refractive index distribution (by operating only upon this in real-space), and samples the refractive index distribution within a unit-cell on a high number of grid points. This has the additional advantage that the method is just as applicable to arbitrary periodic structures as to ideal structures (to be presented in the preceding chapters). As demonstrated by Meade *et al.*, the number of sampling points should preferably be similar to the number of plane waves used for the expansion of electro-magnetic fields and an appropriate interpolation of the refractive index distribution is necessary [2.94]. The interpolation that has been used for the present implementation is based on

$$\varepsilon_{ij} = \varepsilon_{\perp} n_y n_x + \varepsilon_{\parallel}(-n_y n_x) \tag{2.9}$$

where $\overline{n} = n_x \hat{x} + n_y \hat{y}$ is the unit-vector normal to the dielectric interface and $\varepsilon_{\perp}$ and $\varepsilon_{\parallel}$ are the Wiener limits [2.98]:

$$\varepsilon_{\parallel} = f\varepsilon_1 + (f-1)\varepsilon_2 \tag{2.10}$$

$$\varepsilon_{\perp} = \frac{1}{f/\varepsilon_1} + \frac{1}{(f-1)/\varepsilon_2} \tag{2.11}$$

The variational method yields very fast computations for full-periodic structures, and compared to the basic plane-wave method, it provides a much more efficient scaling to allow accurate modelling of large computational tasks such as for PBG-fibres.

Since plane-wave methods utilize the periodicity of the photonic crystals for the expansion of the electromagnetic fields and the refractive index distributions, they are not directly suited for simulating spatial defects. However, the plane-wave methods may still be applied, if the smallest region describing the structure is enlarged to include the defect and several periods of the photonic crystal that surround it. In this way, an artificial super periodicity is introduced, where also the defect is repeated periodically. This is known as a supercell approximation. To accurately determine the properties of the defect region, the size of the supercell must be large enough to ensure that neighbouring defects are uncoupled. Figure 2-8 illustrates the supercell utilized for the simulation of a honeycomb PBG-fibre.

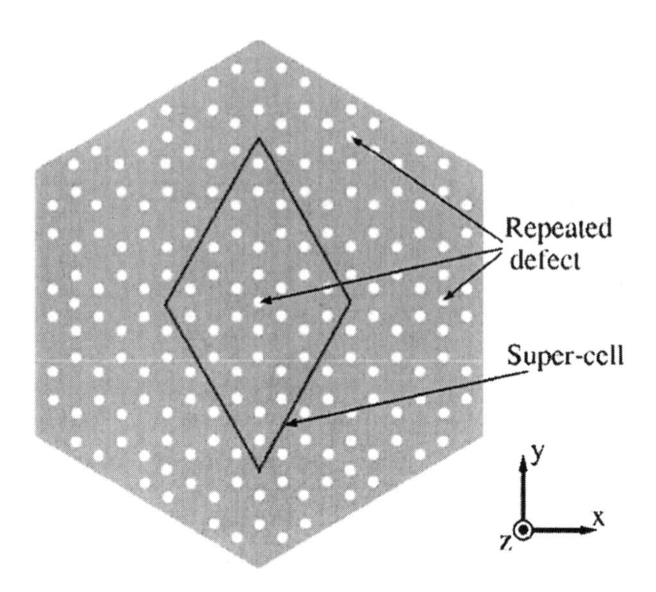

Figure 2-8. Illustration of a supercell with a size of 4 x 4 simple honeycomb cells. The large supercell is required in order to simulate the central defect region as isolated even though it is, in fact, repeated periodically.

To perform a first analysis of honeycomb PBG-fibres, and hereby indicate the properties of these fibre structures, a silica-air honeycomb photonic crystal with a relatively low air-filling fraction f = 10% and a single, similar sized, extra air hole introduced is chosen. Figure 2-9 (a) shows the band diagram for the cladding structure of the fibre (i.e., for the full-periodic honeycomb photonic crystal) at a fixed normalized propagation constant of $\beta\Lambda = 7.0$. A calculation of the band diagram for the photonic crystal including the defect is illustrated in Figure 2-9 (b). A relatively small supercell corresponding to 3 ^{x} 3 simple honeycomb cells was used for the calculation. As seen from the figure, the super-cell causes a folding of the bands in the Brillouin zone. Due to the folding, a PBG appearing above band n for the full periodic structure will correspond a PBG above band nm^2, when using using a m ^{x} m supercell approximation. As seen from the figure, the lower PBG boundary is re-found, when using the super-cell approximation. This boundary is - in this case - defined by the maximum frequency of band number 18. The upper PBG boundary is also re-found for the super-cell calculation, but in addition two bands is found within the PBG region for the super-cell calculation. These two bands are introduced solely by the defect region and correspond to a doubly-degenerate mode.

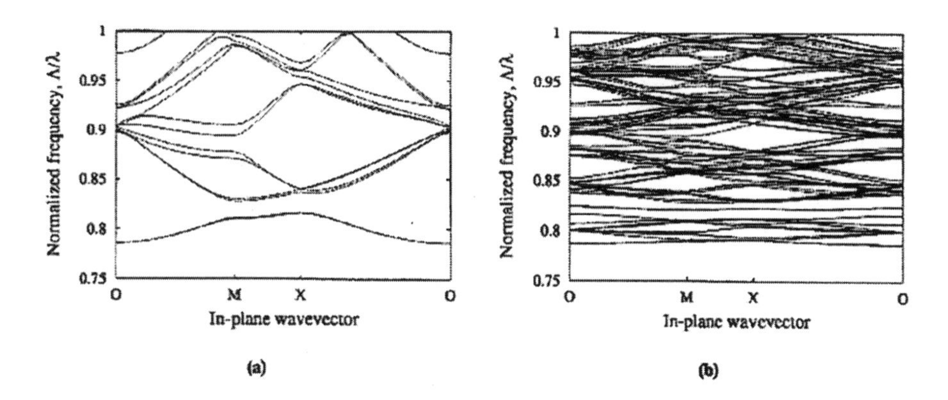

Figure 2-9. (a) Photonic band diagram for a full-periodic honeycomb lattice of circular air holes in a silica background material (f = 10%, $\beta\Lambda$ = 7.0). (b) Photonic band diagram for a similar photonic crystal but with a single, extra air hole introduced to form a defect. The super-cell calculation predicts the same frequencies of the PBG boundaries. However, within the PBG, two additional bands occur, which correspond to a doubly-degenerate defect mode. A super-cell with a size of 3 $^{\times}$ 3 simple cells was used for the calculation. For accurate simulation, the defect bands should be flat across the Brillouin zone, but due to the relatively small super-cell some variation may be observed.

For accurately determination of the properties of the defect modes, the super-cell should ideally be infinitely large. Hence, in reciprocal space, the unit cell used for the approximation would become reduced to a single point and an infinite number of flat, folded bands should occur across the Brillouin zone in the allowed frequency intervals. Only a single defect should, however, exist within the super-cell (irrespectively of its size) and no folding of the bands corresponding to the defect will occur. Following these arguments, an important evaluation of the super-cell approximation and, therefore, of the accuracy on the determination of the defect modes is to determine the variation of the defect bands across the Brillouin zone. For calculations using a super-cell size of 4 $^{\times}$ 4 simple honeycomb cells, 16384 plane waves for the expansion of the electromagnetic fields, and the same number of expansion terms for the refractive index distribution, an accuracy on the frequency of the defect modes of better than 0.01% is achieved. This accuracy is sufficient for accurate determination of most of the PBG-fibre properties that are going to be investigated in most cases. Generally, only for studies of the birefringence properties of PBG-fibres, where the mode-index must be determined very accurately ($\Delta\beta/k \sim 10^{-5}$), will additional evaluation of the numerical inaccuracies of the variational method be needed (birefringence in honeycomb PBG-fibres will be analysed in Section 6.5.2).

2.5 SUMMARY

In this chapter, an introduction to the concept of photonic crystals has been presented. Using heuristic arguments, the operation of 2D and 3D photonic crystals has been explained by taking basis in wellknown lD periodic structures. Issues concerning the fabrication of photonic crystals for use in the optical domain have briefly been discussed and various application-oriented aspects have been addressed.

The chapter further documented how 2D photonic crystals may be utilized to reflect light in the out-of-plane case, i.e., for waves propagating at an oblique angle to the holes/rods of the photonic crystal. This treatment started out by visualizing the behaviour for the complete 2D PBGs from the in-plane case to the out-of-plane case for a high-index contrast photonic crystal.

REFERENCES

[2.1] J. Joannopoulos, I. Winn, and R. Meade, "Photonic Crystals: Molding the Flow of Light", *Princeton University Press*, 1995. ISBN:0-691-03744-2

[2.2] S. G. Johnson, and J. Joannopoulos, "Photonic Crystals: The Road from Theory to Practice", *Kluwer Academic Publishers*, 2002. ISBN: 0-7923-7609-9.

[2.3] J. Joannopoulos, P. Villeneuve, and S. Fan, "Photonic crystals: putting a new twist on light," *Nature,* vol. 386, pp. 143-149, March 1997.

[2.4] E. Yablonovitch, " Photonic band-gap structures", *Journal of the Optical Society of America B*, vol. 10, pp. 283-295, Feb. 1993.

[2.5] C. Soukoulis, ed., "Photonic band gaps and localization", *Proceedings of the NATO advanced research workshop*, Heraklion 1992, vol. 308 of NATO ASI series. Series B, Physics, Dordrecht: Kluwer, 1993.

[2.6] E. Burstein and C. Weisbuch, eds., "Confined electrons and photons New physics and applications", *Proceedings of a NATO Advanced Study Institute*, Erice 1993, vol. 340 of NATO ASI series. Series B, Physics, New York, N.Y. : Plenum Press, 1995.

[2.7] C. Soukoulis, ed.,
"Photonic band gap materials",
Proceedings of the NATO advanced study institute, Elounda 1995, vol. 315 of NATO ASI series. Series E, Applied sciences, Dordrecht : Kluwer, 1996.

[2.8] J. Rarity and C. Weisbuch, eds.,
"Microcavities and photonic band gaps Physics and applications",
NATO ASI series. Vol. 324, Series E, Applied sciences, Dordrecht : Kluwer, 1996.

[2.9] C. M. Bowden, J. P. Dowling, and H. O. Everitt (Editors),
"Development and Applications of Materials Exhibiting Photonic Band Gaps",
Journal of the Optical Society of America B, vol. 10, Feb. 1993, special issue.

[2.10] Special issue on: " Photonic Crystals and Photonic Microstructures", Vol. 145, IEE *Proceedings - Optoelectronics*, Dec. 1998.

[2.11] J. Dowling, H. Everitt, and E. Yablonovitch,
"Photonic & sonic band-gap bibliography",
http://home.earthlink.net/ jpdowling/pbgbib.html.

[2.12] Y. Vlasov,
"The ultimate collection of photonic band gap research links",
http://www.pbglink.com.

[2.13] S. Ramo, J. Whinnery, and T.Van Duzer,
"Fields and waves in communications electronics",
Wiley, 1994

[2.14] G. Agrawal,
"Fiber-optic communications systems",
Wiley Interscience, second ed., 1997.

[2.15] E. Yablonovitch,
"Inhibited spontaneous emission in solid-state physics and electronics",
Physical Review Letters, Vol.58, pp.2059-62, May 1987.

[2.16] S. John,
"Strong localization of photons in certain disordered dielectric superlattices",
Physical Review Letters, Vol. 58, no. 23, pp. 2486-9, 1987.

[2.17] D. Chigrin, A. Lavrinenko, D. Yarotsky, and S. Gaponenko,
"Observation of total omni-directional reflection from a one-dimensional dielectric lattice",
Appl. Phys. A, vol. 68, pp. 25-28, 1999.

[2.18] Y. Fink, J. Winn, S. Fan, C. Chen, J. Michel, J. Joannopoulos, and E. Thomas,
"A dielectric omnidirectional reflector",
Science, vol. 282, pp. 1679-1682, Nov. 1999.

[2.19] T. Søndergaard, J. Broeng, A. Bjarklev, K. Dridi, and S. Barkou,
"Suppression of spontaneous emission for a two-dimensional honeycomb photonic band gap structure estimated using a new effective-index model",
IEEE Journal of Quantum Electronics, vol. 34, pp. 2308-2313, Dec. 1998.

[2.20] T. Krauss, R. De La Rue, and S. Brand,
"Two-dimensional photonic-bandgap structures operating at near-infrared wavelengths",
Nature, vol. 383, pp. 699-702, 24 Oct. 1996.

[2.21] M. Plihal and A. Maradudin,
"Photonic band structure of two-dimensional systems: The triangular lattice,"
Physical Review B, vol. 44, pp. 8565-8571, Oct. 1991.

[2.22] S. McCall, P. Platzman, R. Dalichaouch, D. Smith, and S. Schultz,
"Microwave propagation in two-dimensional dielectric lattices",
Physical Review Letters, vol. 67, pp. 2017-2020, Oct. 1991.

[2.23] R. Meade, K. Brommer, A. Rappe, and J. Joannopoulos,
"Existence of a photonic band gap in two dimensions",
Applied Physics Letters, vol. 61, pp. 495-497, July 1992.

[2.24] P. Villeneuve and M. Piche,
"Photonic band gaps in two-dimensional square and hexagonal lattices",
Physical Review B, vol. 46, pp. 4969-4972, Aug. 1992.

[2.25] A. Maradudin and A. McGurn,
"Photonic band structure of a truncated, two-dimensional, periodic dielectric medium",
Journal of the Optical Society of America B, vol. 10, pp. 307- 313, Feb. 1993.

[2.26] J. Gerard, A. 1zrael, J. Marzin, R. Padjen, and F. Ladan,
"Photonic bandgap of two- dimensional dielectric crystals",
Solid-State Electronics, vol. 37, pp. 1341-1344, April-June 1994.

[2.27] H.- B. Lin, R. Tonucci, and A. Campillo,
"Observation of two-dimensional photonic band behavior in the visible",
Applied Physics Letters, vol. 68, pp. 2927-2929, May 1996.

[2.28] C. Anderson and K. Giapis,
"Larger two-dimensional photonic band gaps",
Physical Review Letters, Vol. 77, pp. 2949-2952, Sept. 1996.

[2.29] A. Rosenberg, R. Tonucci, H.-B. Lin, and A. Campillo,
"Near-infrared two-dimensional photonic band-gap materials",
Optics Letters, Vol. 21, pp. 830-832, June 1996.

[2.30] K. Inoue, M. Wada, K. Sakoda, M. Hayashi, T. Fukushima, and A. Yamanaka, "Near- infrared photonic band gap of two-dimensional triangular air-rod lattices as revealed by transmittance measurement", *Physical Review B,* Vol. 53, pp. 1010-1013, Jan. 1996.

[2.31] V. Kuzmiak, A. A. Maradudin, and A. R. McGurn, "Photonic band structures of two-dimensional systems fabricated from rods of a cubic polar crystal", *Physical Review - Section B - Condensed Matter*, Vol. 55, pp. 4298-4311, Feb. 1997.

[2.32] V. Kuzmiak and A. A. Maradudin, "Photonic band structures of one- and two-dimensional periodic systems with metallic components in the presence of dissipation", *Physical Review - Section B - Condensed Matter*, Vol. 55, pp.7427-7444, Mar. 1997.

[2.33] D. Labilloy, H. Benisty, C. Weisbuch, T. Krauss, R. Houdre, and U. Oesterle, "Use of guided spontaneous emission of a semiconductor to probe the optical properties of two-dimensional photonic crystals", *Applied Physics Letters,* Vol. 71, pp. 738-740, Aug. 1997.

[2.34] D. Cassagne, C. Jouanin, and D. Bertho, "Optical properties of two-dimensional photonic crystals with graphite structure", *Applied Physics Letters,* Vol. 70, pp. 289-291, Ian. 1997.

[2.35] A. Barra, D. Cassagne, and C. Jouanin, "Existence of two-dimensional absolute photonic band gaps in the visible", *Applied Physics Letters,* Vol. 72, pp. 627-629, Feb. 1998.

[2.36] M. Charlton and G. Parker, "Nanofabrication of advanced waveguide structures incorporating a visible photonic band gap", *J. Micromech. Microeng.,* Vol. 8, pp. 172-176, Jun. 1998.

[2.37] J. Nielsen, T. Søndergaard, S. Barkou, A. Bjarklev, J. Broeng, and M. Nielsen, "The two-dimensional Kagome structure, a new fundamental hexagonal photonic crystal configuration", *IEE Electronics Letters,* Vol. 35, pp. 1736-1737, Sept. 1999.

[2.38] T. Birks, P. Roberts, P. Russell, D. Atkin, and T. Shepherd, "Full 2-d photonic bandgaps in silica/air structures", *IEE Electronics Letters,* Vol. 31, pp. 1941-1943, Oct. 1995.

[2.39] J. Broeng, S. Barkou, A. Bjarklev, J. Knight, T. Birks, and P. Russell, "Highly increased photonic band gaps in silica/air structures", *Optics Communications,* Vol. 156, pp. 240-244, Nov. 1998.

[2.40] A. Maradudin and A. McGurn,
"Out of plane propagation of electromagnetic waves in a two-dimensional periodic dielectric medium",
Journal of Modern Optics, Vol. 41, pp. 275- 284, Feb. 1994.

[2.41] P. Roberts, T. Birks, P. Russell, T. Shepherd, and D. Atkin,
"Two-dimensional photonic band-gap structures as quasi-metals",
Optics Letters, Vol. 21, pp. 507-509, April 1996.

[2.42] X. Feng and Y. Arakawa,
"Off-plane angle dependence of photonic band gap in a two-dimensional photonic crystal",
IEEE Journal of Quantum Electronics, Vol. 32, pp. 535-542, Mar. 1996.

[2.43] A. Mekis, J. Chen, I. Kurland, S. Fan, P. R. Villeneuve, and J. D. Joannopoulos,
"High transmission through sharp bends in photonic crystal waveguides",
Physical Review Letters, Vol. 77, pp. 3787-3790, *act.* 1996.

[2.44] R. Ziolkowski,
"FDTD modelling of photonic nanometer-sized power splitters and switches",
Integrated Photonics Research, Vol. 4 of Technical Digest Series, p. IThA2, OSA, March 30 -April1 1998.

[2.45] S. Fan, P. Villeneuve, J. Joannopoulos, and H. Hails,
"Channel drop tunneling through localized states",
Physical Review Letters, Vol. 80, pp. 960-963, Feb. 1998.

[2.46] O. Painter, R. Lee, A. Scherer, A. Yariv, P. O'Brien, J. D. Dapkus, and I. Kim,
"Two-dimensional photonic band-gap defect mode laser",
Science, Vol. 284, pp. 1819-1821, June 1999.

[2.47] P. Russell, J. Pottage, J. Broeng, D. Mogilevtsev, and P. Philips,
"Leak-free bound modes in photonic crystal films",
CLEO'99: Conference on Lasers and Electro-Optics, p. CWF65, May 1999.

[2.48] S. Johnson, S. Fan, P. Villeneuve, J. Joannopoulos, and L. Kolodziejski,
"Guided modes in photonic crystal slabs",
Physical Review B, Vol. 60, pp. 5751-5758, Aug. 1999.

[2.49] O. Martin, C. Girard, D. Smith, and S. Schultz,
"Generalized field propagator for arbitrary finite-size photonic band gap structures",
Phys. Rev. Lett., Vol. 82, pp. 315-318, Jan. 1999.

[2.50] S. Lin and G. Arjavalingam,
"Photonic bound states in two-dimensional photonic crystals probed by coherent-microwave transient spectroscopy,"
Journal of the Optical Society of America B, vol. 11, pp. 2124-7, Oct. 1994.

[2.51] E. Yablonovitch, T. Gmitter, and K. Leung,
"Photonic band structure: the face-centered-cubic case employing nonspherical atoms",
Physical Review Letters, Vol. 67, pp. 2295-2298, Oct. 1991.

[2.52] E. Ozbay, E. Michel, G. Tuttle, R. Biswas, M. Sigalas, and K.-M. Ho,
"Micromachined millimeter-wave photonic band-gap crystals",
Applied Physics Letters, Vol. 64, pp. 2059- 2061, April 1994.

[2.53] E. Brown, C. Parker, and E. Yablonovitch,
"Radiation properties of a planar antenna on a photonic-crystal substrate",
Journal of the Optical Society of America B, Vol. 10, pp. 404-407, Feb. 1993.

[2.54] S. Cheng, R. Biswas, E. Ozbay, S. McCalmont, G. Tilttle, and K-M. Ho,
"Optimized dipole antennas on photonic band gap crystals",
Applied Physics Letters, Vol. 67, pp. 3399-3401, Dec. 1995.

[2.55] L. Jasper and G. Tran,
"Photonic band gap (pbg) technology for antennas",
SPIE, vol. 7, Iuly 1996.

[2.56] M. Kesler, J. Maloney, B. Shirley, and G. Smith,
"Antenna design with the use of photonic band-gap materials as all-dielectric planar reflectors",
Microwave and Optical Technology Letters, Vol. II, pp. 169-174, March 1996.

[2.57] Y. Qian, R. Coccioli, D. Sievenpiper, V. Radisic, E. Yablonovitch, and T. Itoh,
"A microstrip patch antenna using novel photonic band-gap structures",
MICROWAVE J., Vol. 42, p. 66, Jan. 1999.

[2.58] U. Gruning and V. Lehmann,
"Two-dimensional infrared photonic crystal based on macroporous silicon",
Thin Solid Films, Vol. 276, no. 151, 1996.

[2.59] J. Knight, T. Birks, P. Russell, and D. Atkin,
"All-silica single- mode optical fiber with photonic crystal cladding",
Optics Letters, Vol. 21, pp. 1547-1549, Oct. 1996.

[2.60] S. Rowson, A. Chelnokov, and *J.*-M. Lourtioz,
"Macroporous silicon photonic crystals at 1.55 micrometers"
IEE Electronics Letters, Vol. 35, No. 9, pp. 753-755, Apr. 1999.

[2.61] S. Lin, J. Fleming, D. Hetherington, B. Smith, R. Biswas, K. Ho, M. Sigalas, W. Zubrzycki, S. Kurtz, and J. Bur,
"A three-dimensional photonic crystal operating at infrared wavelengths",
Nature, Vol. 394, pp. 251-253, *July* 1998.

[2.62] A. A. Zakhidov, R. Baughman, Z. Iqbal, C. Cui, I. Khayrullin, S. Dantas, J. Marti, and V. Ralchenko,
"Carbon structures with three-dimensional periodicity at optical wavelengths",
Science, Vol. 5390, pp. 897-901, Oct. 1998.

[2.63] S. Kawakami,
"Fabrication of submicrometere 3d periodic structures composed of Si/Si02",
IEE Electronics Letters, Vol. 33, p. 1260, 1997.

[2.64] J. Fleming and S. Lin,
"Three-dimensional photonic crystal with a stop band from 1.35 to 1.95mu m,"
Optics Letters, Vol. 24, pp. 49-51, Jan. 1999.

[2.65] P R. Fan, S. Villeneuve, and J. D. Joannopoulos,
"Theoretical investigation of fabrication-related disorder on the properties of photonic crystals",
J. Appl. Phys., Vol. 78, p. 1415, 1995.

[2.66] A. Chutinan and S. Noda,
"Effects of structural fluctuations on the photonic bandgap during fabrication of a photonic crystal",
J. Opt. Soc. Am. B, Vol. 16, pp. 240-244, Feb. 1999.

[2.67] A. Yariv,
"Propagation, modulation and oscillation in optical dielectric waveguides",
Saunders Clooege Publishing, 1991.

[2.68] P. Villeneuve, S. Fan, J. Joannopoulos, K.-Y. Lim, G. Petrich, L. Kolodziejski, and R. Reif,
"Air-bridge microcavities",
Applied Physics Letters, Vol. 67, pp. 167-169, July 1995.

[2.69] A. Yariv,
"Quantum Electronics",
Wiley, 1988.

[2.70] R. Coccioli, M. Boroditsky, K. Kim, Y. Rahmat-Samii, and E. Yablonovitch,
"Smallest possible electromagnetic mode volume in a dielectric cavity",
lEE Proc. -Optoelectron. , Vol. 145 , Dec. 1998.

[2.71] B. D'Urso, 0. Painter, J. O'Brien, T. Tombrello, A. Yariv, and A. Scherer,
"Modal reflectivity in finite-depth two-dimensional photonic-crystal microcavities",
J. Opt. Soc. Am. B, Vol. 15, pp. 1155-1159, Mar. 1998.

[2.72] M. Boroditsky and E. Coccioli, R. Yablonovitch,
"Analysis of photonic crystals for light emitting diodes using the finite difference time domain technique",
SPIE-Int. Soc. Opt. Eng., Vol. 3283, pp. 184-190, 1998.

[2.73] H. Hirayama, T. Hamano, and Y. Aoyagi,
"Novel spontaneous emission control using 3-dimensional photonic bandgap crystal cavity",
Mat. Sci. Eng. B-Solid, vol. 51, pp. 99-102, Feb. 1998.

[2.74] K. Yoshino, S. Tatsuhara, Y. Kawagishi, M. Ozaki, A. Zakhidov, and Z. Vardeny,
"Amplified spontaneous emission and lasing in conducting polymers and fluorescent dyes in opals as photonic crystals",
Appl. Phys. Lett., Vol. 74, pp. 2590-2592, May 1998.

[2.75] K. Yoshino, S. Lee, S. Tatsuhara, Y. Kawagishi, M. Ozaki, and A. Zakhidov,
"Observation of inhibited spontaneous emission and stimulated emission of rhodamine 6G in polymer replica of synthetic opal",
Appl. Phys. Lett., Vol. 73, pp. 3506-3508, Dec. 1998.

[2.76] C. Smith, H. Benisty, D. Labilloy, U. Oesterle, R. Houdre, T. Krauss, R. De la Rue, and C. Weisbuch,
"Near-infrared micro cavities confined by two-dimensional photonic bandgap crystals" ,
IEE Electronics Letters, Vol. 35, pp. 228-230, Feb. 1999.

[2.77] R. Lee, 0. Painter, B. D'Urso, A. Scherer, and A. Yariv,
"Measurement of spontaneous emission from a two-dimensional photonic band gap defined micro cavity at near-infrared wavelengths",
Appl. Phys. Lett., Vol. 74, pp. 1522-1524, Mar. 1999.

[2.78] T. Søndergaard, J. Broeng, and A. Bjarklev,
"Suppression of spontaneous emission for two-dimensional GaAs photonic crystal mirocavities",
CLEO'99: Conference on Lasers and Electro Optics, p. 525, May 1999. Paper JFA2.

[2.79] H. Benisty,
"Modal analysis of optical guides with two-dimensional photonic band-gap boundaries",
Journal of Applied Physics, Vol. 79, pp. 7483-7492, May 1996.

[2.80] S.-Y. Lin, E. Chow, V. Hietala, P. Villeneuve, and J. Joannopoulos,
"Experimental demonstration of guiding and bending of electromagnetic waves in a photonic crystal",
Science, Vol. 282, pp. 274-276, *Oct.* 1998.

[2.81] O. Hanazumi, Y. Ohtera, T. Sato, and S. Kawakami,
"Propagation of light beams along line defects formed in *a-Si/Si02* three-dimensional photonic crystals: Fabrication and observation",
Applied Physics Letters, Vol. 74, No.6, pp. 777-779, 1999.

[2.82] S. Enoch, G. Tayeb, and D. Maystre,
"Numerical evidence of ultrarefractive optics in photonic crystals",
Optics Communcations, Vol. 161, No.4-6, pp. 171-176, 1999.

[2.83] S. Lin, V. Hietala, L. Wang, and E. Jones,
"Highly dispersive photonic band-gap prism",
Optics Letters, Vol. 21, pp. 1771-1773, 1996.

[2.84] H. Kosaka, T. Kawashima, T. Tomita, M. Notomi, T. Tamamura, T. Sato, and S. Kawakami,
"Superprism phenomena in photonic crystals",
Phys. Rev. B, Vol. 58, pp. 10096-10099, 1998.

[2.85] J. Mathews, and R. Walker,
"Mathematical methods of physics",
Addison-Wesley, Redwood City, California, 1964.

[2.86] C. Kittel,
"Solid state physics",
John Wiley & Sons, New York, 1986.

[2.87] S. I. Bozhevolnyi, J. Erland, K. Leosson K, et al.
"Waveguiding in surface plasmon polariton band gap structures",
Phys. Rev. Lett., Vol. 86, No.14, pp. 3008-3011, Apr. 2001.

[2.88] S. G. Romanov, T. Maka, C. M. S. Torres,
"Suppression of spontaneous emission in incomplete opaline photonic crystal",
J. Appl. Phys., Vol. 91, No.11, pp. 9426-9428, Jun. 2002.

[2.89] H. Notomi, A. Shinya, K. Yamada,
"Singlemode transmission within photonic bandgap of width-varied single-line-defect photonic crystal waveguides on SOI substrates",
IEE Electronics Letters, Vol. 37, No.5, pp. 293-295, Mar 2001.

[2.90] T. Søndergaard, A. Bjarklev, M. Kristensen, J.Erland Østergaard, and J.Broeng,
"Designing finite-height two-dimensional photonic crystal waveguides",
Appl. Phys. Lett., Vol. 77, No. 6, pp. 785-787, Aug. 2000.

[2.91] Sakoda K, Sasada M, Fukushima T, Yamanaka A, Kawai N, Inoue K,
"Detailed analysis of transmission spectra and Bragg-reflection spectra of a two-dimensional photonic crystal with a lattice constant of 1.15 microns",
J. Opt. Soc. Am. B, Vol. 16, No.3, pp. 361-365, Mar. 1999.

[2.92] C. S. Kee, J. E. Kim, H. Y. Park, K. J. Chang, H. Lim,
"Essential role of impedance in the formation of acoustic band gaps"
J. Appl. Phys., Vol. 87, No.4, pp. 1593-1596, Feb. 2000.

[2.93] P. Villeneuve, and M. Piche,
"Photonic bandgaps: What is the best numerical representation of periodic structures?,"
Journal of Modern Optics, vol. 41, pp. 241-256, Feb. 1994.

[2.94] R. Meade, A. Rappe, K. Brommer, and J. Joannopoulos, "Nature of the photonic band gap: some insights from a field analysis," *Journal of the Optical Society of America B,* vol. 10, pp. 328-332, Feb. 1993.

[2.95] R. Meade, A. Rappe, K. Brommer, J. Joannopoulos, and 0. Alerhand, "Accurate theoretical analysis of photonic band-gap materials", *Physical Review B,* vol. 48, pp. 8434-8437, Sept. 1993.

[2.96] H. Sözüer, J. Haus, and R. Inguva, "Photonic bands: convergence problems with the plane-wave method", *Physical Review B,* vol. 45, pp. 13962-13972, June 1992.

[2.97] M. Teter, M. Payne, and D. AlIan, "Solution of Schrödinger's equation for large systems," *Physical Review B,* 1989.

[2.98] D. Aspnes, "Local-field effects and effective- medium theory: A microscopic perspective," *Am.]. Phys.,* vol. 50, pp. 704-709, Aug.1981.

Chapter 3

THEORY AND MODELLING OF MICROSTRUCTURED FIBRES

3.1 INTRODUCTION

In the development of a new research field, it is often very fruitful to have a close interaction between theory and experimental work, and the area of photonic crystal fibres is in no way different on this issue. It is, therefore, an integrated part of the research development to study the theoretical treatment of the problems and challenges leading to the understanding of these new fibres, and as a key element of this work, the numerical modelling of the complex fibre structures become important.

When we wish to address the issue of modelling, it would be natural to consider the methods known from standard waveguide or fibre analysis. However, the first key point is here to realize that the complex nature of the cladding (and sometimes also core) structure of the photonic crystal fibres (PCFs) does generally not allow for the direct use of methods from traditional fibre theory. Especially for the novel PCF, operating by the photonic bandgap (PBG) effect, the full vectorial nature of the electromagnetic waves has to be taken into account, and is most commonly treated by a method closely related to the plane-wave methods used for calculating electronic bandgaps in semiconductors.

Keeping this overall picture in mind, it is on the other hand still possible to use some elements from standard-fibre modelling, and in order to initiate the description of models for PCFs in a manner as close to the generally known approach applied on optical fibres, we have chosen first to look at an approximate method. For the index-guiding PCFs, it is namely possible to apply a simpler scalar model, based on an effective-index of the cladding. This method has proven to give a good qualitative description of the operation of the optical photonic crystal fibre [3.1-3.2]. We will exemplify the use of this model on the high-index core triangular PCFs.

It is, however, also clear that the approximate approaches has a limited application range, and in order to be able to design and accurately predict the properties of new PCF structures, more powerful tools must be applied. The second method to be described in this chapter is the method of localized basis functions [3.13]. In this approach, the mathematical formulation takes point of reference in the spatially localized nature of the modes guided along the core of the optical photonic crystal fibre. The localized-function approach has been mostly applied for calculating the properties of index-guiding PCFs, and analysis of many different structures have been reported.

The third section of this chapter contains a presentation of one of the most widely used tools for the analysis of photonic crystal fibres, namely the full-vectorial plane-wave expansion method [3.24-3.25]. This method is highly suited for the analysis of periodic structures (see also [3.37-3.40]), and it has as such been one of the major tools for the analysis and understanding of photonic bandgap fibres (and planar PBG structures as well). We have, consequently, in this book chosen to place the major effort concerning numerical methods on the plane-wave approach. This choice is also made, because much of the mathematical and numerical considerations related to the different methods contain some of the same fundamental elements, and they may, therefore, be clarified through the description of the full-vectorial plane-wave method.

Another highly accurate numerical method within the discipline of photonic crystal fibres is the biorthonormal-basis method [3.43-3.45], as described in Section 3.5. This approach has its roots in physics, where it has been applied in the solutions of the Schrödinger equation, and it has been extensively used for accurate analysis of photonic crystal fibre properties. Especially has it been applied in accurate calculations of numerically accurate dispersion properties of PCFs.

Among the most recent developments concerning the modelling of photonic crystal fibres has been the multipole method [3.20-3.22]. This method is of the type, which is well known from analysis of electromagnetic scattering problems. The key idea of the method is to treat every single sub-element (or hole) as a scattering element around which the surrounding electromagnetic field may be expressed. Applying rules from Maxwell´s equations to combine the individual field distributions, a system of equations is formed, which leads to the final overall solution. This method is briefly outlined in Section 3.6. A significant advantage of the multipole method is its applicability in the prediction of leakage losses in PCFs.

Section 3.7 describes the Fourier-decomposition method, which is another recently developed approach [3.23] used for the prediction of the confinement loss of silica-air structured optical fibres. In short, the method is based on the solution of a system of equations, which couples field solutions in the central part of the fibre to outward radiating waves in the outer parts of the fibre.

After this presentation of the – to date – most extensively used numerical methods for analysis of photonic crystal fibres, the following sections contain short descriptions of what may be seen as methods, which are well established in the general area of optics, but only has been used to a limited extent in the analysis of PCFs. The first class of methods is the so called finite-difference methods, as outlined in Section 3.8. This approach is well established in the field of integrated optics, and may be formulated either as a time-domain or a frequency-domain method. Section 3.9 describes the so called Finite-Element Method (FEM) which is a widely used method within the area of mechanical design, but which also have been extensively used for the solution of large electromagnetic problems. The third of the "more classical" numerical methods is the Beam-Propagation Method (BPM), which is very briefly outlined in Section 3.10. In this method, which more simplistically may be seen as a step-wise evaluation of the electromagnetic field through concatenated sections consisting of equivalent lenses and homogeneous media, also longitudinal variations - such as gratings – may be described.

The final method, which is shortly mentioned in this chapter, is the recently developed qualitative approach called the Equivalent-Averaged-Index Method (EAIM). As the name indicates, the index profile is averaged in a manner so that each radial point represents an azimuthal averaging of the refractive index. This method is discussed at this point in the chapter, because it so far has not been applied on many structures, and it is, furthermore, a qualitative approach with a number of limitations to its applicability.

The chapter, finally, contains an overview of the key points of the discussed methods – presented on the form of an overview table.

3.2 THE EFFECTIVE-INDEX APPROACH

In order to establish a relatively simple numerical tool that could provide qualitative mode-propagation properties of the high-index core triangular PCFs, Birks *et al.* [3.2] proposed a method in which sequential use of well-established fibre tools was applied. The fundamental idea behind this work was to first evaluate the periodically repeated hole-in-silica structure of the cladding and then (based on the approximate waveguiding properties of this cladding structure) replace the cladding by a properly chosen effective index. In Figure 3-1, a schematic illustration of the replacement of the index guiding photonic crystal fibre structure and an equivalent index profile is shown. In this model, the resulting waveguide then consists of a core and a cladding region that have refractive indices n_{co} and n_{cl}, respectively. The core is pure silica, but the definition of the refractive index of the microstructured cladding region is given in terms of the propagation constant of the lowest-order mode that could propagate in the infinite cladding material. We will now briefly review this scalar effective-index method.

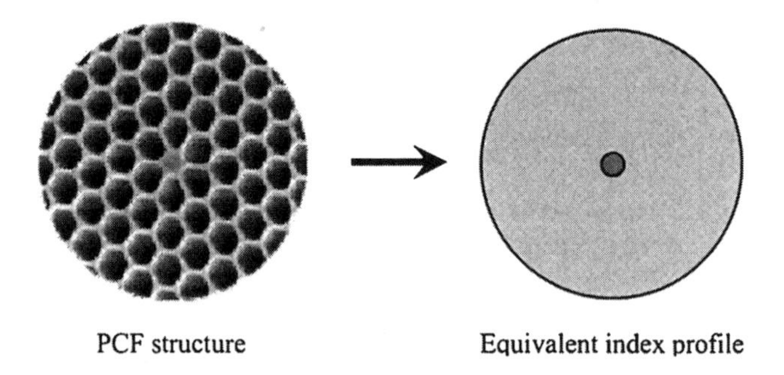

Figure 3-1. Schematic representation of the replacement of the transversal cross-section of a PCF structure with an equivalent effective-index profile.

The first step of the effective-index method is to determine the cladding mode field, Ψ, by solving the scalar-wave equation within a simple cell centred on one of the holes. The diameter of these cells equals the pitch, Λ, between the holes of the cladding structure, and as illustrated in Figure 3-2, their hexagonal shapes are approximated by circular areas in order to make a general circular symmetric mode solution possible. By reflection symmetry, the boundary condition at the cell edge (at radius $\Lambda/2$) is $d\Psi/ds = 0$, where s is the coordinate normal to the edge. This means that in the corners of the original hexagonal-shaped cells, we have to imagine constant fields with values equal to those at the circular cell boundary.

The propagation constant of the resulting fundamental space-filling mode, β_{fsm} is used to define the effective index of the cladding as

$$n_{eff} = \frac{\beta_{fsm}}{k} \tag{3.1}$$

where k is the free-space propagation constant of light with wavelength λ. Figure 3-2 further shows an example of the resulting field distribution in the cladding cell. It should also be noted that in the calculation of this cladding field (together with the effective-index value), we have used the normal weakly guiding field assumption [3.4], although the index step between the central hole (having a refractive index of 1.0) and the surrounding silica (with refractive index around 1.45) actually is considerable. However, the hereby introduced inaccuracy is considered to be less significant than the approximation of the guided-mode field in the effective-index fibre compared to the actual field in the PCF. Note also that the primary aim of the effective-index method (EIM) is to form a qualitative description of the guiding properties, which makes the outlined assumptions acceptable.

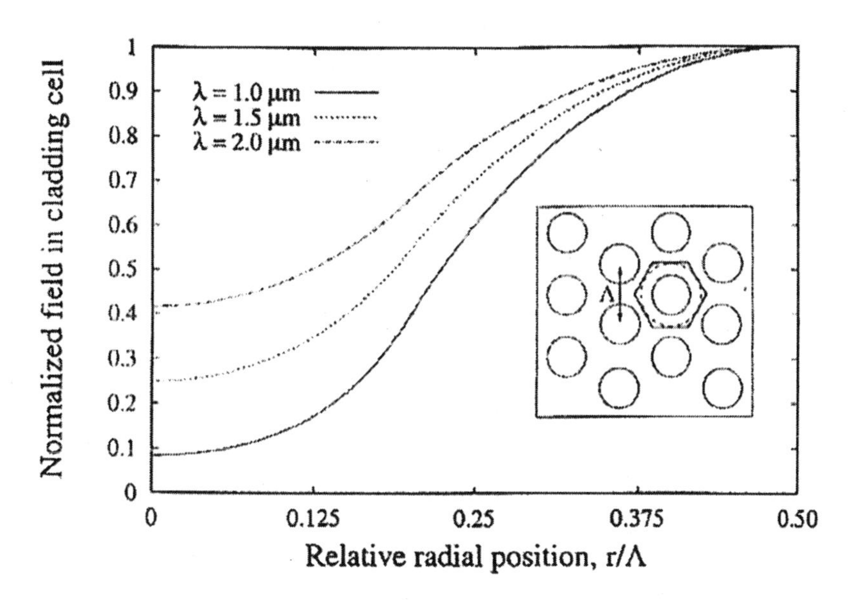

Figure 3-2. Mode distributions in cladding cell for a silica PCF with a pitch of 2.3 μm, and an air-hole size given by $d/\Lambda = 0.40$ (fibre parameters comparable to those presented by Birks *et al.* in [3.5]). The modes are calculated at wavelengths 1.0 μm, 1.5 μm, and 2.0 μm. The inset shows the hexagonally shaped cladding cell with its inscribed circular approximation. A slightly different circular approximation was used in [3.5].

Now having determined the cladding- and core-index values, we may calculate the approximate propagation properties of the PCF as for a step-index fibre with core index n_{co}, core radius $\Lambda/2$, and cladding index $n_{cl} = n_{eff}$. As an extension to the cladding-mode model originally described in [3.5], it was in [3.3] added that the refractive index for silica may be described as being wavelength dependent. This was done through the introduction of the generally applied Sellmeier formula [3.6]:

$$n^2 = 1 + \sum_{i=1}^{3} \frac{A_i \lambda^2}{\lambda^2 - B_i} \tag{3.2}$$

where n is the refractive index, and A_i and B_i denote the Sellmeier coefficients. The values for the Sellmeier coefficients of pure silica are shown in Table 3-1.

	Pure SiO2
A_1	0.69675
A_2	0.408218
A_3	0.890815
B_1	0.0047701
B_2	0.0133777
B_3	98.02107

Table 3-1. Sellmeier coefficients for pure silica.

The described procedure generally provides an effective cladding index of the fundamental cladding mode, which is strongly wavelength-dependant. The core of the equivalent step-index fibre is normally assumed to have the refractive index, which is constant, or only varies with wavelength according to the Sellmeier expression [3.6]. The core is often considered to be of pure silica, and as illustrated in Figure 3-3, the wavelength dependency of the refractive index of silica (corresponding to very small relative hole sizes) is much weaker than that of the effective cladding index of the microstructured part of the waveguide, when the relative hole sizes becomes significant.

It is here relevant to refer to the work of Midrio *et al.* [3.83], who in 2000 derived the dispersion relations of the modes guided by an infinite self-similar air-hole lattice. Midrio *et al.* focussed in particular on the fundamental mode (the so-called space-filling mode), and showed that previous numerical results based on vector methods are accurate, but scalar ones are not. To be a little more specific, Midrio *et al.* [3.83] showed that the solutions to characteristic equations formulated from Maxwell's equations for the fundamental mode may be divided into two sets, analogous to the EH and HE modes of step-index fibres. The dispersion relations derived by Midrio *et al.* [3.83] for these modes may be expressed as follows:

$$\frac{I_{l+1}(w)}{I_l(w)} = -\frac{l}{w} - \frac{w}{2}\left(A_l' + \frac{\varepsilon_{r2}}{\varepsilon_{r1}} B_l'\right) - w\sqrt{\frac{1}{4}\left(A_l' - \frac{\varepsilon_{r2}}{\varepsilon_{r1}} B_l'\right)^2 + \frac{f^2(u,w,l)}{\varepsilon_{r1}}} \tag{3.3a}$$

and

$$\frac{I_{l-1}(w)}{I_l(w)} = \frac{l}{w} - \frac{w}{2}\left(A_l' + \frac{\varepsilon_{r2}}{\varepsilon_{r1}} B_l'\right) + w\sqrt{\frac{1}{4}\left(A_l' - \frac{\varepsilon_{r2}}{\varepsilon_{r1}} B_l'\right)^2 + \frac{f^2(u,w,l)}{\varepsilon_{r1}}} \quad (3.3b)$$

where $w^2 = \omega^2\left(n_{eff}^2 - n_1^2\right)a^2/c^2$, and the quantities called A_l' and B_l' are functions of the parameter u, which is generally known as the normalized attenuation coefficient of standard optical fibres and given as $u^2 = \omega^2\left(n_{e2f}^2 - n_{eff}^2\right)a^2/c^2$. Here the effective index n_{eff} is given from the propagation constant $\beta = (\omega/c)n_{eff}$, ω is the angular frequency, and c is the velocity of light in vacuum. Furthermore, l is the azimuthal mode index, a is the radius of an air hole (e.g., $a = d/2$), and the function $f^2(u,w,l) = l^2\left(\frac{1}{u^2} + \frac{1}{w^2}\right)\left(\frac{\varepsilon_{r1}}{w^2} + \frac{\varepsilon_{r2}}{u^2}\right)$. It should be noted that w in standard fibre theory describing circular symmetric waveguides is known as the normalized transverse propagation constant, and it is related to the normalized frequency v as $v^2 = u^2 + w^2$. The permittivities of the air hole ε_{r1} and of the cladding base material ε_{r2} are also an element of Eqn. (3.3) as well as the function $I_l(\ )$ defines the l-th order modified Bessel function of the first kind. For further details the reader should consult [3.83].

It should be noted that by analogy with step-index fibres Midrio *et al.* [3.83] have names the modes corresponding to solutions of Eqn. (3.3a) as "EH" and those corresponding to solutions of Eqn. (3.3b) was named "HE". Midrio *et al.* analysed the electromagnetic nature of the fundamental space-filling mode (FSFM), and provided a quantitative estimate of the accuracy of the plane-wave expansion technique on this basis. The results indicated that – in a silica-air lattice having a pitch-to-wavelength ratio of about 1.5 – filed expansions over roughly 1000 plane waves are required to get a $5 \cdot 10^{-4}$ accuracy on the estimation of the mode effective index.

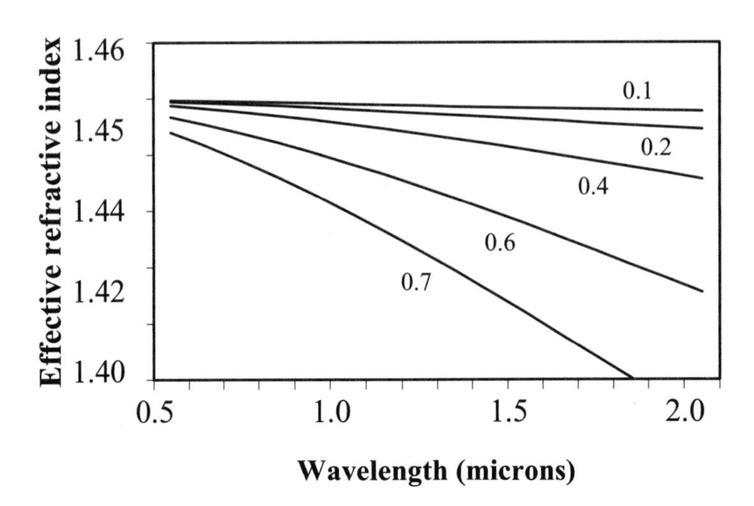

Figure 3-3. Illustration of the wavelength dependency of the effective refractive index of the cladding for different relative hole sizes (values of d/Λ are indicated) for a cladding cell radius of 2.65 μm. In all examples, the triangular (or close-packed) cladding structure is considered.

One difficulty that arises, when using the effective-index approach is to answer the question of how to define the properties of the equivalent step-index fibre, and more specifically, the correct equivalent core radius. A widely applied approach is to choose the core radius of the equivalent structure to be 0.62Λ as found by Birks *et al.* [3.82]. This is done in order for the results obtained using the effective index method to agree well with full simulations via appropriate choice of this constant of proportionality. However, for different (i.e., not closely packed) structures, different choices of equivalent core radius can become necessary. This restricts the usefulness of the effective-index model, since it is typically necessary to determine the best choice of the equivalent structure by referring to results from a more complete numerical model. In [3.8] Riishede *et al.* have explored the possibility of choosing the step-index fibre parameters in a more general fashion by allowing a wavelength dependent core or cladding index. However, to date no entirely satisfactory method for ascribing parameters to the equivalent fibre has been found.

The discussion concerning the equivalent core radius generally assumes that the core is created by the omission of one cladding air hole. Note, however, further that the core could be fabricated of doped glass, which then would call for a correspondingly modified refractive index value, or it could even consist of a structure with air holes, in which case an effective core index could be defined along the lines of the already described procedure.

Despite ignoring the detailed spatial distribution of the refractive index profiles within photonic crystal fibres, the effective-index method can provide useful insight into the qualitative properties of PCF operation. For example, the effective method correctly predicts the endlessly single-mode guidance regime in small-hole PCFs [3.2]. This method has also been used as a basis for the approximate dispersion and bending analysis presented in [3.3]. We will return to these properties in detail in a later chapter. However, this reduced (and scalar) model cannot accurately predict modal properties such as dispersion, birefringence or other polarisation properties that depend critically on the whole configuration within the cladding (or core).

It should, finally, be noted that recent developments in the fabrication of fibre structures with relatively large air holes has made it relevant to approximate the fibre by an isolated strand of silica surrounded by air [3.7], and in such cases approximate methods may provide useful insight concerning the fibre properties.

3.3 THE METHOD OF LOCALIZED BASIS FUNCTIONS

Despite the good qualitative information provided by the simple effective-index model, advanced numerical methods must be used for highly accurate modelling of the PCFs. In 1998, Mogilevtsev *et al.* [3.13] developed a method based on the direct solution of Maxwell's equations, using a representation of the refractive index and the field distributions as sums of localized basis functions. This was initially developed for the modelling of triangular PCFs, but it has been extended and further developed by other scientists (see e.g., [3.15]), who also has applied it for numerous photonic crystal fibre structures since then. We will in this section shortly review the basic elements of the method of localized basis functions.

The guided modes of the PCFs are localized in the close vicinity of the area forming the core. Note that this core area - in literature on photonic bandgap waveguides - often is referred to as the defect, indicating that in this place, the otherwise perfectly periodic structure is broken, i.e., if a hole is replaced by a solid material or visa versa. Because of the mode field localization, it is possible to model the guided modes by representing the fields as sums of functions localized in the vicinity of the core. A clear advantage of such an approach is that for an appropriately chosen set of basis functions, only a modest number of functions are required to accurately describe the bound mode, thereby significantly lowering the demand for computational resources. To implement the method, Maxwell's equations are reformulated for a medium, which is translationally invariant along the z-axis (the length axis of the fibre) as an eigenvalue problem for the propagation constant [3.14],

$$\left(\nabla_{\perp}^{2}+k^{2}\varepsilon\right)\bar{h}_{\perp}+\left(\nabla_{\perp}\ln(\varepsilon)\right)\times\left(\nabla_{\perp}\times\bar{h}_{\perp}\right)=\beta^{2}\bar{h}_{\perp} \tag{3.4}$$

where $\nabla_{\perp}$ denotes the gradient in the periodic *xy*-plane (note that $\nabla_{\perp}^{2}=\partial^{2}/\partial x^{2}+\partial^{2}/\partial y^{2}$), k is the free-space wave number, and the components of the vector $\bar{h}_{\perp} = \left[h_x h_y\right]^T$ correspond to the transversal components of the magnetic field $\bar{H}$, which may be expressed as

$$H_i = h_i \exp\left[j(ckt-\beta z)\right], \qquad i = x, y \tag{3.5}$$

where β is the propagation constant, z is the coordinate indicating the position along the fibre, c is the velocity of light in vacuum, and t is the time.

For the system of basis functions, a set of Hermite-Gaussian functions [3.13] were used, and they may be written as follows:

$$\phi_{m,n} = \exp\left[-\frac{\left(x^2+y^2\right)}{2\Lambda^2}\right]H_m\left(\frac{x}{\Lambda}\right)H_n\left(\frac{y}{\Lambda}\right) \tag{3.6}$$

where H_m is the Hermite polynomial of the order m and Λ denotes the period of the lattice. The functions ϕ_{mn} are mutually orthogonal, and they form a complete system in the *xy* plane. They are localized in the vicinity of the point $(x,y) = (0,0)$, which then indicate that the coordinate system is placed with (0,0) in the central part of the waveguide core. Note that the Hermite-Gaussian functions not only are an advantageous choice due to their localization, but also because their overlap integrals may be evaluated analytically. In the basis of functions ϕ_{mn}, Eqn. (3.4) becomes the algebraic eigenvalue problem:

$$\sum_{k,l} L_{k,l}^{m,n}\bar{h}_{\perp}^{m,n} = \beta^2\bar{h}_{\perp}^{m,n} \tag{3.7}$$

For the vector of coefficients $\bar{h}_{\perp}^{m,n}$ representing the transversal magnetic field in the Hermite-Gaussian basis. $L_{k,l}^{m,n}$ are the matrix coefficients of the operator on the left-hand side of Eqn. (3.4) in the Hermite-Gaussian basis. They are real and may be found analytically for a wide range of lattices.

The implementation of the method becomes especially simple in the high-frequency regime, where the coupling between the orthogonal components of the field in the transversal plane becomes negligible. In this regime, a scalar approximation holds, and the eigenvalue problem Eqn.(3.7) becomes Hermitian. For calculations of the guided modes of the PCFs, the third term in Eqn.(3.4), describing the coupling between the orthogonal components of the field in the transversal plane, scales with the air-filling fraction, and for small holes the high-frequency limit is reached very quickly. For example, triangular PCFs with an air-filling fraction less than 10% are in the high-frequency regime for wavelengths less than the pitch Λ.

This technique, which originally was developed by Mogilevtsev *et al.* [3.13], takes as explained advantage of mode localization, and it is - for well confined and simple mode distributions - often more efficient than the widely used plane-wave methods, which we will review in the following section. The localized-function method, however, cannot be accurate unless the refractive index is also represented well. This is an important issue in the formulation of the model for the following reason; It may be tempting to represent the refractive index structures using localised functions (like Hermite-Gaussians) due to the feasibility of analytical evaluations of overlap integrals (see the work of Knudsen *et al.* [3.34] on this approach). However, in the work by Monro *et al.* [3.15], the transverse refractive index profile is separated into two parts; the periodic lattice of holes is described using periodic functions (cosines), and the central index defect (i.e., the core) is described using the localized Hermite-Gaussian functions. The way in which this decomposition is achieved is according to Monro *et al.* [3.15] sketched in Figure 3-4. By choosing the most natural basis sets to represent each component, the index can be reproduced both accurately and efficiently. Note that in [3.15], the squared refractive index (n^2, given by the relative dielectric constant) is decomposed rather than n as it is the physical quantity of interest, and this choice leads to much simpler forms for the overlap integrals.

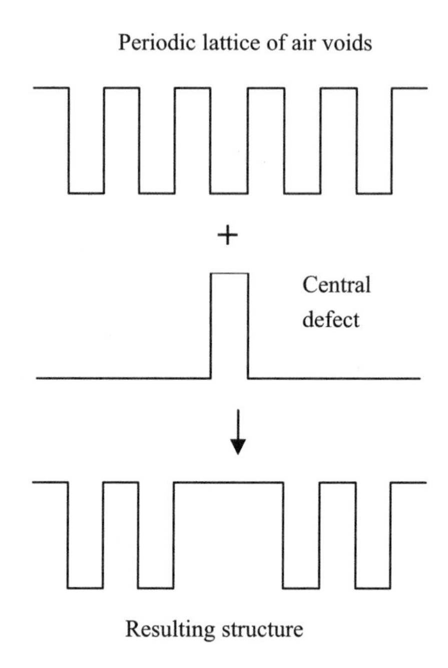

Figure 3-4. Schematic illustration of the way in which the transverse refractive index profile is decomposed in the work by Monro *et al.* [3.15].

The central index defect could be described by the same basis functions used for the mode, as they form a complete basis set. However, it has been found more efficient to choose functions with a characteristic width *(ω_d)*, which is proportional to the defect diameter (*d*) rather than the lattice spacing (Λ), as then fewer terms are required for an accurate description. The constant of proportionality is chosen empirically; Monro *et al.* [3.15], find that when $\omega_d = 0.26d$, the defect can be reconstructed the most accurately with few terms.

Hence, the squared index distribution is decomposed in the following way:

$$n^2(x,y) = \sum_{a,b=0}^{C-1} c_{ab}\psi_a^d(x)\psi_b^d(y) + \sum_{a,b=0}^{P-1} p_{ab}\cos\left(\frac{2a\pi x}{\ell}\right)\cos\left(\frac{2b\pi y}{\ell}\right) \quad (3.8)$$

where C and P indicate the number of terms that are used to represent the defect and holes, respectively, and ℓ is the transverse extent of the structure. The function ψ_a^d represent elements of the following orthonormal set of Hermite-Gaussian basis functions as follows:

$$\psi_a^d(x) = \frac{2^{-a}\pi^{-1/4}}{\sqrt{(2a)!\,\omega_d}} \exp\left(\frac{-x^2}{2\omega_d^2}\right) H_{2a}\left(\frac{x}{\omega_d}\right) \tag{3.9}$$

where H_i is the ith-order Hermite polynomial (i is the leading order of the polynomial). For any given photonic crystal fibre structure, the coefficients c_{ab} and p_{ab} in the index decomposition in Eqn.(3.8) can be evaluated by performing overlap integrals of the cosine functions with the periodic cladding, and the Hermite-Gaussians with the central defect. These coefficients only need to be calculated once for any structure; once they are known, the modal properties of the fibre at any wavelength can be determined.

In the cases, where localized functions are used to represent the air holes, the high-index contrasts may become a significant difficulty, since their resolution and the effectiveness of the model may become hard to obtain.

The hybrid approach, which combines some of the best features of the localized function and plane-wave techniques described above is formulated in a number of publications by Monro *et al.* (see [3.15], [3.16], and [3.17]), and some recent extensions to this approach are outlined briefly here. In [3.15] and [3.16], the air-hole lattice was described using a plane-wave decomposition, as in the plane-wave techniques described in Section 3.4, and the solid core and the modal fields are described using localized functions. This approach allows for an efficient description, particularly for idealised periodic structures, since only symmetric terms need to be used in the expansions. In order to model microstructured fibres with asymmetric profiles or to obtain accurate predictions for higher-order modes, it is necessary to extend this approach to use a complete basis set, as described by Monro *et al.* in [3.17].

When more complex fibre profiles are considered, the advantages of describing the localized core separately from air holes is diminished, and the best combination of efficiency and accuracy is obtained by describing the entire refractive index distribution using a plane-wave expansion, while using localized functions only for the modal fields. This general implementation of the hybrid approach can be used to explore the full range of photonic crystal fibre structures and modes, and it can predict the properties of actual PCFs by using SEM photographs to define the refractive index profile in the model (see for example the work by Bennett *et al.* in [3.18]). This allows the deviations in optical properties that are caused by the subtle changes in structure to be explored. Note, however, at this time that some of the later methods also have been used to determine the properties of photonic crystal fibres based on accurate SEM picture representations.

In the hybrid implementation, the entire transverse refractive index profile is described using a plane-wave expansion, and the Fourier coefficients are evaluated by performing overlap integrals, which only need to be calculated once for any given structure. The modal electric field is expanded into orthonormal Hermite-Gaussian functions (both even and odd functions are included). These decompositions can be used to convert the vector-wave equation into a simple eigenvalue problem (as in the plane-wave method) that can be solved for the modal propagation constants and fields. In order to solve the system, a number of overlap integrals between the various basis functions need to be evaluated. For this choice of decompositions, these overlaps can be performed analytically, which is a significant advantage of this approach.

Most of the modelling done to date has considered ideal hexagonal arrangements of air holes. Group theory arguments can be used to show that all symmetric structures with higher than 2-fold symmetry are not birefringent [3.19]. However, as the techniques described thus far in this Section perform calculations based on a Cartesian grid, they typically predict a small degree of birefringence that can be reduced (but not eliminated) by using a finer grid. However, when modelling asymmetric structures with a form birefringence that is significantly larger than this false birefringence, it is possible to make reasonably accurate predictions for fibre birefringence.

In order to provide a better feeling for some of the key issues of the localized-function method, we will here shortly draw forward results by Knudsen *et al.* [3.34]. The first key point is the ability of the method to accurately represent the refractive index structure under consideration. In order to evaluate this property, Knudsen *et al.* has determined the integral deviation between an ideal step index profile (representing an air hole in silica) and the approximate profile using localized functions to describe this specific index profile. In Figure 3-5, this integral representation error is shown for three different choices of expansion terms (corresponding to the parameter P in Eqn. (3.7)). The integral deviation from the unit step function is given as a function of the width parameter, and it is clear from Figure 3-5 that very significant deviations may be expected, if a non-optimal choice of the width parameter is chosen. This is exactly the reason for representing the refractive index profile by a Fourier series as indicated in Eqn. (3.7). Note that the inset of Figure 3-5 shows an ideal step function and the approximate representation given for a width parameter of 0.1 and using 12 expansion functions. Further details of this method may be found in [3.34].

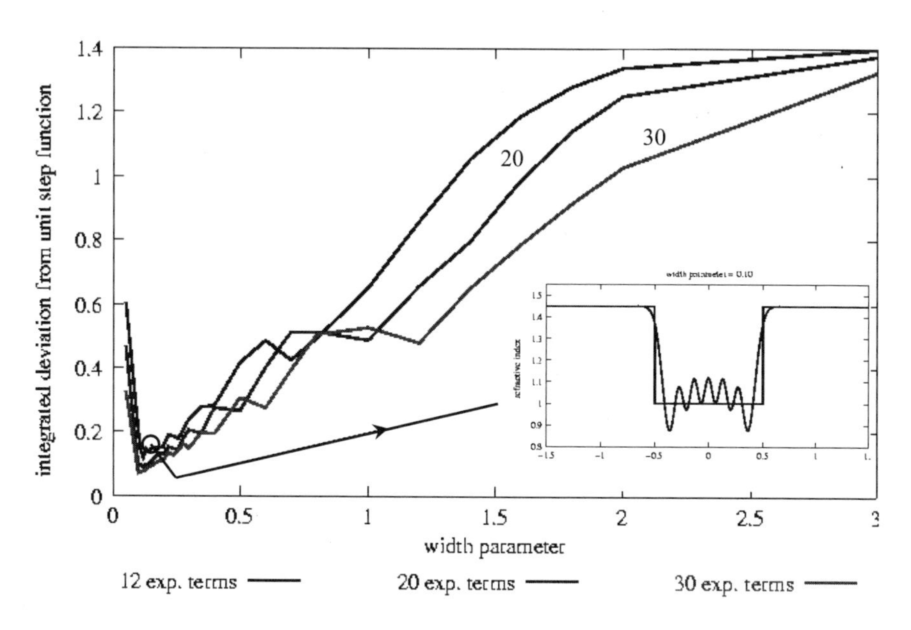

Figure 3-5. Integral deviation from unit step function by a localized-function representation of the refractive index profile of an air hole in a silica-base material. Results are shown as function of the width parameter used in the representation of the hole, and the inset illustrates the deviation between ideal and approximate index profile representation. The illustration origins from work by Knudsen *et al.* [3.34].

3.4 FULL-VECTORIAL PLANE-WAVE EXPANSION METHOD

After now having presented a simple scalar approach to the modelling of photonic crystal fibres and a hybrid approach taking the modal localisation into account, we will now turn to another widely used method for PCF analysis, namely the full vectorial plane-wave expansion method. As we will discuss in the following text, this method is particularly well-suited for the analysis of fully-periodic structures, such as those relevant for photonic bandgap fibres.

3.4.1 Solving Maxwell's equations using a plane-wave expansion method

The plane-wave method was the first theoretical method to accurately analyse photonic crystals [3.24-3.25] and is probably the most applied of all methods. In its basic form, the method is relatively simple, and it utilizes the periodicity of photonic crystals in much the same manner as the periodicity of semiconductors and other solid-state crystals is utilized in methods for solving the Schrödinger equation to predict electronic bandgaps [3.33]. The plane-wave method provides a fast tool for frequency-domain analysis of full-periodic photonic crystals (i.e., crystals without defects), but has a worse scaling with large computational problems, but using advanced implementation, it may be used for large tasks and real-structure PBG fibres. Efficient solution of large photonic crystal problems, as e.g., for the PBG-fibres, may, however, be obtained by an advanced implementation, as will be presented in the following text. Using this implementation, arbitrary fibre designs may be analysed, and due to its relative simplicity and wide applicability, the plane-wave method has been the preferred method for analysis of structures exhibiting PBG effects.

Throughout this book, only photonic crystals realized in dielectric materials with no free charges or currents will be considered unless specifically stated. For such materials, the macroscopic Maxwell's equations on differential form are expressed as [3.32]:

$$\nabla \times \overline{E} = \frac{\partial \overline{B}}{\partial t}, \tag{3.10}$$

$$\nabla \times \overline{H} = \frac{\partial \overline{D}}{\partial t}, \tag{3.11}$$

$$\nabla \cdot \overline{D} = 0, \tag{3.12}$$

$$\nabla \cdot \overline{B} = 0, \tag{3.13}$$

where $\overline{E}$ and $\overline{H}$ are the electric and magnetic fields, respectively, and $\overline{D}$ and $\overline{B}$ are the electric and magnetic flux densities, respectively. The photonic crystals will all be assumed to be constituted of linear, isotropic and non-magnetic materials, hence, the constitutive equations are:

$$\overline{D} = \varepsilon_r \varepsilon_0 \overline{E}, \tag{3.14}$$

$$\overline{B} = \mu_r \mu_0 \overline{H}, \qquad (3.15)$$

where μ_0 is the free-space permeability, μ_r is the relative permittivity (which generally is set to the value $\mu_r = 1$ for dielectric waveguides described in this text), ε_0 is the free-space permittivity and ε_r is the relative dielectric constant. As a final restriction, only non-absorbing materials shall be considered, hence ε_r is real. In the time harmonic case, the position-dependent $\overline{E}$- and $\overline{H}$-fields of angular frequency, ω, may be expressed as:

$$\overline{E}(\overline{r},t) = \overline{E}(\overline{r})e^{j\omega t}, \qquad (3.16)$$

$$\overline{H}(\overline{r},t) = \overline{H}(\overline{r})e^{j\omega t}, \qquad (3.17)$$

Using the above equations, a single equation, which fully describes the propagation of electromagnetic waves, may be deduced. This is known as the wave equation and it may be casted in the form of either the $\overline{E}$ - or the $\overline{H}$-field. For the $\overline{E}$ and the $\overline{H}$-field, respectively, it reads [3.33]:

$$\nabla \times \nabla \times \overline{E}(\overline{r}) = \frac{\omega^2}{c^2} \varepsilon(\overline{r}) \overline{E}(\overline{r}), \qquad (3.18)$$

$$\nabla \times \left[\frac{1}{\varepsilon(\overline{r})} \nabla \times \overline{H}(\overline{r}) \right] = \frac{\omega^2}{c^2} \overline{H}(\overline{r}), \qquad (3.19)$$

where $c = 1/\sqrt{\varepsilon_0 \mu_0}$ is the speed of light in vacuum.

Due to the periodic nature of photonic crystals, a solution to one of the wave equations may according to Bloch's theorem be expressed as a plane wave modulated by a function with the same periodicity as the photonic crystal:

$$\overline{E}(\overline{r}) = \overline{V}_k(\overline{r}) e^{-j\overline{k}\cdot\overline{r}} \qquad (3.20)$$

$$\overline{H}(\overline{r}) = \overline{U}_k(\overline{r}) e^{-j\overline{k}\cdot\overline{r}} \qquad (3.21)$$

where $\overline{V}_k(\overline{r}), \overline{U}_k(\overline{r})$ are the periodic functions, and $\overline{k}$ is the wave vector of the solution. In order to determine the solutions, it is advantageous to operate in reciprocal space and utilize that the periodic functions may be expressed as a Fourier-series expansion in terms of the reciprocal lattice vectors, $\overline{G}$ to give the following expression for the $\overline{E}$ - and $\overline{H}$ - fields:

$$\overline{E}(\overline{r}) = \sum_{G} \overline{E}_{\overline{k}}\left(\overline{G}\right)e^{-j\left(\overline{k}+\overline{G}\right)\overline{r}}, \tag{3.22}$$

$$\overline{H}(\overline{r}) = \sum_{G} \overline{H}_{\overline{k}}\left(\overline{G}\right)e^{-j\left(\overline{k}+\overline{G}\right)\overline{r}}, \tag{3.23}$$

In reciprocal space, the corresponding wave equations are found by Fourier transformations:

$$-\left(\overline{k}+\overline{G}\right)\times\left[\left(\overline{k}+\overline{G}\right)\times\overline{E}_{\overline{k}}\left(\overline{G}\right)\right] = \frac{\omega^2}{c^2}\sum_{\overline{G}'}\varepsilon_r\left(\overline{G}-\overline{G}'\right)\overline{E}_{\overline{k}}\left(\overline{G}'\right), \tag{3.24}$$

$$-\left(\overline{k}+\overline{G}\right)\times\left[\sum_{\overline{G}'}\varepsilon_r^{-1}\left(\overline{G}-\overline{G}'\right)\left(\overline{k}+\overline{G}'\right)\times\overline{H}_{\overline{k}}\left(\overline{G}'\right)\right] = \frac{\omega^2}{c^2}\overline{H}_{\overline{k}}\left(\overline{G}\right), \tag{3.25}$$

where $\varepsilon_r\left(\overline{E}\right)$ are the Fourier coefficients of $\varepsilon_r\left(\overline{r}\right)$, and $\varepsilon_r^{-1}\left(\overline{G}\right)$ are the Fourier coefficients of $\varepsilon_r^{-1}\left(\overline{r}\right)$. In the proceeding section, these will be given for a specific type of 2D photonic crystals. The equations (3.24) and (3.25) may be expressed on matrix form and solved using standard numerical routines as eigenvalue problems. The subscript $\overline{k}$ has been used to indicate that the eigenvalue problem is solved for a fixed wave vector, $\overline{k}$, to find the angular frequencies, ω, of allowed modes. By truncating the summations to N reciprocal lattice vectors, the matrices are of dimensions $3N \times 3N$. However, as demonstrated in literature, e.g., by Ho *et al.* [3.9], the matrix equation involving the $\overline{H}$ -field may be reduced to $2N \times 2N$ dimension. This follows from the transverse condition on the $\overline{H}$ -fields (i.e., $\nabla\cdot\overline{H}(\overline{r}) = 0$), wherefore $\overline{H}_{\overline{k}}\left(\overline{G}\right)$ may be written as a sum over two vectors orthogonal to the relevant $\overline{k}+\overline{G}$:

$$\overline{H}_{\overline{k}}\left(\overline{G}\right) = h_{\overline{k},\overline{G},1}\hat{e}_1 + h_{\overline{k},\overline{G},2}\hat{e}_2, \tag{3.26}$$

where $\hat{e}_1$ and $\hat{e}_2$ are unit vectors perpendicular to $\overline{k}+\overline{G}$. Hence, Eqn. (3.23) may be expressed as:

$$\overline{H}(\overline{r}) = \sum_{G}\sum_{\lambda=1} h_{\overline{k},\overline{G},\lambda}\hat{e}_\lambda e^{-j(\overline{k}+\overline{G})\cdot\overline{r}}, \tag{3.27}$$

and the final matrix equation for the $\overline{H}$-field may (after some steps of deduction) be written as [3.9]:

$$\sum_{\overline{G},\lambda} \overline{H}^{\lambda,\lambda'}_{\overline{k},\overline{G},\overline{G}'} h_{\overline{k},\overline{G},\lambda'} = \frac{\omega^2}{c^2} h_{\overline{k},\overline{G},\lambda}, \tag{3.28}$$

where

$$\overline{H}_{\overline{G},\overline{G}'} = \left|\overline{k}+\overline{G}\right|\left|\overline{k}+\overline{G}'\right| \varepsilon_r^{-1}\left(\overline{G}-\overline{G}'\right) \begin{bmatrix} \hat{e}_2 \cdot \hat{e}'_2 & -\hat{e}_2 \cdot \hat{e}'_1 \\ -\hat{e}_1 \cdot \hat{e}'_2 & \hat{e}_1 \cdot \hat{e}'_1 \end{bmatrix}, \tag{3.29}$$

The reduction from a *3N* x *3N* to a *2N* x *2N* - dimension matrix equation is naturally of high importance with respect to computation times and for this reason, the $\overline{H}$-field version of the wave equation is the preferred implementation. The involved matrix is, furthermore, Hermitian and positive definite, which allow employment of faster numerical routines than when solving for general eigenvalue problems. Although the wave equation is only solved for the $\overline{H}$-field, the corresponding $\overline{E}$-field solution may readily be obtained using Eqn. (3.11), (3.15) and (3.17):

$$\overline{E}(\overline{r}) = \frac{1}{j\omega\varepsilon_r(\overline{r})\varepsilon_0} \nabla \times \overline{H}(\overline{r}), \tag{3.30}$$

The matrix equation (3.28)) is on a general form applicable to all periodic structures. However, in the following section, we shall focus on 2D photonic crystals and see how the equation may be further reduced for a specific type of 2D photonic crystals.

3.4.2 Two-dimensional photonic crystals with hexagonal symmetry

In this section, we shall perform a more accurate analysis of 2D photonic crystal with a hexagonal symmetry. An example of such a structure was previously presented in Section 2.2.2 (see Figure 2-3) and for convenience the structure is repeated in Figure 3-6 (a).

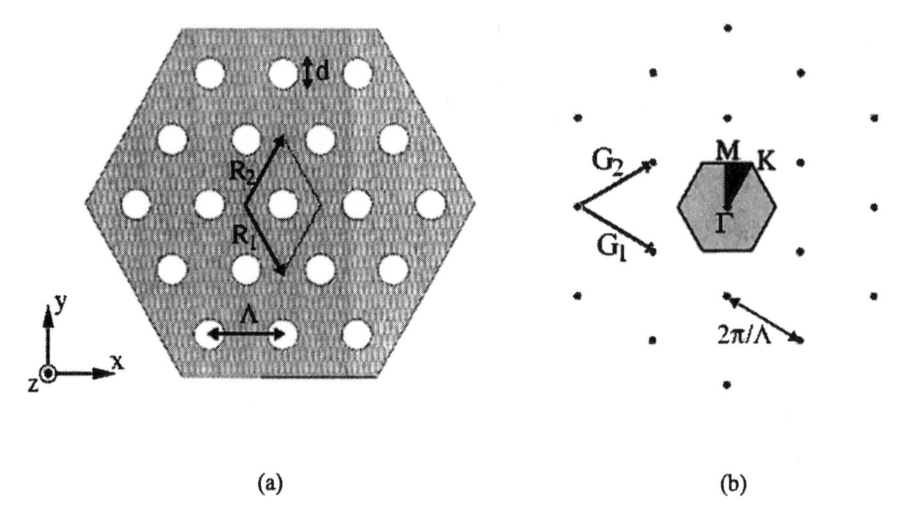

Figure 3-6. (a) Illustration of the real-space primitive lattice vectors $\overline{R}_1$ and $\overline{R}_2$ and unit cell of a 2D photonic crystal with a hexagonal symmetry. (b) Illustration in reciprocal space of the 2D hexagonal lattice structure. The reciprocal lattice vectors are $\overline{G}_1$ and $\overline{G}_2$. The light shaded area indicate the 1. Brillouin zone and the dark shaded area indicate the irreducible 1. Brillouin zone. The symmetry points of the hexagonal structure are indicated by Γ, M and K.

Figure 3-6 (a) further illustrates the two real-space primitive lattice vectors, $\overline{R}_1$ and $\overline{R}_2$, and the unit cell defined as the smallest area, which by mere translation may fully represent the structure. 2D photonic crystals that are characterized by a single hole/rod for each unit cell are also generally referred to as triangular structures [3.33]. Other types of 2D photonic crystals with hexagonal symmetry will be presented in the proceeding chapters. For triangular photonic crystals the primitive lattice vectors are given by:

$$\overline{R}_1 = \frac{\Lambda}{2}\left(\overline{x} - \overline{y}\sqrt{3}\right), \tag{3.31}$$

$$\overline{R}_2 = \frac{\Lambda}{2}\left(\overline{x} + \overline{y}\sqrt{3}\right), \tag{3.32}$$

Hence, the area of the unit cell is well determined as $\left|\overline{R}_1 \times \overline{R}_2\right|$, and an important factor, when discussing photonic crystal fibres, is the total hole/rod volume fraction which, for triangular structures is defined as:

$$f = \frac{\pi}{2\sqrt{3}} \frac{d^2}{\Lambda^2}, \tag{3.33}$$

where d is the hole/rod diameter and f shall be named the filling fraction (when we talk about an air-silica structure, the term air-filling fraction is generally used).

The plane-wave method operates in reciprocal space, and to analyse a given photonic crystal and determine the possible existence of any PBGs, it is in principle required to calculate the frequencies of allowed modes for all possible $\overline{k}$-vectors. However, by utilizing the translational symmetries of the crystals, it is possible to determine any solution (in terms of frequencies) to the eigenvalue problem by only considering $\overline{k}$-vectors restricted to the 1. Brillouin zone. The 1. Brillouin zone is defined as the smallest possible region in reciprocal space that only through translation may be used to represent the full periodic structure. For a triangular structure, the 1. Brillouin zone is illustrated in Figure 3-6 (b), where also the reciprocal lattice vectors $\overline{G}_1$ and $\overline{G}_2$ are shown. These may be constructed from the real-space lattice vectors such that the following equation is fulfilled:

$$\overline{R}_i \cdot \overline{G}_j = 2\pi\delta_{ij}, \quad i, j = 1,2 \tag{3.34}$$

For the triangular structure, this is satisfied for the reciprocal lattice vectors $\overline{G}_1$ and $\overline{G}_2$, where:

$$\overline{G}_1 = \frac{2\pi}{\Lambda}\left(\overline{x} + \overline{y}\frac{\sqrt{3}}{3}\right), \tag{3.35}$$

$$\overline{G}_2 = \frac{2\pi}{\Lambda}\left(\overline{x} - \overline{y}\frac{\sqrt{3}}{3}\right), \tag{3.36}$$

As demonstrated in literature [3.33], mirror- and inversion symmetries allow a further restriction of the $\overline{k}$-vectors that need to be considered. The restricted reciprocal region is known as the irreducible Brillouin zone. This may be demonstrated by analysing the corresponding operations on the vector operator of the $\overline{H}$-field wave equation:

$$\Theta(\bar{r}) = \nabla \times \frac{1}{\varepsilon(\bar{r})} \nabla \times \quad , \tag{3.37}$$

(see Eqn. (3.19)), and the form of the $\overline{H}$ -field solutions (see Eqn.(3.21)). The irreducible Brillouin zone is also illustrated in Figure 3-6 (b), and the symmetry points of the reciprocal lattice are labeled Γ, M, and K. Finally, as for studies of electronic bandgaps in solid-state physics, it is generally accepted that it is sufficient only to vary the $\bar{k}$ -vectors along the boundary of the irreducible Brillouin zone in order to accurately determine the existence of possible PBGs.

The final step required to perform analysis of a given photonic crystal is to determine the coefficients, $\varepsilon_r^{-1}(\overline{G})$, which are required for setting up the matrix equation (3.28). As before the periodicity of the photonic crystal is utilized and a Fourier transformation is employed to go from real-space to reciprocal space. Hence, the real-space dielectric distribution is expanded into a Fourier series:

$$\frac{1}{\varepsilon(\bar{r}_{\|})} = \sum_{\overline{G}_{\|}} \varepsilon_r^{-1}(\overline{G}_{\|}) \bar{e}^{\,j\overline{G}_{\|}\cdot\bar{r}_{\|}}, \tag{3.38}$$

where the Fourier coefficients are given by:

$$\varepsilon_r^{-1}(\overline{G}_{\|}) = \frac{1}{\left|\overline{R}_1 \times \overline{R}_2\right|} \int_{S_{cell}} \frac{1}{\varepsilon(\bar{r}_{\|})} e^{-j\overline{G}_{\|}\bar{r}_{\|}} d^2\bar{r}_{\|}, \tag{3.39}$$

and the surface integral is taken over the unit cell. In Eqn. (3.38) and (3.39), the subscript $\|$ indicate that the vectors are two-dimensional and have components only in the periodic plane. For circular holes/rods with a dielectric constant ε_1 and a background material with a dielectric constant ε_2, the Fourier coefficients are [3.35]:

$$\varepsilon^{-1}(\overline{G}_{\|}) = \begin{cases} \dfrac{1}{\varepsilon_1} f + \dfrac{1}{\varepsilon_2}(1-f), & \overline{G}_{\|} = 0 \\ \left[\dfrac{1}{\varepsilon_1} f - \dfrac{1}{\varepsilon_2} f\right] f \dfrac{2J_1(G_{\|}\Lambda/2)}{(G_{\|}\Lambda/2)}, & G_{\|} \neq 0 \end{cases}, \tag{3.40}$$

where J_1 is the 1.order Bessel function. Although analytic expressions for the dielectric distribution in reciprocal space may be found in the specific case of circular holes/rods in a triangular lattice, this is not the case for arbitrary photonic crystal structures. Hence, for analysis of real photonic crystals, where some distortion from an "ideal" structure may occur during fabrication, the Fourier coefficients must be found numerically. This can readily be done by discretising the real-space distribution and transform it to reciprocal space (e.g., using fast Fourier transformation).

3.5 THE BIORTHONORMAL-BASIS METHOD

In this section, we will describe another numerical method, which successfully has been used to model and solve the transverse 2D structure of photonic crystal fibres. This approach, which first was developed by Silvestre *et al.* [3.43] and further developed by Ferrando *et al.* [3.44], takes into account the full vector character of light propagation in PCFs, and it is generally denoted the biorthonormal-basis method. The background for using this approach is described in detail in the thesis of Miret [3.45], and we will in the following text shortly review these considerations.

One of the major reasons for choosing an approach different from the previously described models is (if we e.g., are looking at the plane-wave method) that the frequency appears as an eigenvalue of the matrix equation (see Eqn. (3.27)) dependent of the refractive index. This makes the direct inclusion of the chromatic dispersion of the material problematic, since it makes it impossible to consider the new equations as an eigenvalue problem. Miret [3.45] points to this as an important limitation for the precise calculation of key fibre parameters such as the group velocity dispersion (GVD). This means that in the case of the previously described methods, it is only possible to get acceptable results by using a less efficient interactive process, and Miret claims that this is only fully feasible, when the material dispersion has a soft dependency with the frequency.

On the other hand, in the 2D vectorial wave equations, the chromatic dispersion of the material can be directly included in the calculations without any trouble, because this formulation leads - in contrast to the 3D formulation - to eigenvalues that are propagation constants (and not in the frequency). Therefore, for the study of the photonic crystal fibres, it will be advantageous to start from the 2D equations rather than from 3D wave equations. To solve the 2D problem, two kinds of methods have been proposed - until now - the linked solutions approach and the modal representation.

To the first kind belongs the so-called multipole method [3.46], which is based on developing the electromagnetic field in the interior of each air hole, as well as in the uniform region between holes. In present formulations, the holes are supposed to be circular, and the fields may be described in harmonic cylinders or by Bessel functions. The imposition of boundary conditions around each of the air holes, allows the formulation of a homogeneous equations system, in which the zeros of the determinant provide the values of the propagation constant. When a realistic simulation of a PCF is faced, the number of terms in the field representations and air holes, which are necessary, make the size of the matrix large. This in term, makes the effort of calculating the zeros of the determinant difficult. Beside this limitation, it has to be pointed out that nowadays this technique only can be used, when the holes have a circular shape. We will return to more details on this method later in this chapter.

Concerning the modal-representation methods in 2D, they will – in agreement with those formulated in 3D – be expressed using a matrix representation with a specific base. A possible selection of such a base is the one discussed in Section 3.3 using localized functions. Having this focus, the key idea is that the guided modes are localized basically in the proximities of the defect, and because of that, it is possible to make a model by developing a function series localized in the surroundings of the fibre core. Miret [3.45], however, points out that the localized-function approach is not very efficient, since it requires an excessive computer calculation time in order to have an acceptable precision. As a matter of fact, almost all calculations reported until now, which use the localized-functions method, has been developed using the scalar approximation.

In the work of Silvestre *et al.* [3.43], Ferrando *et al.* [3.44], and Miret [3.45], the problem is tackled by using, a modal formulation in 2D of the propagation of the electromagnetic fields in the fibre complemented with the introduction of periodic boundary conditions. This has made it possible to develop the electric and magnetic transversal fields in Fourier series, i.e., in linear imaginary exponentials terms. The advantage of utilizing this base is that it posses some very simple transformation properties under translations, which has led the researchers to easily relate the matrix elements representing a given air hole in an arbitrary position, with those that represents one placed at the coordinates in the origin. This results in considerable calculation simplifications for a structure with multiple holes. Using these properties, it is possible to obtain all the matrix elements in an analytical way, obtaining efficiently the complicated spatial structure of the fibre. In the following subsection, we will describe the approach in further detail.

3.5.1 The basic equations of the biorthonormal-basis method

The following description will follow the derivation by Silvestre *et al.* [3.43]. First, we consider a non-magnetic dielectric medium with no free charges or currents. Furthermore, we assume that this medium is invariable to translations along the z direction (the length direction of the fibre). This corresponds to the normally used assumptions concerning propagation through dielectric waveguides, and in this case, the most general solution to Maxwell's equations is a combination of linear fields of the form

$$\overline{E}\left(\overline{\mathrm{x}}_{\mathrm{t}},z,t\right)=\overline{\mathrm{e}}\left(\overline{\mathrm{x}}_{\mathrm{t}}\right)\exp(-j\beta \mathrm{z})\exp(\mathrm{j}\omega t) \tag{3.41}$$

$$\overline{H}\left(\overline{\mathrm{x}}_{\mathrm{t}},z,t\right)=\overline{\mathrm{h}}\left(\overline{\mathrm{x}}_{\mathrm{t}}\right)\exp(-\mathrm{j}\beta \mathrm{z})\exp(\mathrm{j}\omega t) \tag{3.42}$$

where β is the propagation constant, and $\overline{e}$ and $\overline{h}$ contain the dependency of the field in the transverse coordinates $\overline{\mathrm{x}}_{\mathrm{t}} = (x,y)$. These harmonic fields in time *t* and longitudinal position *z* are the modes of the waveguide, in which terms, the propagation of the electromagnetic field can be described. The transversal components of these fields, $\overline{h}_t$ and $\overline{e}_t$, satisfies the wave equations [3.54]:

$$\left[\nabla_t^2 + k_0^2 n^2 + \left(\frac{\nabla_t n^2}{n^2}\right) \times \left(\nabla_t \times o\right)\right]\overline{h}_t = \beta^2 \overline{h}_t, \tag{3.43}$$

$$\left[\nabla_t^2 + k_0^2 n^2 + \nabla_t\left(\left(\frac{\nabla_t n^2}{n^2}\right) \cdot o\right)\right]\overline{e}_t = \beta^2 \overline{e}_t, \tag{3.44}$$

where $n = n(\overline{x}_t)$ is the refractive index profile and $k_0 = 2\pi/\lambda$ is the free-space wave number. ∇_t^2 and ∇_t are, respectively, the Laplacian and the transversal gradient. In Eqn. (3.43-3.44), the square brackets identify the operators responsible for the evolution of the transverse components along the *z* axis. Once the matrix representation is provided, the modes of the system may be obtained by diagonalizing these operators, together with the appropriate boundary conditions and the constraints between field components given by Maxwell's equations.

Note, in relation to Eqn. (3.43) and (3.44) and considering real refractive indices that in the vector theory, these operators are nonself-adjoint, whereas in the scalar theory both operators are self-adjoint. This is the key question pointed out by Silvestre *et al.* [3.43] that distinguishes both cases and prevents blind application of the techniques used in the scalar approximation to the vector case.

In general, a basic property satisfied by the eigenvectors $\overline{\theta}_i$ of a nonself-adjoint operator, L, and the eigenvectors $\overline{\chi}_i$ of its adjoint, $L^\dagger$, i.e., the eigenvectors that fullfill the eigensystems

$$L\overline{\theta}_i = l_i \overline{\theta}_i \tag{3.45}$$

$$L^\dagger \overline{\chi}_i = l_i^* \overline{\chi}_i \tag{3.46}$$

where l_i and l_i^* are the corresponding eigenvalues and * denotes the complex conjungate. The property, which is expressed in Eqn. (3.45-3.46), states that both sets of eigenvectors satisfy a relation named biorthogonality [3.56-3.57]:

$$\left\langle \overline{\chi}_i, \overline{\theta}_j \right\rangle = \delta_{ij} \tag{3.47}$$

where $\langle o, o \rangle$ is the ordinary scalar product in the Hilbert space of the square-integrable complex functions on $R^2, L^2(R^2, C)$. We note that this biorthogonality approach is a well-established method in the field of theoretical physics, but it is due to the work of Silvestre *et al.* [3.43] that it has been adapted to the area of optical waveguides - and more specifically to photonic crystal fibres.

The next step is to identify those eigenvectors with the transverse components of the magnetic field and the electric field, respectively

$$\overline{\theta} = \begin{pmatrix} h_r \\ h_\phi \end{pmatrix}, \quad \overline{\chi} = \begin{pmatrix} -e_\phi^* \\ e_r^* \end{pmatrix} \tag{3.48}$$

and their respective eigenvalues, $l_i = \beta_I^2$. We can recognize the biorthogonality relation as the well-known "orthogonality" relation satisfied by the electromagnetic fields [3.54]

$$\left\langle \overline{\chi}_i, \overline{\theta}_j \right\rangle = \int_{R^2} \left(\overline{\chi}_i^* \cdot \overline{\theta}_j \right) ds = \int_{R^2} \left(\overline{e}_i \times \overline{h}_j \right) \cdot \overline{z} ds \tag{3.49}$$

Because of the symmetry of several examples, Silvestre *et al.* [3.43] choose, with no loss of generality, polar coordinates for representing electric and magnetic field components.

The next step is then to note that in order to perform a modal expansion, only a complete basis and the corresponding projector onto the one-dimensional subspace generated by each eigenvector is required. The scalar product of a field $\overline{\theta}$ and the eigenmodes $\overline{\chi}_i$ of the adjoint problem selects the different components and gives the modal expansion of $\overline{\theta}$ in terms of the eigenmodes $\overline{\theta}_i$, i.e., $\overline{\theta} = \sum_i \langle \overline{\chi}_i, \overline{\theta} \rangle \overline{\theta}_i$. In other words, the projector onto the subspace generated by $\overline{\theta}_i$, $P_i(o)$, is given by $\langle \overline{\chi}_i, o \rangle$. This means that, when we have a nonself-adjoint problem, there is a biorthogonality relation given by Eqn. (3.47) that allows us to obtain the coefficients of the modal expansion in terms of the eigenvector basis of such a nonself-adjoint operator. In the case of L being self -adjoint, the eigenvectors $\overline{\theta}_i$ and $\overline{\chi}_i$ are the same, and, therefore, the biorthogonality relation (3.47) simplifies into the usual orthogonality relation $\langle \overline{\theta}_i, \overline{\theta}_j \rangle = \delta_{ij}$, and the eigenvalues; β_i^2 become real numbers.

Once the relation between the properties of nonself-adjoint operators and the vector wave equations of the electromagnetic field have been shown, it is possible to represent (3.43) and (3.44) in matrix form in the basis provided by an auxiliary system. With this aim, we need an appropriate auxiliary system, described by an index profile *n,* that provides the auxiliary basis

$$\widetilde{L}\widetilde{\theta}_i = \widetilde{\beta}_i^2 \widetilde{\theta}_i, \tag{3.50}$$

and

$$\widetilde{L}^\dagger \widetilde{\chi}_j = \left(\widetilde{\beta}_j^2\right)^* \widetilde{\chi}_j \tag{3.51}$$

The eigenvectors of $\widetilde{L}$ and $\widetilde{L}^\dagger$ satisfy the biortogonality relation Eqn. (3.47), $\langle \widetilde{\chi}_i, \widetilde{\theta}_j \rangle = \delta_{ij}$, and they are the modes of the auxiliary system, when this system is regarded to be a waveguide.

The problem is in the next step developed by expanding the eigenvectors and representing the vector wave equation in matrix form. Each element L_{ij} of the matrix may be calculated by means of the expression:

$$L_{ij} = \langle \widetilde{\chi}_i, L\widetilde{\theta}_j \rangle \tag{3.52}$$

and hereafter Eqn. (3.43) may be written for an arbitrary eigenvector $\overline{\theta}_i$ as the algebraic eigensystem:

$$\sum_k L_{jk}\overline{c}_{ik} = \beta_i^2 \overline{c}_{ij} \tag{3.53}$$

The diagonalisation of Eqn. (3.53) provides the eigenvalues β_i^2 and the eigenvectors $\overline{\theta}_i = \sum_k c_{ik}\widetilde{\theta}_k$, i.e., the propagation constant and the modes are determined.

3.5.2 Biorthonormal-basis method with periodical boundary conditions

The functionality of photonic crystal fibres, and especially photonic bandgap fibres, relies on the periodicity of the waveguide cladding structure. In a modeling context, this has the implication that it is highly advantageous to apply periodic boundary conditions as described in connection with the plane-wave method in Section 3.4.2. However, a similar approach has been described for the biorthonormal-basis method by Ferrando *et al.* [3.44] and by Miret [3.45].

In the previous section, we have seen that the main goal of the biorthonormal-basis method is to transform the problem of solving the system of differential equations for the magnetic and electric fields into an algebraic problem involving the diagonalisation of the L-matrix (Eqn. (3.45)). In this procedure, the choice of an appropriate auxiliary basis is very important for an efficient implementation of the method. In the particular case of a PCF, realistic simulations can contemplate as much as nearly 100 2D step-index individual structures (the air holes of the photonic crystal fibre). Therefore, a brute-force computation of matrix elements can become useless in practice due to losses in numerical precision [3.58]. This loss of precision may especially be critical in dispersion calculations, where results are extremely sensitive to error accumulation, and, consequently, a very high accuracy in the numerical procedure is required. However, the calculation of the matrix elements corresponding to a single hole can be worked out analytically. Moreover, in this basis, the sum over all the holes, and, consequently, the L-matrix of the whole dielectric structure, can also be analytically calculated due to symmetry properties of the often analysed structures (e.g., the hexagonally-centered configuration of the PCF).

An example of an accurately determined mode field calculated by the biorthonormal-basis method for a photonic crystal fibre having a triangular structure of air-holes in the cladding is shown in Figure 3-7.

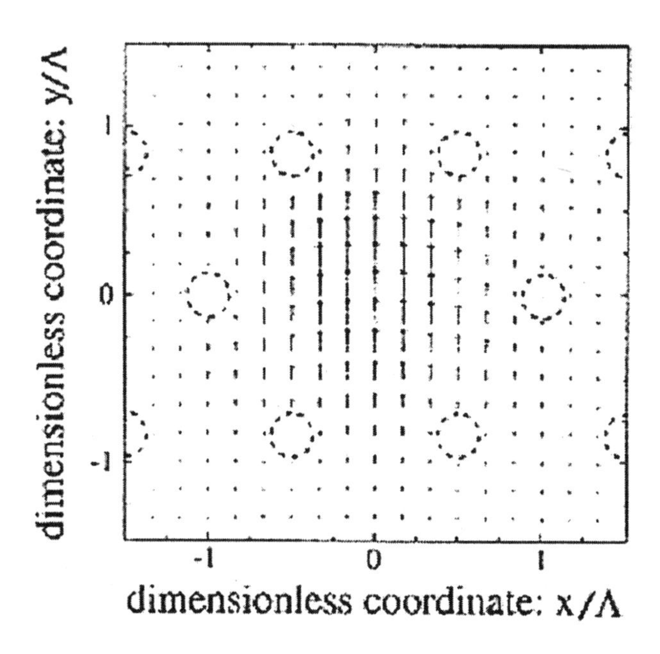

Figure 3-7. Illustration of the magnetic field calculated by the biorthonormal-basis method. The illustration is kindly provided by Dr. Miguel Andrés, University of Valencia, Spain.

3.6 THE MULTIPOLE METHOD

In the proceeding part of this chapter, various approaches have been taken in the modelling of microstructured optical fibres. As we have seen, these methods may be formulated to represent the refractive index distribution in a highly accurate manner, and providing powerful algorithms for the calculation of the vectorial properties of the guided modes in the fibres. However, one crucial draw-back of each of these accurate vectorial methods is that they use periodic boundary conditions, and, therefore, cannot be used to calculate the confinement losses of the structures. These losses may be seen as intrinsic to fibre structures, which solely consists of pure silica with air holes, although changes in the structure (for example increasing the number of rings) can make these losses arbitrarily small. One method that does allow the calculation of confinement losses is the multipole method as described by White *et al.* [3.21-3.22]. This method has also been applied by Monro *et al.* [3.20], and a detailed description may be found in the recent paper by Kuhlmey *et al.* [3.46].

The multipole method is similar to other expansion methods but uses many expansions, one based on each of the holes of the structure. Modes are found by combining the expansions of each element and adjusting the expansion coefficients to meet the boundary conditions. The nature of the expansion does not require the artificial periodicity of the plane-wave method. Furthermore, the multipole method may provide the calculation of a complex propagation constant, and so it can allow confinement losses to be determined. As PCF structures are inherently leaky, this is a real requirement.

In this section, we will shortly outline the main elements of the multipole method. This will be done with point of reference in the work by White *et al.* [3.22], who describes a full vector multipole formulation in which the modes have generally complex propagation constants. The modal fields are expanded in cylindrical harmonic functions about the air holes, which are assumed to be circular. This is of course a limitation for certain PCF structures, but in contrast to the complete lack of predicted loss values, the circular hole assumption is not that significant a problem, and it is chosen in order to improve the efficiency of the method.

In the neighborhood of a cylinder l, the longitudinal electric field component E_z is expanded by use of local polar coordinates (r_l, ϕ_l) centered on cylinder l as follows:

$$E_z = \sum_{m=-\infty}^{\infty} \left[a_m^{(l)} J_m\left(k_\perp^e r_l\right) + b_m^{(l)} H_m^{(1)}\left(k_\perp^e r_l\right) \right] \exp(-jm\phi_l) \exp(-j\beta z) \quad (3.54)$$

where $k_\perp^e = \sqrt{k_0^2 n_e^2 - \beta^2}$ is the transverse wave number in the silica background of refractive index n_e. The transverse wave number $k_\perp^e$ may be complex. β is the propagation constant of the guided mode, and $k_0 = 2\pi/\lambda$ is the free-space wave number, and λ is the wavelength. In agreement with conventional fibre models, a time dependence defined by $\exp(j\omega t)$ is assumed. Inside the cylinder l, where the refractive index is assumed to have the value $n_i = 1$, the field is expressed in regular Bessel functions as

$$E_z = \sum_{m=-\infty}^{\infty} \left[c_m^{(l)} J_m \left(k_\perp^i r_l \right) \right] \exp\left(-jm\phi_l \right) \exp\left(-j\beta z \right) \tag{3.55}$$

where $k_\perp^i = \sqrt{\beta^2 - k_0^2 n_i^2}$. Similar expressions are used for the magnetic field H_z. Now, by applying electric and magnetic boundary conditions on the hole surface, relations that relate the coefficients $a_m^{(l)}$, $b_m^{(l)}$, and $c_m^{(l)}$ on that cylinder is obtained.

The values of the coefficients are determined collectively for the cylinders, which form the full PCF structure, by observing that the regular (J_m) part of the field close to cylinder l must be due to the outgoing ($H_m^{(1)}$) part of the wave field from all other cylinders $j \neq l$. Mathematically, White *et al.* derives these coefficient values by using Graf's addition theorem [3.59], which resulted in a relation among the coefficients of the different cylinders. This relation then lead to an infinite set of homogeneous algebraic equations, depending on β that may be truncated such that $-M \leq m \leq M$ in Eqns. (3.54-3.55), and solved numerically.

The hereby outlined approach will then account for the inner part of the microstructured fibre, i.e., the core and inner cladding. However, in order to describe the effect of a leaky mode, all holes of the cladding is surrounded by a jacketing material having a complex refractive index $n_J = n_e - j\delta$, where $\delta << 1$, ensuring that the fields and their energy density are integrable in transverse plane. Without the jacket, the expansions lead to fields that diverge far away from the core, because the modes are not completely bound. White *et al.* [3.22] has reported that although one may see the inclusion of the jacket as a step to obtain mathematical rigor, the effects of its radius R_J and the attenuation parameter δ on the propagation constants were found to be negligible over a wide range of parameters, provided that the modes are well confined to the fibre core region. Note that the multipole method generally is computationally intensive.

In Figure 3-8, a sketch of the fibre structure is shown with the inner part of the fibre indicated. The refractive indices (with real and imaginary parts) related to the different sections are also shown. The matching of mode solutions are done at the radius indicated by the dashed line.

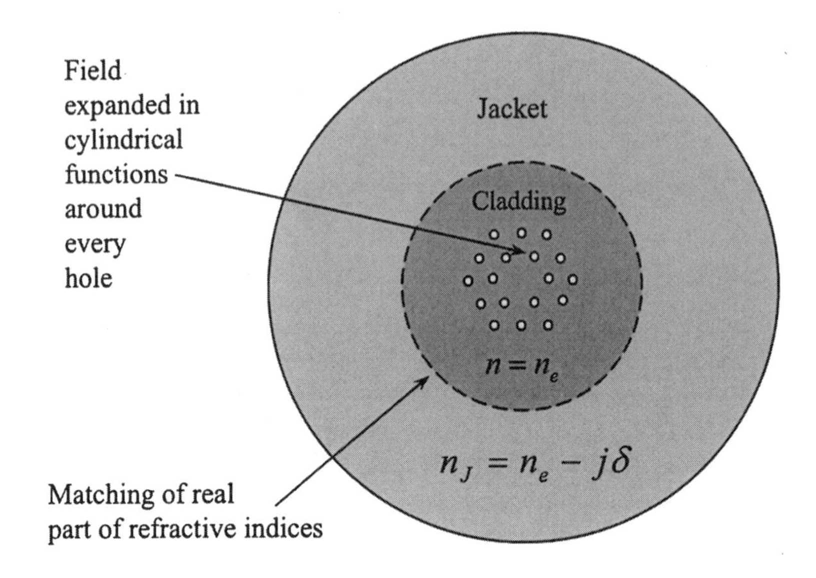

Figure 3-8. Sketch of the refractive index distribution considered in the cylindrical field expansion used in the multipole method.

In Figure 3-9, we have shown some calculated values for confinement loss of a PCF as a function of the number of air-hole rings. For the considered structure the air-holes are placed in a triangular structure, and as shown from the figure, rather high values are determined. However, the calculated values are strongly dependent of the air-hole size, and the confinement loss is predicted to drop several orders of magnitude for increasing number of rings in the fibre structure.

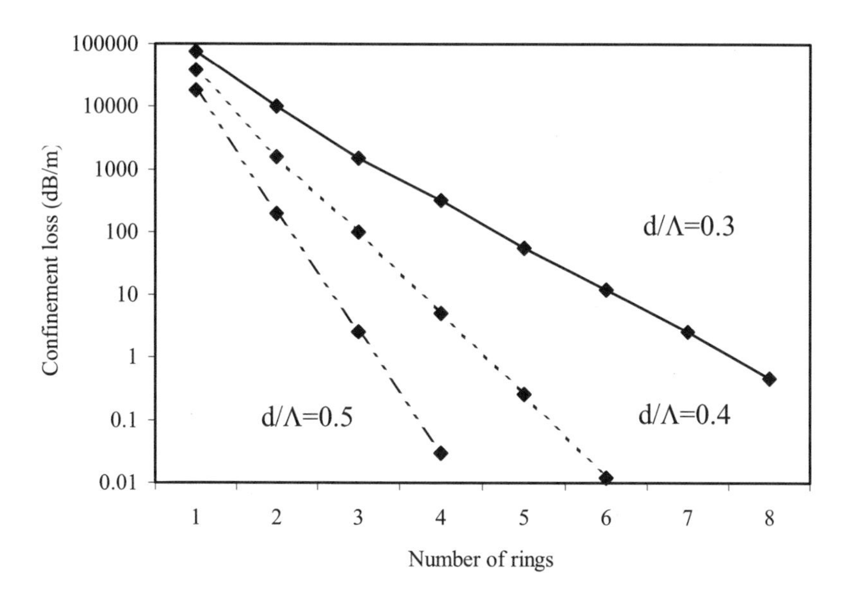

Figure 3-9. Confinement loss determined by the multipole method [3.22] for photonic crystal fibre structures having triangular cladding-hole distributions. The confinement loss values are shown as a function of the number of index rings of the considered structures for three different relative sizes of air-hole diameter d/Λ. The pitch is $\Lambda = 2.3$ μm, and for the specific case, the wavelength $\lambda = 1.55$ μm.

3.7 THE FOURIER DECOMPOSITION METHOD

In this chapter, various approaches have been taken in the modelling of microstructured optical fibres, and their major characteristics, advantages and disadvantages have been listed. Among the discussed methods, it is only the multipole method, which has been used for a reasonable prediction of the confinement loss of photonic crystal fibres. The reason for this draw-back of the majority of the methods is that they either use periodic or zero boundary conditions, and, therefore, cannot be used to calculate the confinement losses of the structures. These losses may be viewed as intrinsic to un-doped microstructured fibres, because the effective index of the guided mode generally is below that of the outer cladding layer (See Figure 3-10). Note, however, that changes in the structure (for example increasing the number of rings) can make these losses arbitrarily small.

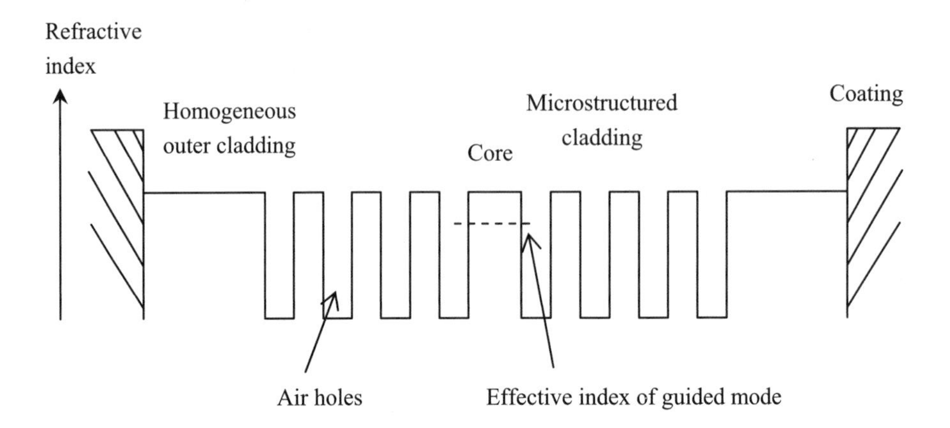

Figure 3-10. The effective-index profile of an index-guiding photonic crystal fibre. The effective index of the guided mode is indicated.

The schematic illustration in Figure 3-10 indicates that the confinement of light in the microstructured fibre is done by the presence of air holes in the inner cladding structure. However, the confinement may also be done very effectively by air-suspension by fine support filaments, as shown in Figure 3-11. The photonic crystal fibres having fine support filaments may either be fabricated (as it is the case for the example illustrated in Figure 3-11 (b)) by stacking of capillary tubes, or it may appear using alternative fabrication techniques, such as the extrusion method, which was recently demonstrated for low-temperature melting point glasses [3.20]. Regardless of the fabrication method, however, un-doped silica-air fibres will not have a raised index core with respect to the outer cladding, and the basic situation will be as indicated in Figure 3-10. In this case, a leaky-mode description becomes relevant, if accurate prediction of leakage losses is desired.

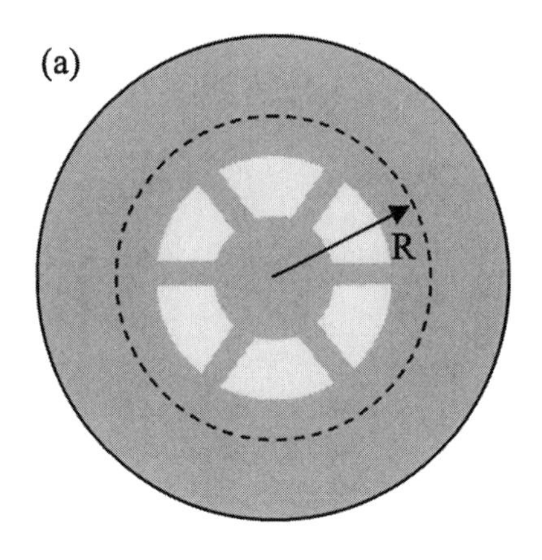

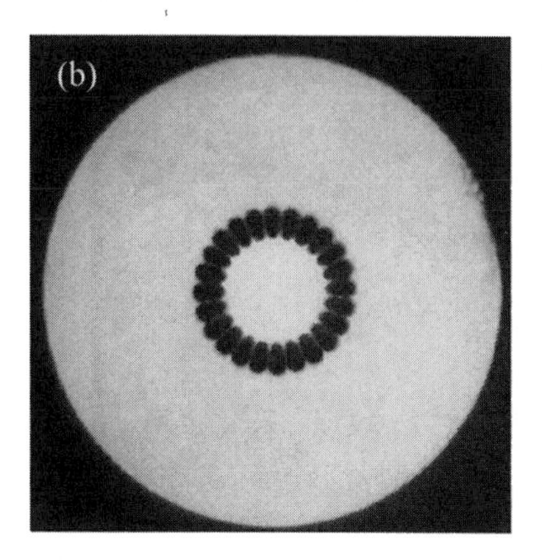

Figure 3-11. (a) Schematic illustration of a PCF structure having an air-suspended central core region. The dashed line indicates the radius, at which mode matching in the Fourier decomposition method is performed. (b) Photonic crystal fibre structure having the central part of the fibre suspended in air by fine support filaments. The photograph is kindly supplied by Crystal Fibre A/S.

As described in the previous section, the multipole method developed by White *et al.* [3.22] is so far limited to the description of confinement losses for structures only having circular air holes. However, it is obvious from the example shown in Figure 3-11 (b) that photonic crystal fibres do not only have circular holes. For this reason, the alternative approach developed by Poladian *et al.* [3.23] becomes very interesting. The method by Poladian *et al.* [3.23] takes point of reference in a basis function expansion technique, which may be used to determine the confinement loss by correctly determining the field outside the computational domain satisfying the outward radiating boundary condition. In this section, we will shortly review the key elements of this method.

Poladian *et al.* [3.23] relates the confinement loss per unit length to the imaginary part of the complex effective mode index, n_{eff}, by

$$\alpha_{rad} = -20\log_{10}(e)k \cdot \mathrm{Im}\lfloor n_{eff} \rfloor \tag{3.56}$$

The effective mode index is determined from a solution to the scalar wave equation. In this approach, a circular computational domain D of radius R is used to encapsulate the central part of the waveguide structure (see Figure 3-11 (a)), and the index is uniform and equal to n_{cl} in the infinite region outside the domain. The scalar wave equation is in the work of Poladian *et al.* [3.23] expressed in dimensionless coordinates $x = r / R$ as follows:

$$\nabla^2\psi(x,\theta)+V^2(x,\theta)\psi(x,\theta)=W^2\psi(x,\theta) \tag{3.57}$$

where $V^2(x,\theta)=k^2R^2\left[n^2(x,\theta)-n_{cl}^2\right]$ depends on the index profile $n(x,\theta)$ of the fibre and $W^2=k^2R^2\left[n_{eff}^2-n_{cl}^2\right]$ is a constant that depends on the effective index of the mode n_{eff}. It should be noted that $V(x,\theta)=0$ outside the central area D.

The field outside the computational domain D can be expressed exactly on the form

$$\psi(x,\theta)=\sum_m B_m e^{jm\theta}K_m(Wx) \tag{3.58}$$

where K_m are modified Bessel functions. For bound modes, W is real and the fields are in this case evanescent. For radiating and leaky modes, the parameter W is complex, and the fields express an outward propagating field. In either case, the expansion outside D has the correct physical behavior at infinity. The field inside D is expanded in some complete set of basis functions

$$\psi(x,\theta)=\sum_{m,n} A_{mn}\psi_{mn}(x,\theta) \tag{3.59}$$

In accordance with the approach chosen by Fletcher [3.60], Poladian *et al.* [3.23] choose a set of weighting functions $\phi_{mn}(x,\theta)$, and the following generalized eigenvalue problem for the unknown coefficients A_{mn} and the eigenvalue W^2 is obtained:

$$M_{\mu\nu}^{mn}A_{mn}=W^2N_{\mu\nu}^{mn}A_{mn} \tag{3.60}$$

where the matrix elements are given by

$$M_{\mu\nu}^{mn}=\left\langle\phi_{\mu\nu}\middle|\nabla^2\psi_{mn}\right\rangle+\left\langle\phi_{\mu\nu}\middle|V^2(x,\theta)\psi_{mn}\right\rangle \tag{3.61}$$

$$N_{\mu\nu}^{mn}=\left\langle\phi_{\mu\nu}\middle|\psi_{mn}\right\rangle \tag{3.62}$$

in which the standard inner product is defined as

$$\langle f|g\rangle=\int_0^{2\pi}\int_0^1 f(x,\theta)^*g(x,\theta)dxd\theta \tag{3.63}$$

Further information on this specific formulation may be found in [3.23].

The complex eigenvalues of leaky modes arise from the non-self-adjoint boundary conditions applied to the system, which forces the matrix operators of Eqn. (3.60) to be non-Hermitian. Consequently, will techniques, which exploit Hermitian operator properties and variational theorems, be unavailable in the solution, and Poladian *et al.* [3.23] choose to use inverse iteration to find the eigenvalues of interest. The essential properties of the expansion functions ψ_{mn} are that they have the correct behavior at the origin (such that the coordinate singularity at the origin is not problematic). The expansion functions should, furthermore, contain adjustable parameters that allow continuity with the expansion outside the domain (allowing adjustable boundary conditions), and, finally, they should allow computationally efficient evaluation of the matrix elements. The choice of Poladian *et al.* was a polar-coordinate harmonic Fourier decomposition on the form

$$\psi_{mn}(x,\theta) = e^{jm\theta}\left[\frac{\sin(n\pi x)}{n\pi} + \alpha_{mn}(W) + \beta_{mn}(W)x\right] \tag{3.64}$$

$$\phi_{mn}(x,\theta) = e^{jm\theta}\sin(n\pi x) \tag{3.65}$$

where

$$\alpha_{mn}(W) = \delta_{m,0}\left[1 + \frac{(-1)^n - 1}{WK_0^{'}(W)}K_0(W)\right] \tag{3.66}$$

$$\beta_{mn}(W) = -\delta_{m,0} + (1-\delta_{m,0})\frac{(-1)^n}{WK_m^{'}(W) - K_m(W)}K_m(W) \tag{3.67}$$

The adjustable parameters α and β provide for the correct behavior at the origin and also satisfy the boundary condition $\psi^{'}(x,0)/\psi(x,0) = WK_m^{'}(Wx)/K_m(W)$ at $x = 1$. The basis functions can be chosen to exploit rotational symmetries of the structure for computational advantage.

According to [3.23], the adjustable-boundary-condition Fourier-decomposition method (ABC-FDM) proceeds as follows: To start the procedure, an initial guess for the value of n_{eff} is used to determine the parameter W appearing in the external field expansion. The adjustable parameters in the basis functions (Eqns. (3.66-3.67)) are chosen to match the external field. The matrix problem is then solved for the eigenvalue and eigenvector of interest. Inverse iteration methods can be used to narrow in on the eigenvalue closest to some chosen value, and additional eigenvalues and eigenvectors can, subsequently, be found by working in the subspace orthogonal to the modes already found. The eigenvalue then gives an improved estimate of n_{eff} for the mode of interest. If required, this new estimate can be used to re-adjust the external field and the adjustable boundary conditions again to obtain an improved estimate and continued iteration can be used to increase the accuracy of the answers. Such re-iteration is reported to converge quickly for any mode with the majority of its guided power within the computational domain. Since the imaginary part is often orders of magnitude smaller than the real part, Poladian *et al.* indicate that - in most cases of interest - a single re-iteration is sufficient to accurately determine the imaginary part of the index unless the loss is extremely large.

Although the algorithm is applied to structures, which only have leaky modes, it can be used directly with structures that have both bound and leaky modes. The algorithm proceeds exactly as above, except that the adjustable parameter W remains real for all bound modes. However, if one is only interested in bound modes, formulations that exploit self-adjointness of bound mode problems are in general considered to be more efficient [3.23].

3.8 THE FINITE-DIFFERENCE METHOD

In this chapter, we have already discussed a number of numerical methods developed for theoretical studies of guided modes in PCFs. Among these are well-established methods such as the plane-wave method, the localized-function method etc. In this section, the focus will be on an even more widely used class of methods, namely the so-called Finite-Difference (FD) methods. As we will describe in the following two sections, this class of methods may be formulated both in time domain (see Section 3.8.1) and in frequency domain (Section 3.8.2).

3.8.1 The Finite-Difference Time-Domain method

The Finite-Difference Time-Domain (FDTD) method was proposed for the full-wave analysis of guided modes in photonic crystal fibres by Qiu [3.61] and Town *et al.* [3.62] in 2001. As described by Qiu [3.61], the FDTD method has been successfully applied to photonic crystals in the computation of band structures [3.63-3.64], for calculation of out-of-plane band structures [3.65], defect modes [3.66], waveguide modes [3.67], and surface modes. For photonic crystal fibres, if one assumes that the propagation constant along the z-direction (propagation direction) is fixed, three-dimensional hybrid guided modes can be calculated using only a 2-dimensional mesh. Furthermore, only real variables are used in the method described by Qiu [3.61]. In this formulation by Qiu, which description we will follow in this section, the computation time grows at a rate of N (order N), where N is the number of discretization points for the FDTD method (notice that the plane wave expansion method used in [3.68] (and described in Section 3.4) is of order N^3). Therefore, the computation time and computer memory are significantly reduced.

We will now describe the major elements of the numerical method. For a linear isotropic material in a source-free region, the time-dependent Maxwell's equations can be written in the following form:

$$\frac{\partial \overline{H}}{\partial t} = -\frac{1}{\mu(\overline{r})}\nabla \times \overline{E} \tag{3.68}$$

$$\frac{\partial \overline{E}}{\partial t} = \frac{1}{\varepsilon(\overline{r})}\nabla \times \overline{H} + \frac{\sigma(\overline{r})}{\varepsilon(\overline{r})}\overline{E} \tag{3.69}$$

where $\varepsilon(\overline{r})$, $\mu(\overline{r})$, and $\sigma(\overline{r})$ are the position-dependent permittivity, permeability, and conductivity of the material, respectively. Note that for silica based optical fibres, the conductivity is zero.

An important element of the FDTD approach is based on the fact that Maxwell's equations can be discretized in space and time by a so-called Yee-cell technique [3.69-3.70] on a discrete three-dimensional mesh. For the guided modes in photonic crystal fibres, it is assumed that the propagation constant along the propagation direction is β. Therefore, each field component has the form $\phi(x,y,z) = \phi(x,y)e^{-j\beta z}$, where ϕ denotes any field component. In Maxwell's equations, therefore, the z-derivatives can be replaced by $-j\beta$, and they may, consequently, be expressed in terms of the transverse variables only [3.71]. Figure 3-12 depicts the unit cell of the two-dimensional mesh, which is to be placed over the cross section of the fibres.

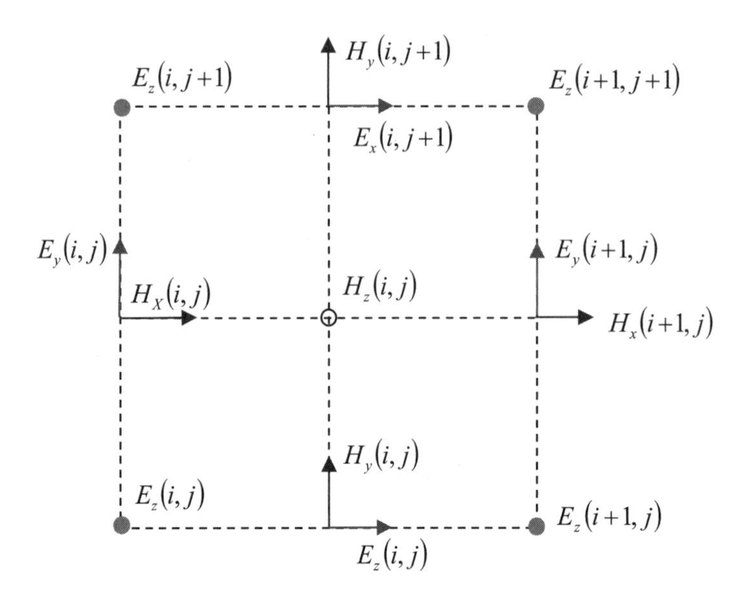

Figure 3-12. Unit cell in the two-dimensional FDTD mesh. The illustration is made from [3.61].

The field components shown on Figure 3-12, could for example be the discrete form of the x-component of Maxwell's first curl equation

$$\frac{\partial H_x}{\partial t} = -\frac{1}{\mu}\left(\frac{\partial E_z}{\partial y} - \frac{\partial E_y}{\partial y}\right) \tag{3.70}$$

which then becomes

$$H_x\Big|_{i,j}^{n+\frac{1}{2}} = H_x\Big|_{i,j}^{n-\frac{1}{2}} - \frac{\Delta t}{\mu_{i,j}}\left(\frac{E_z\Big|_{i,j+1}^{n} - E_z\Big|_{i,j}^{n}}{\Delta y} + j\beta E_y\Big|_{i,j}^{n}\right) \tag{3.71}$$

where the index n denotes the discrete time step, and indices i and j denote the discretized grid point in the x-y plane, respectively. Δt is the time increment, and Δx and Δy are the intervals between two neighboring grid points along the x- and y-directions, respectively. The rest of the equations for other field components can be obtained in a similar manner .

Equations of the type presented in Eqn. (3.71) introduces complex numbers into the computation. However, when computing guided modes in photonic crystal fibres, one wishes to refer only to real numbers, since the computation time and memory will be reduced significantly compared to using complex numbers. One way to eliminate the complex numbers is to assume that E_z, H_x, and H_y have components $\cos(\beta z+\phi)$ (with real amplitudes), and H_z, E_x, and E_y have components $\sin(\beta z+\phi)$ (with real amplitudes) [3.72]. Equation (3.78) becomes

$$H_x\Big|_{i,j}^{n+\frac{1}{2}} = H_x\Big|_{i,j}^{n-\frac{1}{2}} - \frac{\Delta t}{\mu_{i,j}}\left(\frac{E_z\Big|_{i,j+1}^{n} - E_z\Big|_{i,j}^{n}}{\Delta y} + \beta E_y\Big|_{i,j}^{n}\right) \tag{3.72}$$

Now, only real variables remain. For the reader, who is interested in more details on the representation of the other field components, we recommend the consultation of the paper by Qiu [3.61] or the book by Taflove [3.70].

For a fixed total number of time steps, the computation time is proportional to the number of discretization points in the computation domain, i.e., the FDTD algorithm is of order N.

The FDTD time-stepping formulas are stable numerically, if the following condition is satisfied [3.73]:

$$\Delta t \le \frac{1}{c\sqrt{\Delta x^{-2} + \Delta y^{-2} + (\beta/2)^2}} \tag{3.73}$$

Since information out of the computational domain is not available at the boundary of FDTD cells, the fields are updated using special boundary conditions. In the work of Qiu [3.61], the perfectly matched layer (PML) method [3.74] was used for the boundary treatment. In this approach, an artificial initial field distribution is introduced in the FDTD algorithm. The non-physical components in the initial field distribution will disappear in the time evolution, and only the physical components, i.e., guided modes, will remain, if the evolution time is long enough. Since all of the fields are obtained in the time domain in the present method, one needs to transform the calculated fields from the time domain to the frequency domain by a Fourier transform in order to obtain the spectral information. Details for the boundary condition, the initial field distribution, and the Fourier transform can be found in [3.64,3.66].

It should also be noted that the FDTD approach is not a mode-based method, and the propagating field in a multimode waveguide structure may be seen as a summation over excited modes. The FDTD method is flexible, but a significant amount of work must be invested in intelligent design of boundary conditions.

3.8.2 The Finite-Difference Frequency-Domain method

Within the finite-difference methods, it may alternatively be chosen to aim at the solution of Maxwell's equations through a frequency-domain approach. This was first described for photonic crystal fibres by Zhu *et al.* [3.75], and we will here shortly outline the basic elements of this method.

In the Finite-Difference Frequency-Domain (FDFD) mode solvers, two discretization schemes have been used by Zhu *et al.* [3.75]. One is that first proposed by Stern [3.76], in which possible discontinuities lie between two adjacent mesh grids and every grid point corresponds to a unique refractive index. The wave equation in terms of transverse electric field E_t (or magnetic field H_t) may be expressed as follows:

$$\left(\nabla_t^2 + k_0^2\varepsilon_r\right)\overline{E}_t + \nabla\left(\varepsilon_r^{-1}\nabla_t\varepsilon_r \cdot \overline{E}_t\right) = \beta^2\overline{E}_t \tag{3.74}$$

or

$$\left(\nabla_t^2 + k_0^2\varepsilon_r\right)\overline{H}_t + \varepsilon_r^{-1}\nabla_t\varepsilon_r \times \left(\nabla_t \times \overline{H}_t\right) = \beta^2\overline{H}_t \tag{3.75}$$

where $k_0 = 2\pi/\lambda$ is the wave number in free space, ε_r is the waveguide dielectric constant, and β is the propagation constant. These equations are directly discretized by finite difference as presented in [3.77]. This discretization scheme is usually used in the context of the beam propagation method (BPM) [3.77-3.78], which can also be used as a mode solver as further described in Section 3.10. Another discretization scheme is that first proposed by Bierwirth *et al.* [3.79]. In this second scheme, possible discontinuities lie on the mesh grids, so that any grid point can be associated with up to four different refractive indices. The transverse magnetic components are usually used in deriving the discretization matrix [3.80-3.81].

In the paper by Zhu *et al.* [3.75], a full-vector finite-difference mode solver that is based on discretization scheme first proposed by Yee [3.69] is presented. Yee's mesh, which is shown in Figure 3-12, is (as previously mentioned) widely used in the FDTD analysis [3.51]. Zhu *et al.* [3.75] use Yee's two-dimensional mesh in their frequency-domain mode solver for complex optical waveguides. Before we look closer at this method, it should, however, be noted that in order to improve the staircase approximation for curved interface, an index averaging technique is used for the cells across interfaces.

In Yee's mesh, the mesh grids for electric fields lie on possible dielectric discontinuities. Since all the transverse field components are tangential to the unit cell boundaries, the continuity conditions are automatically satisfied.

Zhu *et al.* [3.75] assumes that the fields have a dependence of position z and time t according to $\exp[j(-\beta z+\omega t)]$. From Maxwell's curl equations

$(\nabla\times\overline{E}=-\partial\overline{B}/\partial t,\quad \nabla\times\overline{H}=\partial\overline{D}/\partial t)$, after scaling $\overline{E}$ by the free-space impedance $Z_0=\sqrt{\mu_o/\varepsilon_0}$,we then have

$$jk_oH_x=-\partial E_z/\partial y+j\beta E_y \tag{3.76}$$

$$jk_0H_y=j\beta E_x+\partial E_z/\partial x \tag{3.77}$$

$$jk_0H_z=-\partial E_y/\partial x+\partial E_x/\partial y \tag{3.78}$$

and

$$jk_o\varepsilon_rE_x=\partial H_z/\partial y+j\beta H_y \tag{3.79}$$

$$jk_o\varepsilon_rE_y=-j\beta H_x-\partial H_z/\partial x \tag{3.80}$$

$$jk_o\varepsilon_rE_z=\partial H_y/\partial x-\partial H_x/\partial y \tag{3.81}$$

Equations (3.76-3.81) are now discretized, and Zhu *et al.* obtains the following type of equations:

$$jk_oH_x(i,j)=-[E_z(i,j+1)-E_z(i,j)]/\Delta y+j\beta E_y(i,j) \tag{3.82}$$

$$jk_oH_y(i,j)=j\beta E_x(i,j)+[E_z(i+1,j)-E_z(i,j)]/\Delta x \tag{3.83}$$

$$jk_oH_z(i,j)=-[E_y(i+1,j)-E_y(i,j)]/\Delta x+[E_x(i,j+1)-E_x(i,j)]/\Delta y \tag{3.84}$$

The equations (3.79-3.81) may be written in a similar manner (please consult [3.75] for details on this). We note that the terms *j* in the brackets indicate a number (and this number has nothing to do with the complex term appearing outside the brackets). Note, furthermore, that Zhu *et al.* approximate the refractive indices by averaging the refractive indices of adjacent cells.

Eqs. (3.82-3.84), and the corresponding discretized equations representing Eqs. (3.79-3.81), can be written in matrix form as follows:

$$-jk_0\begin{bmatrix}\overline{Hx}\\ \overline{H}_y\\ \overline{Hz}\end{bmatrix}=\begin{bmatrix}0 & j\beta\overline{I} & \overline{U}_y\\ -j\beta\overline{I} & 0 & -\overline{U}_x\\ -\overline{U}_y & \overline{U}_x & 0\end{bmatrix}\begin{bmatrix}\overline{E}_x\\ \overline{E}_y\\ \overline{E}_z\end{bmatrix} \tag{3.85}$$

$$-jk_0\begin{bmatrix}\varepsilon_{rx} & 0 & 0\\ 0 & \varepsilon_{ry} & 0\\ 0 & 0 & \varepsilon_{rz}\end{bmatrix}\begin{bmatrix}\overline{E}_x\\ \overline{E}_y\\ \overline{E}_z\end{bmatrix}=\begin{bmatrix}0 & j\beta\overline{I} & \overline{V}_y\\ -j\beta\overline{I} & 0 & -\overline{V}_x\\ -\overline{V}_y & \overline{V}_x & 0\end{bmatrix}\begin{bmatrix}\overline{H}_x\\ \overline{H}_y\\ \overline{H}_z\end{bmatrix} \tag{3.86}$$

where $\overline{I}$ is a square identity matrix, and $\overline{\varepsilon}_{rx}$, $\overline{\varepsilon}_{ry}$ and $\overline{\varepsilon}_{rz}$ are diagonal matrices determined by the following equations:

$$\varepsilon_{rx}(i,j)=[\varepsilon_r(i,j)+\varepsilon_r(i,j-1)]/2 \tag{3.87}$$

$$\varepsilon_{ry}(i,j)=[\varepsilon_r(i,j)+\varepsilon_r(i-1,j)]/2 \tag{3.88}$$

$$\varepsilon_{rz}(i,j)=[\varepsilon_r(i,j)+\varepsilon_r(i-1,j-1)+\varepsilon_r(i,j-1)+\varepsilon_r(i-1,j)]/4 \tag{3.89}$$

In Eqns. (3.85-3.86), the matrices $\overline{U}_x, \overline{U}_y, \overline{V}_x$, and $\overline{V}_y$ are square matrices, which depend on the boundary conditions of the rectangular computation window. For further details on the representation of these, please consult [3.75]. Now having established a set of matrix equations including the finite-difference formulation, these may be solved using available numerical eigenvalue routines, which then provide the effective modal index $n_{eff}=\beta/k_0$ and modal fields of the guided modes.

In the finite difference analysis of waveguides with curved interfaces, the staircase approximation has to be used in the rectangular mesh. In order to improve this approximation, Zhu *et al.* [3.75] use averaged refractive indices for the mesh cells across the interface. Similar techniques have been previously used in the plane-wave expansion method and FDTD analysis. The use of the averaged refractive index of interfacial cells can significantly accelerate convergence and improve the modeling accuracy for waveguides with curved interfaces, such as PCFs.

3.9 THE FINITE-ELEMENT METHOD

If we for a moment consider the whole area of numerical modeling of complex physical problems in electromagnetics as well as mechanics, it will soon be clear that one of the strongly employed methods is the Finite-Element method (FEM). This numerical method is described in several textbooks (see e.g., [3.48] for a description of the FEM in microwave engineering). The method has also been used in the analysis and design of standard optical fibres, and relevant for the research described in this book, Brechet *et al.* [3.49] applied the FEM in the analysis of photonic crystal fibres in 2000. It is, however, also obvious that the subject of FEM modeling is not only highly developed, but also an area, which is extensively described in literature. For this reason, we have chosen only to present a short description of the basics of the finite element method at this point. We will here follow the description of Brechet *et al.* [3.49] rather close.

The first step in this description is to consider the representation of the refractive index profile and mode fields of the photonic crystal fibre. In order to obtain a precise description of the field distribution over a fibre cross section, and especially near the holes, the classical Maxwell differential equations must be solved for a large set of properly chosen elementary subspaces. Doing this it is essential that the conditions of continuity of the fields is taken into account. The general approach first consists of a step, where the cross section of the modeled guide is split into distinct homogeneous subspaces. This parceling results in a mesh of simple finite elements: triangles and quadrilaterals in the two-dimensional case as shown in Figure 3-13. This approach allows for the application of different-sized elements, depending on the structure and expected mode-field response. For a better description of the fields, as the distance of the subspaces to the center becomes shorter, their dimensions are generally chosen to be smaller. The Maxwell equations are discretized for each element, leading to a set of elementary matrices [3.48]. The combination of these elementary matrices creates a global matrix system for the entire structure studied. Finally, the effective index and the distributions of the amplitudes and of the polarizations of the modes may be numerically computed. This is done by taking into account the conditions of continuity at the boundary of each subspace.

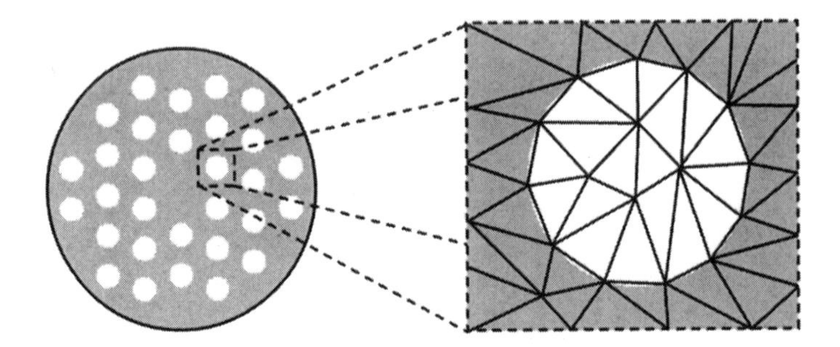

Figure 3-13. The cross-section of a photonic crystal fibre, showing a subsection around one air hole, where discretization in elementary subspaces (finite-elements) has been made.

In the work by Brechet *et al.* [3.49], the calculations were performed using software, which was initially created by Aubourg *et al.* [3.50] for microwave circuits and resonator applications. This model may also be applied to homogeneous and linear materials, isotropic or not, and these materials may be with or without loss. The tools are, therefore, quite general, and the PCF fibres fully comply with these conditions. The FEM may, consequently, be used efficiently for studies of PCFs.

Using the above-sketched FEM, Brechet *et al.* [3.49] studied the polarization of the first electromagnetic modes of a PCF. Their first simulations were performed over a whole section of a PCF, and the results showed that the fields of the modes exhibit at least two symmetries. Thus, in order to reduce the duration of the calculations, it was concluded that it is possible to study only a quarter of the fibres. For this, an electric or magnetic short circuit (named, respectively, ESC and MSC) was applied along each symmetry axis. The idea is that an ESC (respectively MSC) makes the electric field perpendicular (respectively parallel) to it. Then, the polarization of a considered mode over a quarter of the section is preserved - subject to the choice of a suitable combination of the two applied short circuits. An example is given in Figure 3-14 with a sketch of the HE_{11} mode and a vertical-horizontal ESC-MSC combination. This part of our description also follows the work by Brechet *et al.* [3.49]. Due to the various axes of symmetry that some modes exhibit, it is possible to find these modes with different combinations of short circuits. Brechet *et al.* [3.49] for example notes that the HE_{31} mode is found with both the MSC-ESC and ESC-MSC combinations. They further find that the difference between the effective indexes corresponding to each combination for one particular mode is close to 10^{-6} most of the time. This number corresponds to the limit of precision of the software. Nevertheless, for better precision Brechet *et al.* [3.49] chose to study each case and calculate the average value of the computed effective indexes. Note that Brechet *et al.* have validated the FEM method by successfully computing the well-known modes of step index fibres (SIFs).

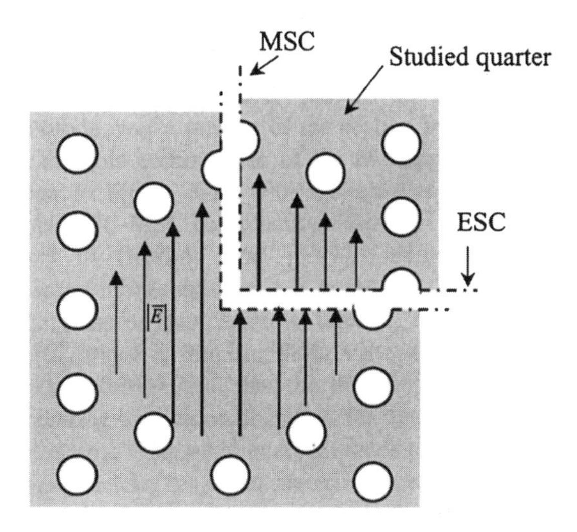

Figure 3-14. A combination of a magnetic short circuit (MSC) and an electric short circuit (ESC), applied on a quarter of a photonic crystal fibre. In the specific example, the finite-element method (FEM) has been used to compute the HE_{11} mode with its right polarization. The illustration is sketched after the idea published by Brechet *et al.* [3.49].

It is also relevant to draw forward a quite recent paper by Cucinotta *et al.* [3.55] describing the analysis of a microstructured fibre with very small effective area analysed by the FEM. This work by Cucinotta *et al.* demonstrates the possibility to design fibres applicable for highly nonlinear applications and dispersion compensation (exemplified with a dispersion value as low as –330 ps/(km·nm) for a specific design) is mentioned as an option.

3.10 THE BEAM-PROPAGATION METHOD

As we discussed in the previous section, also well-established numerical methods – and primarily those, which takes the vectorial nature of the mode fields into account – may be successfully applied in the analysis of photonic crystal fibres. This is also the case for the method, which we shortly should mention in this section, namely the well-known Beam-Propagation Method (BPM). The fact that a method is well established in the field of electromagnetics, should, however, not lead someone to believe that it is without challenges to apply the method on a new problem. The reason is that very often, the computer code is adapted to a specific sub-class of components, and it may be a significant task to describe the appropriate structures, and to obtain the parameters characteristic for a new class of components.

The BPM is of course a very well known tool for analysis of optical waveguides, and it is described in numerous textbooks (see e.g., [3.53]). The method has, however, to our knowledge not formed the basis of many independent research papers, but it has been referred to by well established researchers, in the field of photonic crystal fibres (note that Eggleton *et al.* [3.11] use the term microstructured optical fibres). One of the earliest reported applications of the BPM in the analysis of PCFs were described by Eggleton *et al.* [3.11,3.52].

Eggleton *et al.* [3.52] states that the beam-propagation method provides a simple intuitive method of determining the modal spectrum and modal profiles for complex waveguides. They further notes that the beam-propagation correlation method has been used extensively in the studies of complex waveguides and that it is particularly well suited to compute mode evolution in waveguides that vary in the longitudinal direction and in geometries, where leaky modes are important [3.11]. The particular reason for this line of argumentation is naturally related to the type of problems that Eggleton *et al.* are solving, namely the analysis of grating structures in microstructured optical fibres. Briefly explained, the BPM correlation method – as applied by Eggleton *et al.* – propagates a launched field profile within a waveguide. The propagation of the field along the z-direction through a transverse guide can be written as:

$$E(x,y,z) = \sum_i \alpha_i E_i(x,y) e^{-j\beta_i z} \tag{3.90}$$

where for each mode i, $E_i(x,y)$ is the transverse modal profile, α_i is the amplitude strength of each mode, and $\beta_i = 2kn_i$ is the wave vector in the propagation direction z.

The correlation function computes the initial launched profile, and the profile at each z-value is given by:

$$P(z) = \int E(x,y,z) E^*(x,y,z) dxdy \tag{3.91}$$

This longitudinal profile $P(z)$ is then determined for a given refractive index function, and the coupling constants necessary for a coupled mode analysis may then be derived.

3.11 THE EQUIVALENT AVERAGED INDEX METHOD

The final method, which we have chosen to include in this chapter, represent a recent result, but it also belongs to the class of more qualitative approaches taking point of reference in a simplified system of equations describing the waveguide and mode propagation. More specifically, the method is a scalar approach, which was developed by Peyrilloux *et al.* [3.47] and presented in a comparison between results from the finite-element method (FEM), the localized-function method (LFM), and this more simple approach named the equivalent averaged index method (EAIM). This method is scalar, and it was developed by Peyrilloux *et al.* [3.47] because the aforementioned LFM and FEM methods are demanding on computer time and memory. A novel and more simple method based on an equivalent averaged index profile was, therefore, proposed to reduce computation time.

First, Peyrilloux *et al.* [3.47] considered the 2D-PCF index profile $n(r,\varphi)$ (see Figure 3-15 (a)). r and φ are the radial and azimuthal co-ordinates, respectively. Due to the $2\pi/6$ periodicity of the hexagonal structure considered by Peyrilloux *et al.*, it was possible to develop $n\ (r,\varphi)$ together with the guided mode expression $E = E(r,\varphi)\, e^{j(\omega t-\beta z)}$ into Fourier series as follows:

$$n^2(r,\varphi) = \overline{n^2(r)} + \sum_{m=0} \alpha_m(r)\cos(6m\varphi) \tag{3.92}$$

$$E(r,\varphi) = \overline{E(r)} + \sum_{m=0} a_m(r)\cos(6m\varphi) \tag{3.93}$$

where the real higher-order Fourier terms $\alpha_m(r)$ and $a_m(r)$ are given as:

$$\alpha_m(r) = \frac{2}{\pi}\int_0^{\pi} n^2(r\cdot\varphi)\cos(6m\varphi)\,d\varphi \tag{3.94}$$

$$a_m(r) = \frac{2}{\pi}\int_0^{\pi} E^2(r\cdot\varphi)\cos(6m\varphi)\,d\varphi \tag{3.95}$$

The EAIM approach only considers the mean terms of the above Fourier series:

$$\overline{E(r)} = \frac{3}{\pi}\int_0^{\pi/3} E(r,\varphi)\,d\varphi \tag{3.96}$$

and

$$\overline{n^2(r)} = \frac{3}{\pi}\int_0^{\pi/3} n^2(r,\varphi)\,d\varphi \tag{3.97}$$

While E is a solution to the two-dimensional scalar wave equation, the averaged parameter $\overline{E} = \overline{E(r)}e^{j(\omega t - \beta z)}$ is a solution to the one-dimensional wave equation given as [3.47]:

$$\frac{d^2\overline{E(r)}}{dr^2} + \frac{1}{r}\frac{d\overline{E(r)}}{dr} + \left(k_0^2\overline{n^2(r)} - \beta^2\right)\overline{E(r)} = 0 \tag{3.98}$$

Peyrilloux *et al.* [3.47] take advantage of the symmetry axis shown in Figure 3-15 (a) in order to minimize the computation time. Hence, at each distance $\mathrm{r}, \mathrm{n}^2(\mathrm{r},\varphi)$ is integrated over a $\pi/6$ angular sector providing the equivalent averaged refractive index profile shown in Figure 3-15 (b). A classical resolution of Equation (3.98) is then performed leading to the averaged electric field amplitude (sketched on Figure 3-15 (b)) and to the propagation constant $\beta = k_0 n_e$.

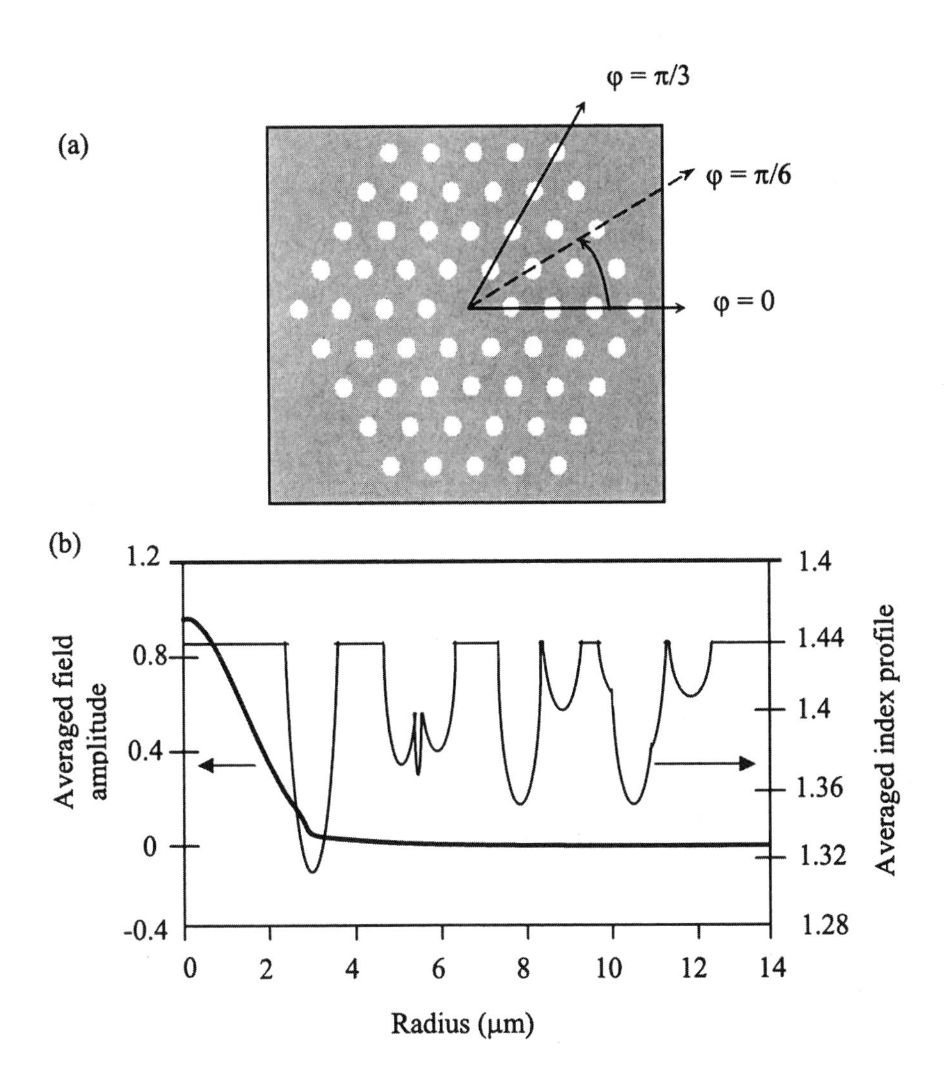

Figure 3-15. (a) Illustration of the two-dimensional refractive index profile $n(r, \varphi)$. (b) The averaged refractive index profile $\overline{n(r)}$ and averaged electric field amplitude $\overline{E(r)}$. The illustration is developed from the idea presented in [3.47].

3.12 SUMMARY

In this chapter, we have shortly reviewed ten different numerical methods, which over the past 7 years have been developed for the analysis and design of photonic crystal fibres. Some of the methods are fundamentally interesting for the understanding of the development of the required waveguide theory (e.g., the plane-wave method), whereas others are of a more general character, and they have not been discussed in much detail here. Yet other methods have their strength in a relative simple formulation, although the same methods generally are insufficient for accurate parameter determination - or even more requiring – for calculation on band gap fibres. However, it has been our ambition under all circumstances to present sufficient information and references so that further studies and developments of numerical tools within the area of PCF research are facilitated. In order to help this process further along, we have included a table containing a few key points for each of the discussed methods, together with a few references, where further information may be sought.

The application of many of these methods will be reflected in several results presented in the proceeding chapters on fibre properties and applications. Finally, it should be noted that considering the very high research activity within the area of photonic crystal fibres, it should be expected that even more numerical methods will be taken in use.

Method	Advantages	Disadvantages	Ref.
Effective-Index Method (EIM)	+ Simple + Low computation time	- No polarisation prop. - Inaccurate modal prop. - No PBG analysis possible	[3.2], [3.3], [3.8]
Localized Functions Method (LFM)	+ Solutions may be adapted to finite size structures + May describe random hole distributions	- Needs great care on result interpretation - Rel. complex method - Inefficient PBG analysis	[3.13]
Plane-Wave Method (PWM)	+ Well suited for PBGs + Good agreement with experiments demonstrated + Widespread approach	- Large supercells needed for complex structures - Very demanding for full fibre analysis	[3.9], [3.10]
Biorthonormal Basis Method (BBM)	+ Vectorial mode repress. + Flexible and efficient + Highly accurate	- Demanding formulation	[3.43] [3.44] [3.45]
Multipole (MPM)	+ Describes effects of finite cladding region + No false birefringence errors + Suited for symmetry stud. + Leakage loss prediction	- Cannot analyse arbitrary cladding configurations	[3.20] [3.21] [3.22] [3.46]
Fourier Decomposition Method (FDM)	+ Leakage loss prediction	- Scalar formulation - Requires adjustable boundary condition	[3.23]
Finite Difference (FD) Methods (FDTD, FDFD)	+ Very general approach + May describe arbitrary structures + Well established and tested	- Non-modal appraoch - Numerically intensive - Requires detailed treatment of boundaries	[3.61] [3.62] [3.75]
Finite Element Method (FEM)	+ Reliable (well-tested) method + Accurate modal descript.	- Complex definition of calculation mesh	[3.12] [3.49]
Beam-Propagation Method (BPM)	+ Reliable (well-tested) method + Commercially available + May use complex prop. constant	- Relatively computationally inefficient	[3.11] [3.52]
Equivalent Averaged Index Method (EAIM)	+ Simple to implement + Numerically efficient	- Scalar approaximation - Only qualitative results	[3.23]

Table 3-1. Overview of advantages and disadvantages concerning different numerical methods used for analysis and design of photonic crystal fibres.

REFERENCES

[3.1] J. Broeng, D. Mogilevtsev, S. E. Barkou, and A. Bjarklev, "Photonic crystal fibres: a new class of optical waveguides", *Optical Fiber Technology,* vol. 5, July 1999, pp. 305-330.

[3.2] T. Birks, J. Knight, and P. St. J. Russell, "Endlessly single-mode photonic crystal fiber", *Optics Letters,* vol. 22, July 1997, pp. 961-963.

[3.3] A. Bjarklev, J. Broeng, S. Barkou, and K. Dridi, "Dispersion properties of photonic crystal fibers", *European Conference on Optical Communications,* pp. 135-6, Madrid, September 20-24, 1998.

[3.4] D. Gloge, "Weakly guiding fibers", *Applied Optics*, Vol.10, 1971, p.2252.

[3.5] T. A. Birks, P. J. Roberts, P. St. J. Russell, D. M. Atkin, and T. J. Shepherd, "Full 2-d photonic band gaps in silica/air structures," *IEE Electronics Letters*, vol. 31, 1995, p.1941.

[3.6] J. W. Fleming, "Material dispersion in lightguide glasses", *IEE Electronics Letters,* vol, 14, 1978, p. 326.

[3.7] J. C. Knight, J. Arriaga, T. A. Birks, A. Ortigosa-Blanch, I. W. Wadsworth, and P. St. J. Russell, "Anomalous Dispersion in Photonic Crystal Fiber", *IEEE Photonics Technology Letters*, Vol.12, 2000, pp. 807-809.

[3.8] J. Riishede, S. B. Libori, A. Bjarklev, J. Broeng, and E. Knudsen, *Proc. 27th European Conference on Optical Communication*, ECOC'2001, Paper Th.A.1.5 (2001).

[3.9] K. M. Ho, C. T. Chan, and C. M. Soukoulis, "Existence of a photonic gap in periodic dielectric structures", *Phys. Rev. Lett.,* Vol.65, 1990, pp. 3152-3155.

[3.10] R. D. Meade, A. M. Rappe. K. D. Brommer, J. D. Joannopoulos, and O. L. Alerhand, "Accurate theoretical analysis of photonic band-gap materials", *Physical Review B*, Vol.48, 1993, pp. 8434-8437.

[3.11] B. J. Eggleton, P. S. Westbrook, R. S. Windeler, S. Spalter, and T. A Strasser, "Grating resonances in air/silica micro structured optical fibers", *Optics Letters*, Vol.24, 1999, pp.1460-1462.

[3.12] F. Brechet, J. Marcou, D. Pagnoux, and P. Roy,
"Complete analysis of the characteristics of propagation into photonic crystal fibers, by the Finite Element Method",
Optical Fiber Technology, Vol.6, 2000, pp.181-191.

[3.13] D. Mogilevtsev, T. A. Birks, P. St. J. Russell,
"Group-velocity dispersion in photonic crystal fibers",
Optics Letters, Vol.23, 1998, pp.1662-1664.

[3.14] A. W. Snyder, and J. D. Love,
"Optical Waveguide Theory",
Chapmann and Hall, London, 1983, pp.595-606.

[3.15] T. M. Monro, D. J. Richardson, N. G. R. Broderick, and P. J. Bennett,
"Holey optical fibers: An efficient modal model",
IEEE Journal of Lightwave Technology, Vol. 17, 1999, pp. 1093-1101.

[3.16] T. M. Monro, D. J. Richardson, N. G. R. Broderick, and P. J. Bennett,
"Modelling large air fraction holey optical fibers",
IEEE Journal of Lightwave Technology, Vol. 18, 2000, pp. 50-56.

[3.17] T. M. Monro, N. G. R. Broderick, and D. J. Richardson,
"Exploring the optical properties of holey fibres",
NATO Summer School on Nanoscale Linear and Nonlinear Optics (Erice, Sicily) Jul.2000.

[3.18] P. J. Bennett, T. M. Monro, D. J. Richardson, '
"Towards practical holey fibre technology: Fabrication Splicing Modeling and Characterization",
Optics Letters, Vol.24, 1999, pp.1203-1205.

[3.19] M. J. Steel, T. P. White, C. M. de Sterke, R. C. McPhedran, and L. C. Botten,
"Symmetry and degeneracy in microstructured optical fibers",
Optics Letters, Vol. 26, 2001, pp. 488- 490.

[3.20] T. M. Monro, K. M. Kiang, J. H. Lee, K. Frampton, Z. Yusoff, R. Moore, J. Tucknott, D. W. Hewak, H. N. Rutt, and D. J. Richardson,
"Highly nonlinear extruded single-mode holey optical fibers",
Proc OFC'2002, OSA Technical Digest 315-317, Anaheim, California, 2002.

[3.21] T. P. White, R. C. McPhedran, L. C. Botten, G. H. Smith, and C. M. de Sterke,
"Calculations of air-guided modes in photonic crystal fibers using the multipole method",
Optics Express, Vol. 11, 2001, pp. 721- 732.

[3.22] T. P. White, R. C. McPhedran, C. M. de Sterke, L. C. Botten, and M. J. Steel,
"Confinement losses in microstructured optical fibres",
Optics Letters, Vol.26, 2001, pp. 1660-1662.

[3.23] L. Poladian, N. A. Issa, and T. M. Monro,
"Fourier decomposition algoritm for leaky modes of fibres with arbitrary geometry",
Optics Express, Vol. 10, No. 10, 2002, pp. 449-454.

[3.24] K. Leung, and Y. Liu,
"Full vector wave calculation of photonic band structures in face-centered-cubic dielectric media",
Physical Review Letters, Vol. 65, Nov. 1990, pp. 2646-2649.

[3.25] Z. Zhang, and S. Satpathy,
"Electromagnetic wave propagation in periodic structures: Bloch wave solution of Maxwell's equations,"
Physical Review Letters, Vol. 65, Nov. 1990, pp. 2650-2653.

[3.26] R. Ziolkowski,
"FDTD modelling of photonic nanometer-sized power splitters and switches",
Integrated Photonics Research, Vol. 4 of *Technical Digest Series,* p. ITuA2, OSA, March 30 – April 1, 1998.

[3.27] S. Fan, P. Villeneuve, J. Joannopoulos, and H. Haus,
"Channel drop tunneling through localized states",
Physical Review Letters, Vol. 80, Feb. 1998, pp. 960-963.

[3.28] A. Taflove,
"Computational electrodynamics: The finite-difference time-domain method",
Artech House Publishers, 1995.

[3.29] J. Pendry,
"Photonic band structures",
Journal of Modern Optics, Vol. 41, Feb. 1994, pp. 209-229.

[3.30] C. Chan, Q. Yu, and K. Ho,
"Order-n spectral method for electromagnetic waves,"
Phys. Rev. B, 1995.

[3.31] D. Maystre,
"Electromagnetic study of photonic band gaps,"
Pure and Applied Optics: Journal Of The European Optical Society, Part A, Vol.3, 1994, p.975.

[3.32] G. Agrawal,
"Nonlinear fiber optics",
Academic Press, Second ed., 1995.

[3.33] J. Joannopoulos, I. Winn, and R. Meade,
"Photonic Crystals: Molding the Flow of Light",
Princeton University Press, 1995.

[3.34] E. Knudsen, A. Bjarklev, J. Broeng, S. B. Libori,
"Modelling photonic crystal fibres with localised functions",
SPIE Photonics West 2002, Optical Fibers and Sensors for Medical Applications II, Vol. 4616, pp. 81-90, 2002.

[3.35] M. Plihal and A. Maradudin,
"Photonic band structure of two-dimensional systems: The triangular lattice,"
Physical Review B, vol. 44, pp. 8565-71, Oct. 1991.

[3.36] E. Yablonovitch,
"Photonic band-gap structures,"
Journal of the Optical Society of America B, vol. 10, pp. 283-95, Feb. 1993.

[3.37] P. Villeneuve, and M. Piche,
"Photonic bandgaps: What is the best numerical representation of periodic structures?,"
Journal of Modern Optics, vol. 41, pp. 241-56, Feb. 1994.

[3.38] R. Meade, A. Rappe, K. Brommer, and J. Joannopoulos,
"Nature of the photonic band gap: some insights from a field analysis,"
Journal of the Optical Society of America B, vol. 10, pp. 328-32, Feb. 1993.

[3.39] R. Meade, A. Rappe, K. Brommer, J. Joannopoulos, and 0. Alerhand,
"Accurate theoretical analysis of photonic band-gap materials",
Physical Review B, vol. 48, pp. 8434-7, Sept. 1993.

[3.40] H. Sözüer, J. Haus, and R. Inguva,
"Photonic bands: convergence problems with the plane-wave method",
Physical Review B, vol. 45, pp. 13962-72, June 1992.

[3.41] M. Teter, M. Payne, and D. Allan,
"Solution of Schrödinger's equation for large systems,"
Physical Review B, 1989.

[3.42] D. Aspnes,
"Local-field effects and effective- medium theory: A microscopic perspective,"
Am.J. Phys., vol. 50, pp. 704-709, Aug.1981.

[3.43] E. Silvestre, M. V. Andrés, and P. Andrés,
"Biorthonormal-basis method for the vector description of optical-fiber modes",
IEEE Journal of Lightwave Technology, Vol.16, pp.923-928, 1998.

[3.44] A. Ferrando, E. Silvestre, J. J. Miret, P. Andrés, and M. V. Andrés,
"Full-vector analysis of a realistic photonic crystal fiber",
Optics Letters, Vol.24, No.5, pp.276-278, 1999.

[3.45] J. J. Mirret Marí,
"Description Vectorial de las Fibras de Cristal Fotónico. Propiedades y Aplicaciones",
Tesis Doctoral, Julio 2002 Valencia España.

[3.46] B. T. Kuhlmey, T. P. White, G. Renversez, D. Maystre, L. C. Botten, C. M. de Sterke. and R. McPhedran,
"Multipole method for microstructured optical fibers II. Implementation and results",
J. Opt. Soc. Am., Vol.19, No.10, pp.2331-2340, 2002.

[3.47] A. Peyrilloux, S. Février, J. Marcou, L. Berthelot, D. Pagnoux, and P. Sansonetti,
"Comparison between the finite element method, the localized function method and a novel equivalent averaged index method for modelling photonic crystal fibres",
Journal of Optics A: Pure and applied optics, Vol. 4, pp.257-262, 2002.

[3.48] T. Itoh, G. Pelosi, and P. Silvester,
"Finite Element Software for Microwave Engineering",
Wiley-Interscience, New York, 1996.

[3.49] F. Brechet, J. Marcou, D. Pagnoux, and P. Roy,
"Complete analysis of the characteristics of propagation into photonic crystal fibers, by the finite element method",
Optical Fiber Technology, Vol. 6, pp.181-191, 2000.

[3.50] M. Aubourg, and P. Guillon,
"A mixed finite element formulation for microwave device problems. Applications to MIS structure",
J. Electromag. Waves Appl., Vol.5, No.45, 1991, p.371.

[3.51] K. S. Kunz, and R. J. Luebbers,
"The finite difference time domain method for electromagnetics",
CRC, Boca Raton, 1993.

[3.52] B. J. Eggleton, C. Kerbage, P. S. Westbrook, R. S. Windeler, and A. Hale,
"Microstructured optical fiber devices",
Optics Express, Vol.9, No.13, Dec.2001, pp.698-713.

[3.53] K. Okamoto,
"Fundamentals of Optical Waveguides",
Academic Press, San Diego, 2000, ISBN 0-12-525095-9

[3.54] A. W. Snyder, and J. D. Love,
"Optical Waveguide Theory",
London, U.K., Chapman and Hall, 1983, pp.595-606.

[3.55] A. Cucinotta, S. selleri, L. Vincetti, and M. Zoboli,
"Holey fiber analysis through the Finite-Element Method",
IEEE Photonics Technology Letters, Vol.14, No.11, Nov.2002, pp.1530-1532.

[3.56] P. M. Morse, and H. Feshbach,
"Methods of Theoretical Physics",
New York,: McGraw-Hill, 1953, pt.1, pp.884-886.

[3.57] F. R. Gantmacher,
"Théorie des Matrices",
Paris, France: Dunod, 1966, vol.1, pp.268-271.

[3.58] A. Ferrando, E. Silvestre, J. J. Miret, J. A. Monsoriu, M. V. Andrés, and P. St. J. Russell,
"Designing a photonic crystal fibre with flattened chromatic dispersion",
IEE Electronics Letters, Vol.35, No.4, Febr. 1999, pp.325-327.

[3.59] M. Abramowitz, and I. A. Stegun,
"Handbook of mathematical functions",
Dover, New York, 1965.

[3.60] C. A. J. Fletcher,
"Computational Galerkin Methods",
Springer Verlag, 1984

[3.61] M. Qiu,
"Analysis of guided modes in photonic crystal fibers using the finite-difference time-domain method",
Microwave and Optical Technology Letters, Vol.30, No.5, Sept.2001, pp.327-330.

[3.62] G. E. Town, and J. T. Lizier,
"Tapered holey fibers for spot size and numerical aperture conversion",
Proc. CLEO'2001, Paper CtuAA3, p.261.

[3.63] C. T. Chan, Q. L. Yu, and K. M. Ho,
"Order-N spectral method for electromagnetic waves",
Phys. Rev. B, Vol.51, 1995, p.16635.

[3.64] M. Qiu, and S. He,
"A non-orthogonal finite-difference time-domain method for computing the band structure of a two-dimensional photonic crystal with dielectric and metallic inclusions",
J. Appl. Phys., Vol.87, 2000, p.8268.

[3.65] M. Qiu, and S. He,
"FDTD algorithm for computing the off-plane band structure in a two-dimensional photonic crystal with dielectric or metallic inclusions",
Phys. Lett. A., Vol.278, 2001, p.348.

[3.66] M. Qiu, and S. He,
"Numerical method for computing defect modes in two-dimensional photonic crystals with dielectric or metallic inclusions",
Phys. Rev. B, Vol.61, 2000, p.12871.

[3.67] M. Qiu, and S. He,
"Guided modes in a two-dimensional metallic photonic crystal waveguide",
Phys. Lett. A., Vol.266, 2000, p.425.

[3.68] S. E. Barkou, J. Broeng, and A. Bjarklev,
"Silica-air photonic crystal fiber design that permits waveguiding by a true photonic bandgap effect",
Optics Letters, Vol.24, 1999, p.46.

[3.69] K. S. Yee,
"Numerical solution of initial boundary value problems involving Maxwell's equations in isotropic media",
Antennas Propagation, Vol.14, 1966, p.302.

[3.70] A. Taflove,
"Computational electrodynamics: The finite-difference time-domain method",
Artech House, Norwood, MA, 1995.

[3.71] A. Asi, and L. Shafai,
"Dispersion analysis of anisotropic inhomogeneous waveguides using compact 2D-FDTD"
IEE Electronics Letters, Vol.28, 1992, p.1451.

[3.72] M. Celuch-Marcysiak, and W. K. Gwarek,
"Spatially looed algorithms for time-domain analysis of periodic structures",
IEEE Trans. Microwave Theory Tech., Vol.43, 1995, p.860.

[3.73] A. C. Cangellaris,
"numerical stability and numerical dispersion of a compact 2-D/FDTD method used for the dispersion analysis of waveguides",
IEEE Microwave Guided Wave Lett., Vol.3, 1993, p.3.

[3.74] J. P. Berenger,
"A perfectly matched layer for the adsorption of electromagnetic waves",
J. Computational Phys., Vol.114, 1994, pp.185-200.

[3.75] Z. Zhu, and T. G. Brown,
"Full-vectorial finite-difference analysis of microstructured optical fibers",
Optics Express, Vol.10, No.17, Aug.2002, pp.853-864.

[3.76] M. S. Stern,
"Semivectorial polarized finite difference method for optical waveguides with arbitrary index profiles",
IEE Proc. J. Optoelectron., Vol.135, 1988, pp.56-63.

[3.77] W. P. Huang, and C. L. Xu,
"Simulation of three-dimensional optical waveguides by a full-vector beam propagation method",
IEEE Journal of Quantum Electronics, Vol.29, 1993, pp.2639-2649.

[3.78] W. P. Huang, C. L. Xu, S. T. Chu, and S. K. Chaudhuri,
"The finite-difference vector beam propagation method. Analysis and assessment",
IEEE Journal of Lightwave Technology, Vol.10, 1992, pp.295-305.

[3.79] K. Bierwirth, N. Schulz, and F. Arndt,
"Finite-difference analysis of rectangular dielectric waveguide structures",
IEEE Trans. Microwave Theory Tech., Vol 34, 1986, pp. 1104-1113.

[3.80] H. Dong, A. Chronopoulos, J. Zou, and A. Gopinath,
"Vectorial integrated finite-difference analysis of dielectric waveguides",
IEEE Journal of Lightwave Technology, Vol.11, 1993, pp.1559-1563.

[3.81] P. Lüsse, P. Stuwe, J. Schüle, and H. G. Unger,
"Analysis of vectorial mode fields in optical waveguides by a new finite difference method",
IEEE Journal of Lightwave Technology, Vol.12, 1994, pp.487-493.

[3.82] T. A. Birks, D. Mogilevtsev, J. C. Knight, and P. St. J. Russell,
"Single material fibres for dispersion compensation",
Proc. OFC′1999, pp. 108-110, paper FG2

[3.83] M. Midrio, M. P. Singh, and C. G. Someda,
"The space filling mode of holey fibers: An analytical vectorial solution",
IEEE Journal of Lightwave Technology, Vol. 18, No. 7, July 2000, pp. 1031-1037.

Chapter 4

FABRICATION OF PHOTONIC CRYSTAL FIBRES

4.1 INTRODUCTION

The idea of producing optical fibres from a single low-loss material with microscopic air holes goes back to the early days of optical fibre technology, and already in 1974 Kaiser *et al.* [4.1] reported the first results on single-material silica optical fibres. In the early days – as well as today – the key issues have been to obtain a desired fibre structure for a given application, and maintain this structure for very long fibre lengths. It will, generally, be needed that the fibre attenuation is kept at a rather low level, and the acceptable attenuation level will be given by the specific application. In this chapter, we will address the fundamental issues of fabrication of photonic crystal fibres, by first discussing the most commonly used preform fabrication method. Secondly, we will report details about the fibre drawing and coating procedure. Furthermore, we will discuss how additional doping techniques are needed for providing hybrid fibre types (such as the hole-assisted lightguide fibre (HALF) [4.6]) combining the approach of micro-structuring with index-raised doped glass or active dopants such as rare-earth ions needed for new amplifiers and lasers. The chapter will also shortly address the issues of photonic crystal fibres in low-melting-point glasses and polymers.

4.2 PRODUCTION OF PHOTONIC CRYSTAL FIBRE PREFORMS

The traditional way of manufacturing optical fibres involves two main steps: fabrication of a fibre-preform and drawing of this using a high-temperature furnace in a tower set-up [4.2]. For conventional silica-based optical fibres, both techniques that have been developed over the past two-to-three decades may today be considered very mature (see e.g., [4.3]). Various vapour deposition techniques have been developed for the fabrication of fibre preforms (including among others; Modified Chemical Vapour Deposition (MCVD), Vapour Axial Deposition (VAD), and Outside Vapour Deposition (OVD) [4.1]). These techniques allow for fabrication of preforms with silica glasses having very low, un-intentional, impurity levels, and very precisely controlled doping levels. A common characteristic of these perform fabrication techniques is, however, that they are tailored for fabrication of circular-symmetric preforms, and although a very accurate control over the deposition is obtainable, this may without significant modifications of the methods only be achieved in radial direction. Therefore, for fabrication of preforms for photonic crystal fibres, where the transversal preform morphology often is realized in a circular symmetric structure and even as it is the case for photonic bandgap fibres with a two dimensional periodic structure, the above-referenced techniques are not directly applicable. Preforms for optical fibres having a non-circular symmetry have, however, been fabricated for several years, e.g., for polarization preserving fibres, where typically a mechanical post-processing of the circular preforms is performed. This may be done, e.g., by drilling a very limited number of holes down the fibre perform and feeding additional material components into these holes (and provide, e.g., stress-induced refractive index changes in the core region) or milling/cutting the outer surface of the perform to provide an outer asymmetry that during drawing gives rise to an asymmetric core [4.4].

Although drilling of several tens to hundreds of holes in a periodic arrangement into one final preform may be utilized for fabrication of photonic crystal fibres, a different - and relatively simple - method has been developed over the past three to four years. The method typically utilizes a hand-stacking technique of silica capillary tubes and solid silica rods and allows relatively fast, low-cost and flexible preform manufacturing. The method was first presented by Birks *et al.* in 1996 [4.5], who described the fabrication of a 2D photonic crystal cladding structure using a method, where a central hole was drilled into a silica rod. Initially, Birks *et al.* started from 25 cm long solid silica rods with a diameter around 3 cm, and used an ultrasonic drill to form a central hole through these. This production step was followed by milling the outside of the tubes to form six flats. After drawing these large tubes to smaller dimensions (diameter around 1 mm), they were stacked in a close-packed manner using a special mechanical supporting unit to provide a fibre preform with an outer flat-to-flat distance of around 2-3 cm. Typically, around 300 silica capillary tubes were used to create a single preform.

Recently, the method has been further developed (or simplified) in the sense that the fibre preforms often are manufactured by direct stacking of silica tubes and rods with a circular outer shape (i.e., the hexagonal milling of the rods have been omitted). Figure 4-1 shows a picture of a photonic crystal fibre preform manufactured by stacking circular elements.

Figure 4-1. A photonic crystal fibre preform formed by manual stacking of silica rods and tubes. The outer diameter of the stack is 2 centimeters. The picture is kindly provided by Crystal Fibre A/S.

It should be noted that the method employing circular elements, however, introduces additional air gaps in the fibre preform. These air gaps are positioned at the interstitial sites mid-between the centres of three adjacent tubes/rods - and they are, therefore, referred to as interstitial holes. It is important to note that the accurate control of fibre structures with regard to control of hole dimensions and location is essential for the manufacturing of photonic crystal fibres with specific properties. In this respect, the inclusion or elimination of interstitial holes may be highly relevant for the bandgap formation (as described elsewhere in this book). A significant effort has, therefore, been put into the accurate control of interstitial holes, and fibres having numerous interstitial holes has been fabricated as shown in Figure 4-2.

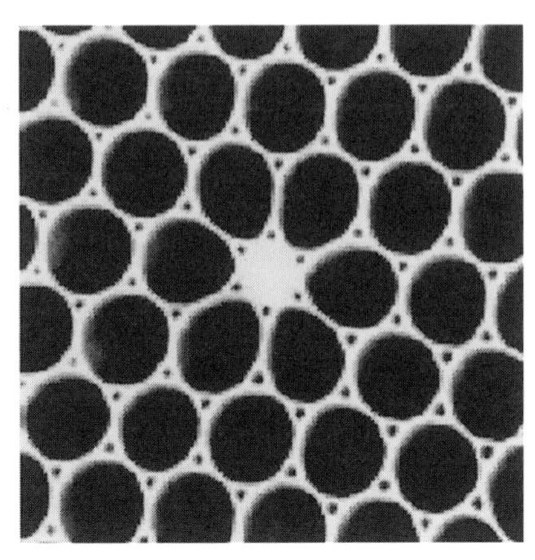

Figure 4-2. Scanning electron micrograph of end-phase of a photonic crystal fibre having numerous interstitial holes maintained open after fibre drawing. The picture is kindly provided by Crystal Fibre A/S.

Several different fibre structures may be obtained using the stacking of rods and tubes with equal outer dimensions. Among some of the most relevant are the close-packed hole structure - or as we choose to name it in this book – the triangular structure (as shown in Figure 4-2), the honeycomb structure (see Figure 4-3) or the Kagomé structure [4.23].

With respect to honeycomb photonic crystals, the interstitial holes were found to have a predominantly positive effect, and the modified preform fabrication method, therefore, is further advantageous by favouring these interstitial holes. To realize fibres with a honeycomb photonic crystal cladding and a single, central periodicity-breaking air hole, a preform with a cross-section as indicated in Figure 4-3 may be designed. This kind of photonic crystal fibre preform is specifically interesting, since it lead to the first PBG guiding fibre demonstrated in 1998 [4.7].

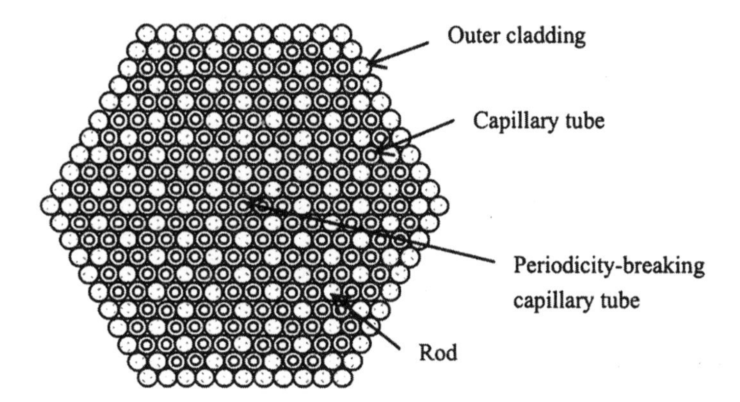

Figure 4-3. Schematic illustration of the arrangement of capillary tubes and rods that may be used as design for the fabrication of a PBG- fibre with a honeycomb cladding structure.

For the PBG-fibre preform illustrated in Figure 4-3, the periodicity-breaking core region is introduced in a very simple manner by replacing a single rod by a tube of similar size as the remainder of tubes forming the fibre preform. As seen from the illustration, the periodic cladding structure is terminated after about four honeycomb cells with a single surrounding layer of solid silica rods. Other types of termination of the periodic cladding structure may be thought of, e.g., utilization of a large silica over-cladding tube as previously discussed in this chapter. Such additional steps have, however, not been taken for the first PBG fibres.

After stacking, the capillaries and rods may be held together by thin wires and fused together during an intermediate drawing process, where the preform is drawn into preform-canes. Such an intermediate step may be introduced to provide a large number of preform-canes for the development and optimization of the later drawing of the PCFs to their final dimensions. During the drawing process, the outer lying tubes/rods may experience some distortion, but the core region and its nearest surroundings generally retain to a large degree the desired morphology.

4.3 DRAWING OF PHOTONIC CRYSTAL FIBRE

The drawing of silica-based photonic crystal fibres is generally performed in a conventional drawing tower operating at a relatively low temperature around 1900°C. The reason for drawing at this temperature level is that surface tension otherwise may be found to collapse the air holes. The lowering of the preform drawing temperature (by a couple of hundred degrees) compared to standard fibre fabrication conditions is reported by several other research groups, see e.g., Eom *et al.* [4.17]. Eom *et al.* further reports that as the drawing time increased, enlarged opening of air holes was seen. This indicates that time dynamics and temperature variations become significant properties in the accurate control of PCFs.

An illustration of the drawing of a PCF is illustrated in Figure 4-4. On the picture we see, how the preform is feed into a circular oven from the top (the preform is lighting up due to the high temperature), and below the oven, the photonic crystal fibre is handled (we see the hands of the operator).

Figure 4-4. Drawing of a photonic crystal fibre from at a temperature around 1900°C. The picture is kindly provided by Crystal Fibre A/S.

The key element in the drawing of the photonic crystal fibre is the ability to maintain the highly regular structure of the preform all the way down to fibre dimensions. The typical outer fibre dimensions of "state-of-the-art" PCFs are 125 micron diameters, although outer dimensions may be chosen for special applications. Examples of this will be given in Chapter 7. To illustrate the transformation from preform to fibre dimensions, we have included a picture of the drawing transition in Figure 4-5 (a).

An example of the cross section of an index-guiding PCF fabricated under accurate control of the drawing conditions is shown in Figure 4-5 (b). We note that the Scanning Electron Micrograph in Figure 4-5 (b) demonstrates a highly regular distribution of cladding holes in a triangular pattern.

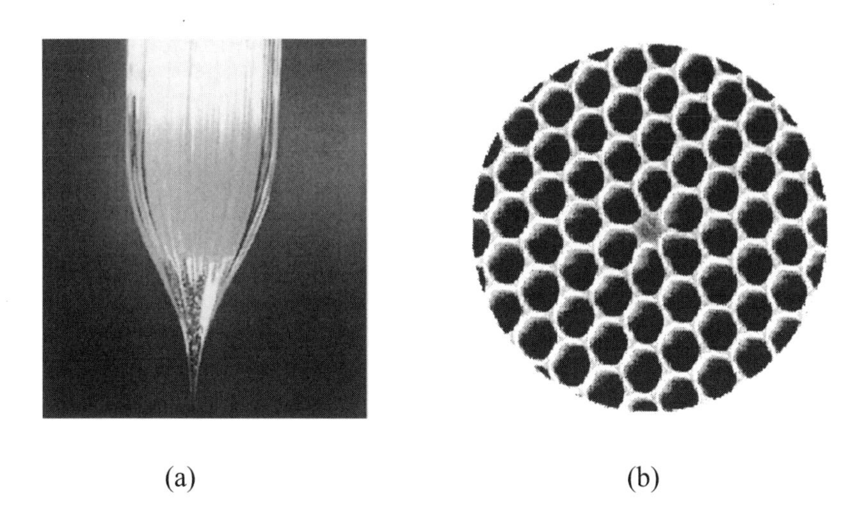

Figure 4-5. a) Picture illustrating how the outer dimensions are reduced from preform to photonic crystal fibre. b) A SEM picture of the cross-section of a highly regular index-guiding PCF structure. Both pictures are kindly provided by Crystal Fibre A/S.

Another example of a photonic crystal fibre is the honeycomb PBG fibre illustrated in Figure 4-6 (a). This type of fibre, which was the first demonstrated waveguide [4.7] operating by the photonic bandgap principle at optical wavelengths, was initially drawn to a centre-to-centre hole spacing of 2.0 μm. Figure 4-6 shows an example of, how the honeycomb air-hole arrangement is well preserved after the drawing process although surface tension forces has reduced the initial air-filling fraction. In the specific example, the preform structure was of the type shown in Figure 4-3. The surface tension forces are also responsible for the collapse of the interstitial holes that owing to their smaller size compared to the honeycomb-arranged holes will collapse at an earlier stage in the drawing process. The surface tension forces have a strongly non-linear dependency on the size of the air holes, and an accurate simulation of the contraction and distortion of the air holes due to surface tension forces requires a fluid-dynamic analysis of the glass structure during the drawing process. Such simulations represent a study in a much different field than electromagnetics and falls beyond the scope of this book. It should, however, be noted that several forces are important in fibre manufacturing, e.g., viscosity of the material, gravity, and surface tension. In a standard optical fibre, the only really important one of these is viscosity. This is not true for a microstructured fibre, because the larger surface area and smaller volume of material makes surface tension relatively much more important. It is also of relevance to note that recent research performed by researchers from the ORC, University of

Southampton, UK, has demonstrated new insight into the flow mechanisms involved in the drawing of a single-hole silica-fibre structure [4.15].

Again considering the example of the first-manufactured honeycomb fibre [4.7], also the core region was seen to be well preserved in the final fibre. In contrast to the cladding region, however, the interstitial holes was seen to remain in the core region. This may partly be understood from considering the surface tension forces from neighbouring tubes/rods that act differently on the interstitial holes in the core and cladding region. In the core region, three adjacent capillaries cause an interstitial hole to form, whereas in the cladding structure two capillaries and a rods form an interstitial hole. Since the surface tension forces that contracts the capillaries will slightly counteract the collapsing of the interstitial holes, the extra adjacent capillary in the core region is believed to contribute to the non-collapsing of the six interstitial holes in the core region. For the honeycomb fibre structure shown in Figure 4-6, this is, [illegible]
interstitial holes are collapsed.

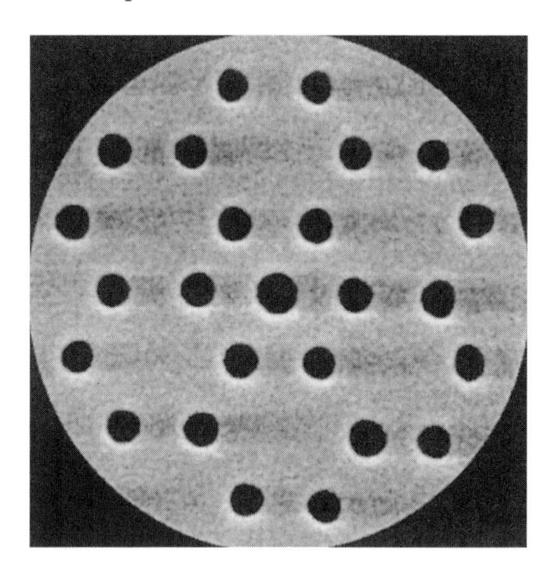

Figure 4-6. Central part of the cross section of a honeycomb fibre structure. The picture is kindly provided by Crystal Fibre A/S.

A very important issue concerning the fabrication of photonic crystal fibres by the preform stacking approach is to compare with the vapour deposition methods generally used for fabrication of standard optical fibres. It is obvious that standard fibre preforms has much less surface area, and, consequently, much less chance of becoming contaminated. Furthermore, the stacking approach requires a very careful handling, and the control of air-hole dimensions, positions, and shapes in PCFs makes the drawing significantly more complex. However, the technology described in this chapter also leads to new possibilities and among the most remarkable is the high degree of design flexibility. An example of this flexibility is to be found in the feasibility of multi core photonic crystal fibres, such as the dual-core PCF structure shown in Figure 4-7. It should be noted that numerous possibilities appears in the application of such designs, since the required number of cores may readily be added, and by controlling the hole dimensions and the core spacing, coupling of optical power between the individual cores may be enhanced or reduced depending on the specific application.

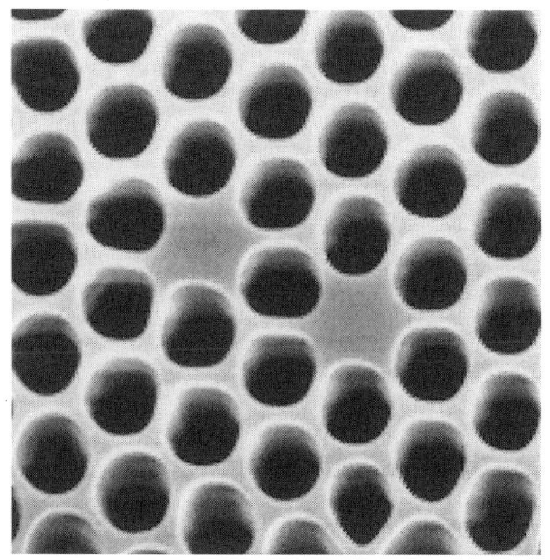

Figure 4-7. Scanning electron micrograph of end-phase of a dual-core high-index photonic crystal fibre having cladding holes placed in a close-packed pattern. The picture is kindly provided by Crystal Fibre A/S.

4.4 PHOTONIC CRYSTAL FIBRES IN NEW MATERIALS OR MATERIAL COMBINATIONS

Over the past 1-2 years, a number of alternatives or additions to the pure-silica PCF technology has emerged. These possibilities include fabrication of PCFs using the air-holes as means of improving waveguiding properties of doped fibres, PCFs in non-silica glass, and microstructured fibres made from polymers. In this Section, we will shortly refer to these possibilities.

4.4.1 Fabrication of hole-assisted lightguide fibres

Early photonic crystal fibres have shown fibre losses, which were significantly higher than those of standard optical fibres, which are typically 0.2 to 0.5 dB/km in the 1550 nm wavelength range. This may be of no significant concern in cases, where the PCFs are used in short lengths (possibly on a meter scale - or shorter), but is naturally of concern for applications that involve long fibre lengths. A solution towards lower loss values has been presented by Hasegawa *et al.* [4.9 - 4.10], who describe a method for realizing novel dispersion characteristics without severe degradation in transmission loss. The solution discussed by Hasegawa *et al.* is to use a hole-assisted lightguide fibre (HALF) [4.9], which is an optical fibre composed of a high-index core, a low-index cladding, and several holes surrounding the core. We will return to the discussion of this structure in section 5.5.

4.4.2 Chalcogenide fibres with microscopic air-hole structures

As an alternative to silica glass fibres, compound glasses offer a range of useful properties. As described by Monro *et al.* [4.12], the value of the compound glasses are that they offer properties, which may not be possible in silica glass, e.g., extended wavelength transmission range etc. Examples of compound glasses include sulphides, halides and heavy metal oxide glasses, as mentioned by Monro *et al.* [4.12]. One example of specific interest is gallium lanthanum sulphide (GLS) glass, which is transparent for wavelengths up to 5 μm,, and which has a high refractive index (in the order of 2.3 to 2.5), and can be doped with over 10% rare-earth ions by weight [4.13]. Despite their promise, compound glass fibres have not yet found widespread application owing to difficulty in fabricating low-loss singlemode compound glass fibres. In [4.9], Monro *et al.* for the first time describes, how microstructured fibre technology may be used as a powerful new technique for producing compound glass fibres. In addition, the photonic crystal fibre geometry dramatically increases the range of possible optical properties.

The fibre described in [4.9] was drawn from a preform in a single step, and this preform was constructed in a manner analogous to silica PCF preforms as described previously in this chapter (see also [4.11]). A solid rod was used to form the core of the fibre, and this rod was surrounded by six capillaries. Subsequently, this structure was placed inside a larger capillary both for structural support and to increase the preform dimensions, and this preform was then drawn into fibre with an outer diameter of approximately

100 μm, and a core diameter of 10 μm. The holes in the cladding capillaries range in diameter from 1.5 μm to 4.0 μm. The drawing temperature for compound glasses is typically lower than for silica, and conventional equipment and experimental parameters cannot be used [4.9]. GLS glass is drawn around 700 – 800 °C in a furnace with precisely controlled heating. In general, the temperature window available for drawing in compound glasses is much smaller than in silica glasses. The excellent structure retention obtained by Monro *et al.* [4.9], however, indicates that the viscosity/temperature relationship in GLS glass is well suited to microstructured (or holey) fibre fabrication.

4.4.3 Microstructured polymer optical fibres

The final approach that we will mention in this chapter also concerns PCFs fabricated in a base material different from silica. More specifically, we will address the work of Argyros *et al.* [4.14]. On ring structured microstructured polymer optical fibres (MPOFs). Argyros *et al.* [4.14] used PMMA of relatively low optical quality for the initial experiments. The material losses (absorption and scattering) were measured to be about 4 dB/m, which was determined in a measurement with polished rods of different lengths. The preform was formed by stacking a 1 mm thick structured cane with capillaries inside a tube. The structured sections of the cane consisted of three equally spaced rings of holes with the same air-filling fraction in each ring.

As reported in [4.14], the preform was drawn down to fibre of about 260 μm diameter on a draw tower with feed speed of 4 mm/min and draw speed of no more than 5 m/min at a temperature of approximately 175 °C. Electron microscope images of the cross section of the fibre shows that the dimensions of the holes in the final mode profile is insensitive to bending of the fibre and changing of the launch conditions. However, when the fibre was cut back to 5 cm, the same core was found to support more than one mode, as indicated by the observation of interference between different modes in the near field pattern, which changed as a function of the launch conditions (by translating the microscope objective that focused the laser beam onto the core).

As described by Eijkelenborg *et al.* [4.18], one of the key motivations for working with polymer PCFs is the possibility to obtain alternative geometries or hole shapes. In this respect, MPOF has a number of advantages. The much lower processing temperatures of polymers and the controllability of the polymerization process allow a variety of ways to produce the polymer preforms. In addition to the capillary stacking technique, polymer preforms can be made using techniques such as extrusion, polymer casting, polymerization in a mould and injection moulding [4.18]. With such techniques available, it becomes straightforward to obtain different cross-sections in the preform, with holes of arbitrary shapes and sizes in any desired arrangement. Creating holes of specific shapes in a glass PCF is complicated, because the balance between viscosity and surface tension tends to modify the hole structure during the draw process, and often the structure in the fibre is significantly different from that in the preform [4.12 – 4.13]. In addition, the drawing temperature is a crucial parameter, when drawing glass microstructured fibre; temperature changes of a few percent can lead to very significant changes in the fibre microstructure. According to Eijkelenborg *et al.* [4.18], this does not seem to be the case for MPOF, which can be drawn over a large temperature range. For a PMMA example, the drawing temperature could be varied from 150 °C to 200 °C without significant change to the fibre structure. There seems to be a much closer relationship between the preform structure and the final structure in the fibre. The reason for this is well understood in polymer processing. The drawing process aligns the polymer chains, so that the final material is anisotropic along the fibre direction, and has enhanced strength properties [4.18]. Annealing after the draw can be used to relieve the internal stresses, while maintaining the microstructure in the fibre.

Another interesting issue is that polymers are intrinsically modifiable [4.18]. This means that it is possible to design and manufacture polymers that include any atomic species, molecular components, dispersed molecules and dispersed phases. Examples of the types of materials that could be used in MPOF are: polymers with enhanced non-linearities, electro or magneto-optic effects, metallic or rare-earth inclusions, birefringent materials such as liquid crystals, photorefractive and photo chromic materials, dyes, polymers used in the detection of particular compounds and porous materials. The polymers can be specifically designed to allow the fabrication of particular fibre-optic components based on MPOF.

4.4.4 Extruded non-silica glass fibres

In the previous sections of this chapter, we have discussed recent results on new base materials aiming at spectral ranges and non-linear properties different from those obtainable using silica. It is exactly this search for properties not possible in silica, which also previously in the development of optical fibres (e.g., in connection with the development of fluoride glass fibres [4.19]) has lead to interesting alternatives. In this section, we will shortly review results on the first microstructured single-mode non-silica fibre, which recently was reported by Monro *et al.* [4.20] and Kiang *et al.* [4.21].

One of the major reasons for the research in non-silica glass fibres is that e.g., compound glasses offer a range of properties such as enhanced non-linearity and relatively low fabrication temperature, which are attractive compared to silica based fibres. However, the development of these fibres has been limited because of high transmission losses in compound glass fibres fabricated using conventional techniques. As discussed previously in this chapter, it has been demonstrated by Monro *et al.* [4.12] that the microstructured fibre technology provides a powerful new technique for producing compound glass fibres, and recently the same group of scientists reported the first single-mode fibre fabricated in non-silica glass [4.20-4.21].

This result is interesting not only because it for the first time demonstrates a photonic crystal fibre made in commercially available Schott glass SF57, but also because it reports a new fabrication approach of extruding compound glass preforms. The specific SF57 glass has a high lead concentration, which results in a relatively high refractive index of 1.83 at a wavelength of 633 nm, and 1.80 at 1530 nm. It is, furthermore, interesting, because the nonlinear refractive index (n_2) is measured at 1060 nm to be $4.1 \cdot 10^{-19}$ W^2/m [4.22], which is more than an order of magnitude larger than that of pure silica glass fibres [4.23]. Since the effective nonlinearity of a fibre is $\gamma = n_2/A_{eff}$, the combination of highly non-linear glass and the potentially small effective area A_{eff} (made possible by the photonic crystal fibre technology) should allow for significant improvements in the achievable non-linearity [4.21]. These improvements should of course be evaluated with respect to other properties such as acceptable fibre attenuation and connection issues.

On the fabrication side, it is relevant to note the low softening temperature of 519 °C for the SF57 glass, which made it possible to extrude the microstructured fibre perform (with an outer diameter of 16 mm) from bulk SF57 glass. The resulting structure comprised a central solid core supported by three long thin membranes. The cane was inserted within an extruded jacketing tube, and the assembly was drawn down to a fibre with an outer diameter of 120 μm, resulting in a core diameter of 2 μm suspended by three 2 μm long supports that were less than 400 nm wide.

Kiang *et al.* [4.21] have reported that robust single-mode guidance was observed in a 50 m long fibre at wavelengths of 633 nm as well as of 1500 nm. The observed fibre loss was 3 dB/m at 633 nm and 10 dB/m at 1550 nm (compared to bulk glass losses of 0.7 dB/m and 0.3 dB/m, respectively). The perspective of the extrusion technique is (among others) that by avoiding the stacking procedure, fewer interfaces are involved, and consequently the extrusion method may, ultimately, offer lower losses than existing techniques. Whether or not this advantage may counteract demands on melting temperature and related material attenuation will be an issue for further research.

4.5 SUMMARY

In this chapter, we have outlined some of the basic elements in the fabrication of photonic crystal fibres. It is obvious that a rapidly developing field of research and development such as the one described in this book will result in numerous new details on fabrication, and it is equally obvious that many of these novelties will not be immediately published, due to the scientific and commercial interests of competition. For the same reason, the description presented in this chapter has been mostly limited to the details generally known in literature.

In the chapter, the production of fibre preforms through stacking of glass rods and tubes has been presented. The issue of the possible appearance of interstitial holes are discussed, and examples of performs structures are shown. Thereafter, the drawing of the fibre performs obtained through the stacking of elements is described, and examples of resulting fibre structures are presented. Finally, some alternative fabrication methods are described, including the possibility of fabricating photonic crystal fibres in polymers and soft-glass compositions. These alternatives do – in contrast to the silica based fibres – suggest possibilities such as perform extrusion etc.

REFERENCES

[4.1] P.V.Kaiser, and H.W.Astle,
"Low-loss single-material fibers made from pure fused silica",
The Bell System Technical Journal, Vol.53, 1974, pp.1021-1039.

[4.2] A.Bjarklev,
"Optical Fiber Amplifiers: Design and System Application",
Artech House, Boston-London, August 1993, ISBN: 0-89 006-659-0.

[4.3] C.K.Kao,
"Optical Fibre",
Peter Peregrinus, London, 1988.

[4.4] *Biomedical Sensors, Fibers, and Optical Delivery Systems,* vol. 3570 of *Proceedings of the SPIE- The International Society for Optical Engineering,* 1999.

[4.5] T. Birks, D. Atkin, G. Wylangowski, P. Russell, and P. Roberts, "2D photonic band gap structures in fibre form,"
Photonic Band Gap Materials (C. Soukoulis, ed.), Kluwer, 1996.

[4.6] T. Hasegawa, E. Sasaoka, M. Onishi, M. Nishimura, Y. Tsuji, and M. Koshiba, "Hole-assisted lightguide fiber for large anomalous dispersion and low optical loss",
Optics Express, Vol.9, No.13, Dec.2001, pp.681-686.

[4.7] J. Knight, J. Broeng, T. Birks, and P. Russell, "Photonic band gap guidance in optical fibers,"
Science, Vol. 282, Nov. 1998, pp.1476-1478.

[4.8] J. Broeng, D. Mogilevtsev, S. Barkou, and A. Bjarklev,
"Photonic crystal fibres: a new class of optical waveguides,"
Optical Fiber Technology, Vol.5, July 1999, pp. 305-30.

[4.9] T. Hasegawa, E. Sasaoka, M. Onishi, M. Nishimura, Y. Tsuji, and M. Koshiba, "Novel hole-assisted lightguide fiber exhibiting large anomalous dispersion and low loss below 1 dB/km",
OFC'2001, 2001, Post deadline paper PD5.

[4.10] T. Hasegawa, E. Sasaoka, M. Onishi, M. .Nishimura, Y. Tsuji, and M. Koshiba, "Hole-assisted lightguide fiber for large anomalous dispersion and low optical loss"
Optics Express, Vol.9, No.13, Dec.2001, pp.681-686.

[4.11] P. J .Bennett, T .M. Monro, and D. J. Richardson,
'Towards practical holey fibre technology: fabrication, splicing, modelling and characterization',
Optics Letters, Vol.24, 1999, pp.1203-1205.

[4.12] T. M. Monro, Y. D. West, D. W. Hewak, N .G. R. Broderick, and D. J. Richardson, "Chalcogenide holey fibres",
IEE Electronics Letters, Vol.36, No.24, Nov.2000.

[4.13] Y. D. West, T. Schweizer, D. J. Brady, and D. W. Hewak,
"Gallium lanthanum sulphide fibers for infrared transmission",
Fiber and Integrated Optics, Vol.19, 2000, pp.229-250.

[4.14] A. Argyros, I. M. Bassett, M. A .van Eijkelenborg, M. C. J. Large, J. Zagari, N. A. P. Nicorovoci, R. C. McPhedran, and C. M. de Sterke,
"Ring structures in microstructured polymer optical fibres",
Optics Express, Vol.9, No.13, Dec.2001, pp.813-820.

[4.15] A. D. Fitt, K. Furusawa, T. M. Monro, and C. P. Please,
"Modelling the fabrication of hollow fibers: Capillary drawing",
IEEE Journal of Lightwave Technology, Vol.19, No.12, Dec.2001, pp.1924-1930.

[4.16] J. Broeng,
"Photonic crystal fibres",
Ph.D. Thesis, Research Center COM, Technical University of Denmark,
September 30, 1999, ISBN:87-90 974-07-7

[4.17] J. B. Eom, K. W. Park, Y .Chung, W-T. Han, U-C. Paek, D. Y. Kim, and B. H. Lee,
"Optical properties measurement of several photonic crystal fibers",
SPIE, Photonics West 2002, San Jose, CA, USA.

[4.18] M. A. van Eijkelenborg, M. C. J. Large, A. Argyros, J. Zagari, S. Manos, N. Issa,
I. Bassett, S. Fleming, R .C. McPhedran, C. M. de Sterke, and N. A. P. Nicorovici,
"Microstructured polymer optical fibre",
Optics Express, Vol.9, 2001, pp.319-327.

[4.19] P.W. France, and M.C. Brierley,
"Progress in Fluoride Fibre Lasers and Amplifiers",
Proceedings of Society of Photo-Optical Instrumentation Engineers Conference, OE/Fibers'90, part: Fiber Laser Sources and Amplifiers II, San Jose, Sepot.1990, Vol.1373, pp.33-39.

[4.20] T.M. Monro, K.M. Kiang, J.H. Lee, K. Frampton, Z. Yusoff, R. Moore, J. Tucknott,
D.W. Hewak, H.N. Rutt, and D.J. Richardson,
"Highly nonlinear extruded single-mode holey optical fibers",
Proc OFC'2002, OSA Technical Digest 315-317, Anaheim, California, 2002.

[4.21] K.M. Kiang, K. Frampton, T.M. Monro, R. Moore, J. Tucknott, D.W. Hewak,
D.J. Richardson, and H.N. Rutt,
"Extruded singlemode non-silica glass holey optical fibres",
IEE Electronics Letters, Vol.38, No.12, June 2002, pp.546-547.

[4.22] S.R.Friberg, and P.W.Smith,
"Nonlinear optical glasses for ultrafast optical switches",
IEEE Journal of Quantum Electronics, Vol.QE-23, 1987, pp.2089-2234.

[4.23] G.P. Agrawal,
"Nonlinear Fibre Optics",
Academic Press, San Diego, 1995.

[4.24] J. B. Nielsen, T. Søndergaard, S. E. Barkou, A. Bjarklev, J. Broeng, M. B. Nielsen,
"Two-dimensional Kagome structure, fundamental hexagonal photonic crystal configuration",
IEE Electronics Letters, Vol. 35, No. 20, pp. 1736-1737, 1999.

Chapter 5

PROPERTIES OF HIGH-INDEX CORE FIBRES

5.1 INTRODUCTION

Photonic crystal fibres may be divided into classes of which the two major ones are the index-guiding or high-index core fibres and the photonic bandgap or low-index core fibres. In this chapter, we will describe the fundamental properties of high-index core fibres, in which the waveguiding principle may be expressed as modified total internal reflection (MTIR). In high-index core PCFs, the core will – as expressed by the name – have a higher effective refractive index value than the surrounding cladding material. However, these effective refractive indices are typically obtained through the introduction of air holes, which may be ordered in different patterns, and which may have different dimensions, shapes etc.

The most typical realisation of high-index core fibres has a solid core region (often fabricated from pure un-doped material) surrounded by a cladding with regularly arranged holes in a periodic arrangement. These holes act to lower the effective refractive index in the cladding region, and so light is confined to the solid core, which has a relatively higher index. High-index core fibres can be made entirely from a single base material with an adequately distributed number of holes, and they are normally fabricated from pure silica [5.1-5.2], although PCFs have also been fabricated in chalcogenide glass [5.3] and in polymers [5.4] (see Chapter 4 for more details on the fabrication methods).

At a first glance, one might think that, since the high-index core PCFs operate by the well-known principle of total internal reflection, these fibres may just represent another (and maybe more difficult) way of obtaining "standard optical fibre" properties. However, as it will be come apparent from this chapter, the index-guiding PCFs have a large number of unique properties that are not obtainable in conventional solid optical fibres, and we will here introduce the most significant of these in the following order:

First, Section 5.2 describes the first fibres made from "a single material" (pure silica) and air as early as 1973. From this short review, we step forward in time to the mid 1990'ies, where fibres of undoped glass with claddings formed from periodically distributed, air-filled voids was suggested. Section 5.3, presents the basic properties of these high-index core fibres, and outlines their unique cutoff properties. The section further describes additional basic properties such as spectral dependency of macrobending loss, as well as the fundamental dispersion properties are analysed. All these basic properties are outlined for structures, which are considered perfect with respect to periodicity and circular shape of the air holes. Waveguides may, however, also be formed from non-ideal hole distribution or using holes with non-circular shapes as described in Section 5.4. Finally, another sub-class of microstructured fibres are reviewed in Section 5.5, namely the so-called hole-assisted lightguiding fibres. These are (as the name indicates) optical fibres, in which air holes play a role in the waveguiding process, but on the other hand are not dominant in determining the fibre properties. This last sub-class of photonic crystal fibres will typically use a combination of standard doping techniques, and some added microstructuring.

5.2 BACKGROUND – "SINGLE-MATERIAL" FIBRES

Very early in the development history of optical fibres, the idea of using pure core materials was formed, and to some extend, the basis of the air-silica fibre technology was founded already in the early 1970′ies. We will in this section look a little closer at these early steps.

In 1973, Kaiser *et al.* [5.5] published the results on a new optical fibre, which formed a viable, handleable transmission medium by a structural form that used only a single low-loss material (of course in combination with air). The motivation for this work was the concern that most of the efforts devoted to optical fibres back in the early 1970′ies focussed on fibres having a central glass core surrounded by a cylindrical glass cladding with a slightly lower refractive index. This in turn required that the chemical composition of the core glass differed from that of the cladding glass (like in any present-day standard optical fibres). These compositional differences were believed to lead to undesired effects at the core-cladding interface (e.g., internal stress etc.), and there was a concern, whether or not the dopant materials would limit the minimum fibre losses achievable.

The fibres proposed by Kaiser *et al.* [5.5] had a schematic shape as illustrated in Figure 5-1 (a), and they were named single-material fibres. The guided energy is concentrated primarily in the central enlargement of the structure shown in Figure 5-1 (a). There is an exponentially decaying field extending outward from the central member in the slab enclosed in the circular outer structure. It is stated [5.5] that by appropriate spacing between the central enlargement and the outer cylinder, the guided-wave field at the outside surface can be made negligibly small, and the fibre could be handled exactly as the conventional core-cladding type fibre. Slab modes are possible on the supporting structure, but these are strongly coupled to the outer shell and are readily lost to the surrounding medium.

The single-material fibre structure developed by Kaiser *et al.* [5.5] can have a single propagating mode for any supporting slab thickness, t, and for any shape of the central enlargement, provided the size of the central enlargement is properly chosen. A very detailed study of the properties of these so-called slab-coupled waveguides was presented by Marcatili [5.6] in 1974.

Kaiser *et al.* [5.5] reported that single-material fibres were drawn from fused quartz tubes containing thin, polished plates and small-diameter rods supported in the centre of the tubes. The resulting single-mode fibre had a height and width of the centre element of 6.5 μm and 5 μm, respectively, and a slab height of 4 μm. At a wavelength of 1.06 μm, the lowest losses were measured to be 55 dB/km. Note also that Kaiser *et al.* [5.5] suggested that active fibre guides could be created by placing an active material in the central core or by putting it in a liquid surrounding the central member.

In a subsequent work by Kaiser *et al.* [5.7] in 1974, a significant loss reduction compared to the just mentioned 55 dB/km was obtained, since single-material fibres with a loss of only 3 dB/km at a wavelength of 1.1 μm was obtained in a 130 m long fibre.

It is also very interesting to note that Kaiser *et al.* [5.7] further describes, how instead of using rod-plate technique for the preform fabrication, longer preform lengths may be achieved by the use of thin-walled tubes as supports. A single-mode, single-material fibre was created at the intersection of two tubes as shown schematically in Figure 5-1 (b). Also the use of three thin-walled tubes in an outer cladding tube is mentioned as means of creating three intersection points.

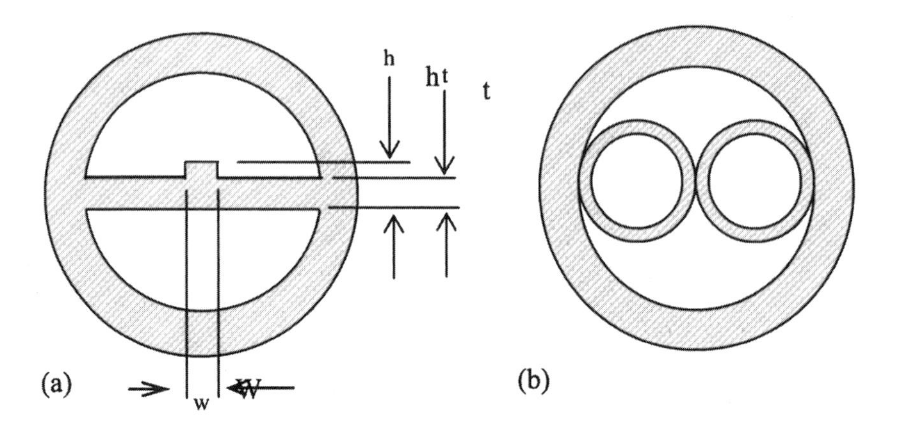

Figure 5-1. (a) Cross-section of a single-material fibre with a rectangular core supported by a slab in the central part of the optical fibre. (b) Schematic illustration of preform for the formation of single-mode, single-material fibres made by fusing two thin-walled tubes. The illustrations are drawn after the original illustrations of [5.5] and [5.7].

Although the single-material fibre technology, therefore, was showing very interesting results almost 30 years ago, the fibre consumers had to wait more than two decades before the ideas was carried further on by Knight *et al.* [5.10], who in their search for the photonic bandgap fibre showed results on single-material fibres in 1996. the reason for the very long period, where almost nothing was published on single-material fibres (few works in the 80'ies by Hicks [5.32] and the early 90'ies by Vali *et al.* [5.33]) was primarily due to the very large success of alternative fabrication techniques using doped fibre material, i.e., methods such as the Modified Chemical Vapour Deposition (MCVD) technique [5.11], the Outside Vapour Deposition (OVD) technique [5.12], and the Vapour Axial Deposition (VAD) technique [5.13].

It should, finally, be pointed out that not all researchers like to apply the name "single-material" fibres, because the fibres strictly speaking are operating as waveguides due to the combination of the base material and the voids or air-filled structures in it. A more correct term could, therefore, be "two-material fibres".

5.3 FIBRES WITH PERIODIC CLADDING STRUCTURES

There are many reasons, why it is practical to fabricate high-index core PCFs having a cladding structure with an-at least partly-periodic hole distribution, and among the most important are issues such as fabrication reproducibility, spectral property control etc. To describe these issues in more detail, we will here address some of the basic spectral properties of PCFs.

5.3.1 Basic properties of high-index core photonic crystal fibres

In order to be a little more specific concerning the fibre structures that we consider in this section, Figure 5-2 shows a scanning electron micrograph of a specific high-index core PCF. The PCF is characterized by triangularly arranged cladding holes. With the present technology, the relative size of the cladding holes d/Λ may range from a few percent and up to around 90%, and the centre-to-centre hole spacing, Λ, may find values from around a micron and up to 20 μm. In the specific case illustrated on Figure 5-2, a single, missing air hole forms the fibre core, and once again it is noteworthy that the dimensions of the fibre core may be anything between 1 μm to 20 μm depending of the fibre design and its desired properties. The PCFs may, consequently, offer a wide variety of potential properties ranging from highly linear performance to fibres with high non-linear coefficients.

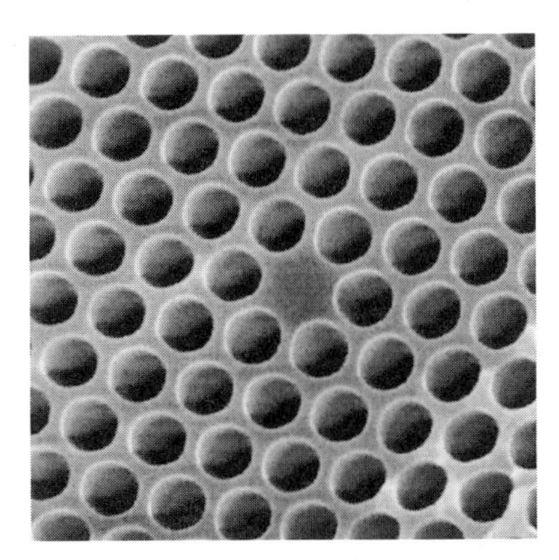

Figure 5-2. Scanning electron micrograph of a typical high-index core PCF. The PCF is characterized by triangularly arranged cladding holes of size $d/\Lambda = 0.7$, a centre-to-centre hole spacing, Λ, of 5 μm, and a single, missing air hole forming the fibre core. The photograph is kindly provided by Crystal Fibre A/S.

To understand the operation of high-index core PCFs in more detail, it is useful to analyse the PCF by plotting the modal-index as a function of the normalized frequency Λ/λ. It should be stressed that this kind of analysis here is made under the assumption that the refractive index of the base material is constant as a function of wavelength. This is, as described in Section 3.2, an approach, which allows us to understand the effective influence of the air holes, independent of specific material properties, and the calculations and analysis may, subsequently, be modified in order to describe the full material influence, e.g., by perturbation methods or by iterative steps in the solutions. Note also that the effective-modal-index illustrations later will be used extensively to describe the properties of PBG-fibres.

The behaviour of the fundamental mode index with respect to normalized frequency is illustrated in Figure 5-3. The figure further shows the behaviour of the effective cladding and core index. The effective cladding index is determined from the properties of the fundamental-space-filling mode as β_{fsm}/k, where β_{fsm} is the propagation constant of the lowest order allowed mode in the cladding structure, and k is the free-space wave number. The cladding index is seen to be strongly wavelength dependent, whereas the core refractive index remains fixed and equal to the refractive index of silica (the illustration is calculated for a fixed silica refractive index value of 1.45). The figure shows that the high-index core design supports confined modes with a β/k-ratio that obeys the relation:

$$n_{cl,eff} < \frac{\beta}{k} < n_{co} \tag{5.1}$$

where $n_{cl,eff}$ is the effective refractive index of the cladding structure and β is the propagation constant of the guided mode. The relation (5.1) is in principle identical to that of conventional fibres. However, as a very important difference to conventional fibres, the specific PCF was not found to exhibit any higher-order guided modes over the studied frequency range. This property is supported by the experimental observations of Knight *et al.*, where an index-guiding PCF was found to support only a single mode over an extraordinary broad wavelength range from at least 337 nm to 1550 nm [5.8].

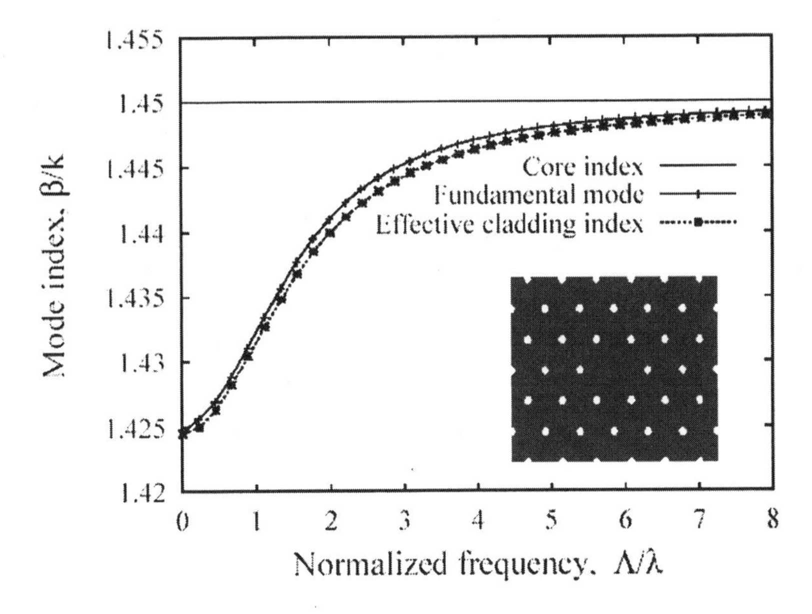

Figure 5-3. Modal-index illustration of the operation of a PCF with a triangular photonic crystal cladding structure and a high-index core formed from a single missing air hole. The cladding air holes have a size of $d/\Lambda = 0.23$. The figure shows the core index, which is considered frequency independent, the mode index of the fundamental mode supported by the PCF, and the effective cladding index of the PCF, which is determined from the fundamental space filling mode of a full periodic photonic crystal corresponding to the cladding structure. The PCF supports only a single mode over the studied frequency range, which is attributed to the strongly frequency dependent cladding index; the cladding index approaches the core index in the high-frequency regime, and the vanishing index contrast between the core and cladding indices suppress any higher-order guided modes. The inset shows a cross-section of the PCF, including the fibre core region.

When looking for an explanation of why the specific PCF does not become multi-moded at short wavelengths, the strongly wavelength dependent cladding index of the PCF must be considered. State-space arguments will be presented in more detail for both standard Step-Index Fibres (SIFs) and PBG-guiding PCFs later in this book (see Section 6.4.2). However, it is at this point relevant to note that for the high-index core PCFs, an expression for the number of index-guided modes, N_{PCF}, could be expressed as [5.9]:

$$N_{PCF} \approx \frac{(k\rho)^2\left(n_{co}^2 - n_{cl,eff}^2\right)}{4} \tag{5.2}$$

where ρ is the effective core radius. From this expression, it may qualitatively be understood that an infinite number of supported modes as for conventional fibres in the high-frequency limit ($\lambda \to 0$) may not necessarily be a feature of the index-guiding PCFs, due to the vanishing core–cladding index difference. For completeness, it should be mentioned that at the other extreme, $\lambda \to \infty$, the mode field will, in principle, be infinitely extending in the cross-section, and for an ideal PCF, having infinitely extending cladding structure in the cross-section, the modal index of the fundamental mode becomes equivalent to the cladding index. Of course, in this situation, the PCF can not be considered as guiding, which is in agreement with $N_{PCF} \to 0$.

The above-discussed PCF showed no confined higher-order modes at the studied frequencies, and using effective index considerations, this indicates that the PCF may potentially be single-moded over an infinite frequency range. The ability of suppressing any higher-order modes is, however, strongly related to the cladding air-filling fraction of the PCFs, as will be demonstrated in the proceeding section, where the cut-off properties of PCFs will be addressed.

5.3.2 Cut-off properties of index-guiding photonic crystal fibres

For comparison with the PCF that was presented in the preceding sections, a modal index analysis of PCF with a similar design but a relatively large air-filling fraction in the cladding structure is illustrated in Figure 5-4. The PCF has cladding air holes of size d/Λ=0.60 corresponding to an air-filling fraction of 33%. Figure 5-4 illustrates that this PCF supports a second-order mode, which has a normalized cut-off frequency of around 1.5.

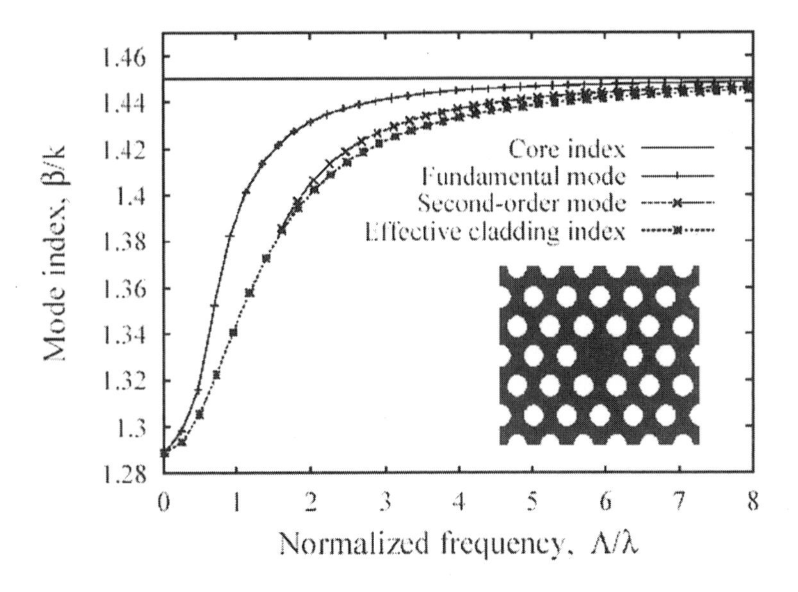

Figure 5-4. Modal index illustration of PCF with similar design as the PCF studied in Figure 5-2, but with a large cladding air-filling fraction ($d/\Lambda = 060$). The PCF supports a second-order mode, which has a normalized cut-off frequency, Λ/λ, around 1.5. The inset shows a cross-section of the PCF, including the fibre core region.

The second-order mode is illustrated in Figure 5-5 at a normalized frequency of 4.0. Apart from these two modes, the PCF was, however, not found to exhibit any additional modes. This is a further indication of the special cut-off properties of the index-guiding PCFs. To provide a more general approach to the analysis of these properties, the PCFs will next be treated using an analogy to SIFs. In traditional fibre theory, a normalized frequency, V, is commonly used for analysis of the cut-off properties of SIF [5.9]:

$$V = k\rho\sqrt{n_{co}^2 - n_{cl}^2} \tag{5.3}$$

where n_{co} and n_{cl} are the largely wavelength independent refractive indices of the solid core and cladding, respectively, and ρ is the fibre core radius. Note that in standard fibre theory, it is common practice to operate with a normalised frequency as opposed to e.g., an expression for the number of guided modes, N, equivalent to the N_{PCF} expression in Eq. (5.2) for analysis of cut-off properties of conventional fibres. The V-parameter is, therefore, chosen for the analysis presented in this section. It can be shown that for a highly multi-moded SIF that $N \approx V^2/2$ [5.9]. In order to use an expression similar to that of Eq. (5.3) for PCFs, the cladding refractive index should be replaced with β_{fsm}/k. Hence, the equivalent expression for an effective V-value for the index-guiding PCFs becomes:

$$V_{eff} = \rho\sqrt{k^2 n_{co}^2 - \beta_{fsm}^2} \tag{5.4}$$

It must, however, be further noticed that a well-defined core radius does not exist for PCFs. An effective V-value for index-guiding PCFs was first presented by Birks *et al.* [5.14]. Birks *et al.* used a core radius of Λ, and a similar effective core radius of the PCFs will be used to start out with for the analysis performed in this section.

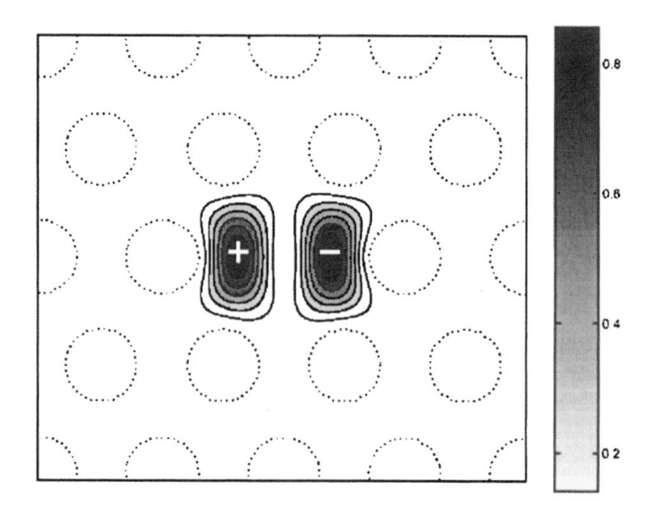

Figure 5-5. Second-order mode of triangular, high-index core PCF with d/Λ = 0.60. The mode was calculated at a normalized frequency Λ/λ =4.0. The '+' and '–' signs indicate the phase reversal between the two lobes.

For a series of triangular, high-index core PCFs with various cladding-hole dimensions, the effective V-values as function of normalized frequency are illustrated in Figure 5-6. A highly unusual feature compared to SIFs is observed from the figure, which is that the effective V-values for the PCFs tend to stationary values in the high frequency limit $(\lambda \rightarrow 0)$. Although it is not observable from Figure 5-6, triangular, high-index core PCFs with any d/Λ-value (including the PCF with d/Λ= 0.95) were found to exhibit a stationary V_{eff} -value in the high-frequency limit. This documents, more generally, what was initially addressed in the preceding section, namely that the PCFs will be characterized by a *finite* number of guided modes at any frequency.

As indicated in Figure 5-6, the stationary V_{eff}-value for a given triangular, high-index core PCF with fixed refractive indices of hole and background material is only dependent on the ratio between the hole diameter d and the period of the lattice Λ, and it increases with the ratio. Thus by designing triangular, high-index core PCFs with d/Λ below a certain value, it is in fact possible that the finite number of modes supported by a given PCF is equal to one, i.e., these PCFs may be classified as being endlessly single-moded. This term was first used by Birks *et al.* [5.14] in 1997. The "endlessly single-mode" property is in strong contrast to the properties of standard optical fibres, where the cladding index is largely wavelength independent, and the finite core–cladding index step results in $V \rightarrow \infty$ for $\lambda \rightarrow 0$ (whereby the fibres become highly multi-moded). From the specific PCF having d/Λ= 0.60 (see Figure 5-4), it is found that the second-order mode cut-off is at a normalized frequency of Λ/λ = 1.5. According to Figure 5-6, this corresponds to a cut-off V_{eff}-value of approximately 4.2.

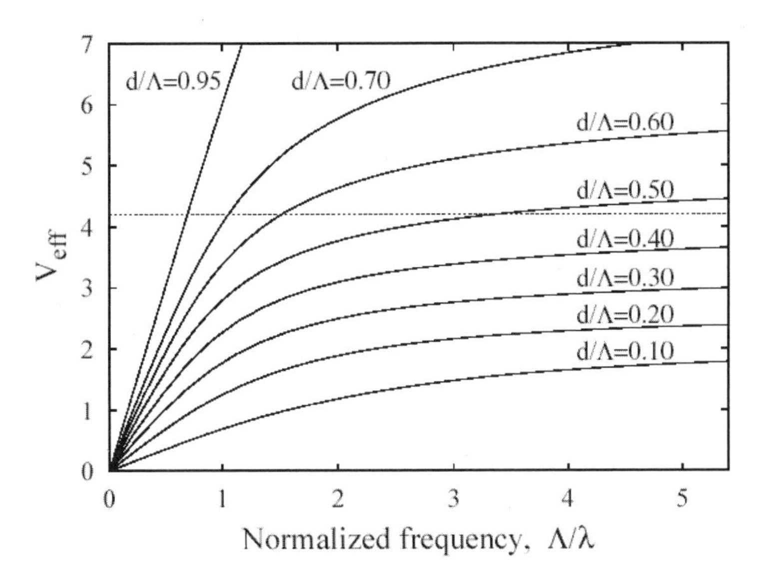

Figure 5-6. Effective V-value for triangular, high-index core PCFs. The effective V-value of any given PCFs will reach a stationary level in the high-frequency limit. The exhibition of a stationary V_{eff} - value in the high-frequency regime is explained as a result of the strong wavelength dependency of the effective photonic crystal cladding index, which tends to the core index material. The horizontal curve (V_{eff}= 4.2) indicates the second-order mode cut-off. PCFs with $d/\Lambda \leq 0.4$ can not exhibit V_{eff}'s above this cut-off value, and will, therefore, only support a single mode at any frequency. The effective V-values have been calculated using an effective PCF core radius, ρ, equal to Λ.

To analyse in further detail the cut-off properties, it is useful also to define an effective normalized core parameter, U_{eff}, for the PCFs, which is equivalent to the U-parameter for conventional fibres:

$$U_{eff} = \rho\sqrt{k^2 n_{co}^2 - \beta^2} \tag{5.5}$$

where β is the propagation constant of a guided mode. Using again ρ=Λ as the effective core radius of the PCFs, the relations between V_{eff} and U_{eff} for triangular, high-index core PCFs with various cladding hole dimensions are illustrated in Figure 5-7. This type of illustration was chosen, since it provides a very clear visualization of the cut-off properties (modes will only be guided for $V_{eff}>U_{eff}$ - according to Eq. (5.1)), as well as it allows a direct comparison between PCFs with different hole dimensions. As for SIFs, the index step between the core and cladding is incorporated into the normalized parameters V_{eff} and U_{eff}, and, therefore, also into the relation between them. Hence, Figure 5-7 should be independent of the *d*/Λ-ratio, and for PCFs with *d*/Λ= 0.35 to 0.75 the V_{eff} - U_{eff} - relations were found to almost coincide (remembering, of course, that the V_{eff}-value that can be reached for various PCFs with different *d*/Λ-values is not identical). The main reason, why a complete coincidence was not found is attributed to the discrepancy in effective core radius for the PCFs with different hole sizes. The figure shows a second-order mode V_{eff} cut-off value of 4.2 for all the PCFs that do exhibit a second-order mode (in agreement with the cut-off value found for the specific PCF with *d*/Λ = 0.60). In combination with Figure 5-6, it may, therefore, be concluded that fibres with the specific triangular, high-index core design may be classified as being endlessly single-moded for cladding hole dimensions *d*/Λ less than approximately 0.45.

The endlessly single-mode property of index-guiding photonic crystal fibres does have practical limits such as loss limits etc. as we will discuss further in section 5.3.3. However, recent studies by Kuhlmey *et al.* [5.22] investigate, how the mode number can be characterised in finite-size PCFs. Kuhlmey *et al.* show that there exist a clear boundary between single- and dual-mode regions, and at this boundary, the second mode changes rapidly from filling the entire cross section to be tightly confined about the defect, thus exhibiting a distinct cutoff. One of the key issues of the work by Kuhlmey *et al.* is that in contrast to conventional fibres, which has a finite set of bound modes with a real propagation constant, all modes of PCFs having a finite set of confining holes have complex propagation constants. Kuhlmey *et al.* [5.22], therefore, search for another criteria for bound modes replacing the requirement of real propagation constants, and in this work an interesting starting point is the effective-area criterion suggested by Mortensen [5.23]. Kuhlmey *et al.* use the Multipole Method [5.24] (see also section 3.6) for their investigations, and they focus on the imaginary part of the effective index and link this to the geometrical loss of PCFs with 4, 6, 8, and 10 rings of hexagonally-packed circular (triangular arranged) holes, respectively. The results show a sharp transition in the ratio of loss versus normalized wavelength, λ/Λ, around a relative hole size of $d/\Lambda = 0.45$. For a more detailed analysis, the effective radius and effective area of the mode field is mapped out as a function of normalized wavelength [5.22], and for $d/\Lambda > 0.45$ a very fine agreement is found with the criteria formulated by Mortensen [5.23]. However, for normalized hole sizes below 0.45, the transition of the imaginary part of the effective index becomes less and less pronounced, until the transition completely ceases to occur at $d/\Lambda = 0.406 \pm 0.003$. Kuhlmey *et al.* [5.22] also provide a best fit for the numerical data determining the single mode – dual mode boundary as

$$\frac{\lambda}{\Lambda} = \alpha\left(\frac{d}{\Lambda} - 0.406\right)^{\gamma} \tag{5.6}$$

with the parameters $\alpha = 2.80 \pm 0.12$ and $\gamma = 0.89 \pm 0.02$.

Having determined that the basic operation of high-index core PCFs and conventional fibres are very much the same and already used simple expressions based on SIF theory, it is at this stage advantageous to consider in more detail, if a closer analogy between SIFs and PCFs may be developed. If PCFs may be treated using such an analogy, well-established fibre tools could be taken advantage of for qualitative analysis of a long range of advanced properties of the index-guiding PCFs – and could, therefore, prove very useful from a practice point of view for first-approach design and development of PCF-based optical systems.

To develop an analogy between SIFs and index-guiding PCFs, it appears appropriate to base this upon the above-discussed V and U parameters, since these incorporate-on a normalized form-the actual operating frequency together with index difference between the guided mode, and the core as well as the cladding.

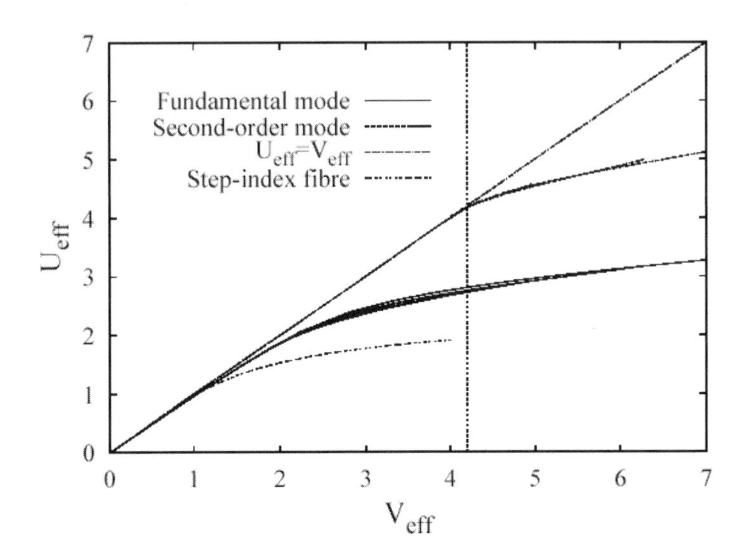

Figure 5-7. Illustrations of the fundamental and second-order mode U_{eff} versus V_{eff} for high-index core, triangular PCFs. The (V_{eff}, U_{eff})-relations for a series of PCFs with d/Λ ranging from 0.35 to 0.75 in steps of 0.1 has been over-layered to illustrate the (almost complete) independency on the d/Λ-ratio. The vertical line shows the second-order mode cut-off (using $\rho=\Lambda$). The (V,U)-relation for an SIF has also been included.

Figure 5-7 also illustrates the fundamental mode U versus V for a step-index fibre (SIF). In order to treat PCFs using an SIF analogy, it is a requirement that the two sets of (V,U)- values for PCFs and SIFs coincide. This is clearly not the case, providing the initial conclusion that it is not feasible to simply substitute the effective V-value into well-established SIF fibre tools. However, as will be demonstrated, a much better correspondence may, in fact, be found by redefining the core radius of the PCF. In order to determine a more appropriate core radius, the effective V and U values for PCFs need to be rescaled. This can be done very simply through the core radius ρ. For a range of ρ-values, the ratio between the effective U-value for a PCF and the U-value for an SIF with the same effective V is illustrated in Figure 5-8 (a) The figure is calculated for a specific PCF with $d/\Lambda = 0.60$, and, as seen, a ratio close to unity is found for $\rho = 0.625\Lambda$. Hence, a close correspondence between the propagation constants (defined through U) in PCFs and SIFs sharing the same V-value does, in fact, exist. It is worth noticing, that the rescaling using $\rho = 0.625\Lambda$ also gives a second-order mode cut-off value for the PCFs of around $V_{eff}= 2.6$, which is much closer to the corresponding value of SIFs ($V = 2.4$). As seen from Figure 5-8 (b), the core radius of 0.625Λ does, however, not appear optimum for all d/Λ-values. For PCFs with lower air-filling fractions, a ρ-value around 0.63Λ was found to result in a U_{eff}/U - ratio closest to unity, whereas the effective core radius for a PCF with cladding air holes of size $d/\Lambda = 1$ should, from geometrical considerations, have $\rho = 1.15\Lambda/2$. Hence, the optimum effective core radius varies slightly with cladding hole dimensions. For the remainder of this chapter, however, an effective core radius of 0.625Λ will be applied, when utilizing the SIF-analogy.

Having determined (using the full-vectorial variational method) the appropriate core dimensions for establishing a relatively accurate analogy between SIFs and PCFs, it is useful to take advantage of this analogy to provide qualitative information about some of the more advanced properties of high-index core PCFs.

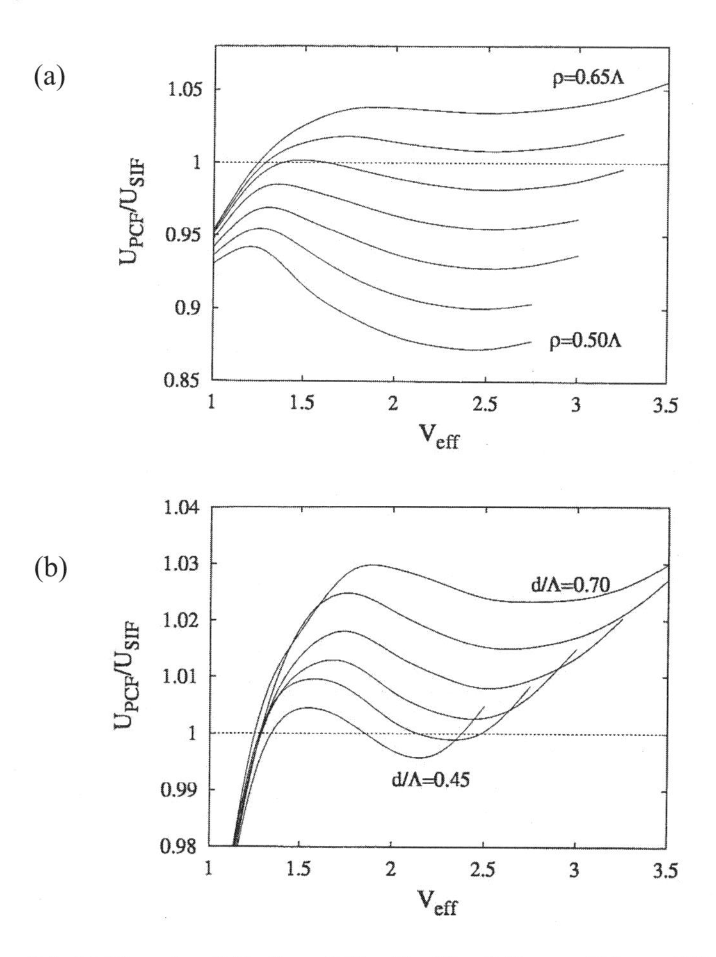

Figure 5-8. (a) The ratio of U_{eff} for a PCF with $d/\Lambda = 0.60$ over U for an SIF. The ratio is illustrated for PCFs with effective core radii ranging from $\rho = 0.50\Lambda$ (lowest curve) to $\rho = 0.65\Lambda$ (top curve) in steps of 0.025Λ. For the analogy between the two types of fibres, an optimum ρ-value of around 0.625Λ is found. (b) Same ratio as in (a), but for PCFs with d/Λ ranging from 0.45 (lowest curve) to 0.70 (top curve) in steps of 0.05.

5.3.3 Macro-bending losses of index-guiding PCFs

The macro-bending losses of optical fibres are very important to address, not only from a practical handling point of view, but also because they play a central role, when defining the spectral window, in which the fibre may be operated. In [5.14], the bending properties of PCFs were described by the introduction of a critical bend radius, i.e., a radius under which a PCF may not be bend in order for the excess bending loss to be below a given limit. However, in order to numerically characterize the PCF bending properties, it is here chosen to utilize the analogy between SIFs and PCFs and apply the bending loss formula described in [5.15]. In this formulation, which have proven to provide very accurate results for standard optical fibres, the power loss coefficient due to macro bending is written as:

$$\alpha = \frac{\sqrt{\pi} A_e^2 \rho \cdot \exp\left(-\frac{4\Delta W^3}{3\rho V^2} R \right)}{4PW\sqrt{\frac{WR}{\rho} + \frac{V^2}{2\Delta W}}} \tag{5.7}$$

where Δ is the relative index difference between the maximum refractive index in the core region and the cladding index, ρ is the core radius, V is the normalized frequency, and W is the normalized decay parameter in the cladding. R denotes the radius of curvature, A_e is the amplitude coefficient of the cladding electric field, and P is the propagation power carried by the fundamental mode. Applying equation (5.7), the bend loss is directly calculated from the Bessel-function coefficients and propagation constant of the effective-index fibre. In order to choose a realistic bend radius in this analysis, a situation where the full length of the fibre is coiled at a radius of 6.0 cm (corresponding to dispersion-compensating fibre coils, or coils in a laboratory) is considered. The bending-loss values for different air-hole dimensions are shown in Figure 5-9 for a silica PCF with $\Lambda = 2.3\ \mu m$. One of the most important observations is that both an upper (with respect to wavelength) and a lower bend edge is found. The explanation of these two bend edges is again the vanishing core–cladding index difference in the high-frequency regime (causing the lower bend edges) and the widely extending mode-field in the low-frequency regime (causing the upper bend edge). While the upper bend edge is well known, and found for all conventional fibres, the lower bend edge is unique to this type of PCFs. The bend edge for a standard SIF, which has a core–cladding refractive-index difference of 0.01 at $\lambda = 1.3\mu m$ and a core radius of $0.625\Lambda = 1.45\mu m$ is, furthermore, illustrated in Figure 5-9.

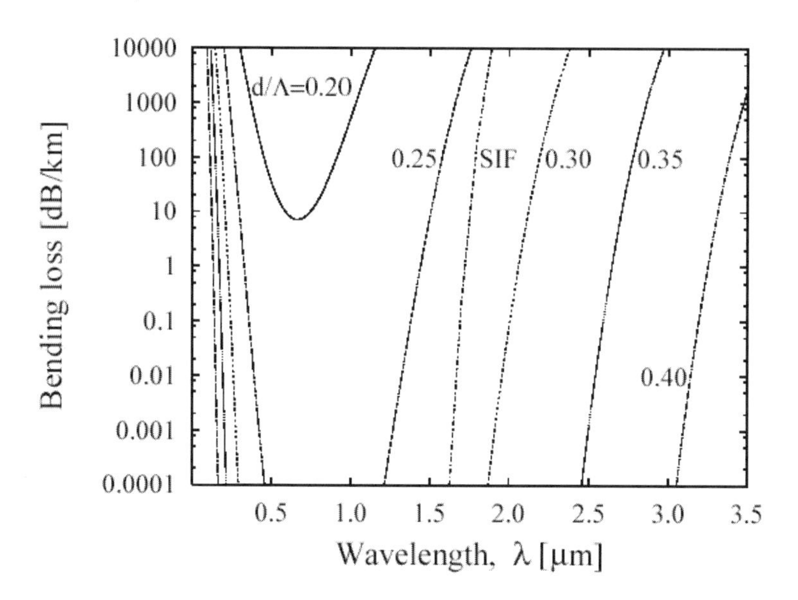

Figure 5-9. Bending properties of triangular, high-index core PCFs for various hole sizes, but a fixed hole spacing Λ of 2.3 μm. The curve indicated 'standard' corresponds to a standard step-index fibre, which has a core radius of 1.45 μm and a refractive index difference of 0.01 at λ = 1.3 μm. The results presented in this figure and in [5.16] are not identical due to the use of two different effective core radii.

The figure also illustrates that for PCFs with small air-hole sizes $d/\Lambda \le 0.20$, the bending losses may be considerable and that for the realization of robust PCFs, the normalized hole size, d/Λ, should be above approximately 0.25 d/Λ. This conclusion is, however, for PCFs with specific hole spacing (of 2.3μm) and to analyse the influence of this parameter on the bending properties, Figure 5-10 provides an illustration the bending loss for PCFs with a fixed normalized hole size of $d/\Lambda = 0.25$ and centre-to-centre hole spacings, Λ, ranging from 1μm to 5μm.

The two figures 5.9 and 5.10 demonstrate that large operational windows at visible and near-infrared wavelengths (of interest to optical communications) are available, if PCFs are designed with a photonic crystal cladding structure period that is approximately twice the operational wavelength as well as with air-hole sizes of around d/Λ= 0.25 or larger.

In Chapter 7, we will return to some of the experimentally characterised macrobending loss properties of practical (large-mode-area) photonic crystal fibres.

Having established an understanding of index-guiding PCFs and determined a basic knowledge of the parameter range, in which they can be explored in order to maintain both single mode operation as well as a certain robustness, a more application-oriented analysis will, finally, be turned to for the remainder of this section. If applications of PCFs are considered within the area of optical-communication systems, then the most interesting question at present will probably be, how the dispersion properties of the PCFs are? This question will be addressed in the proceeding section.

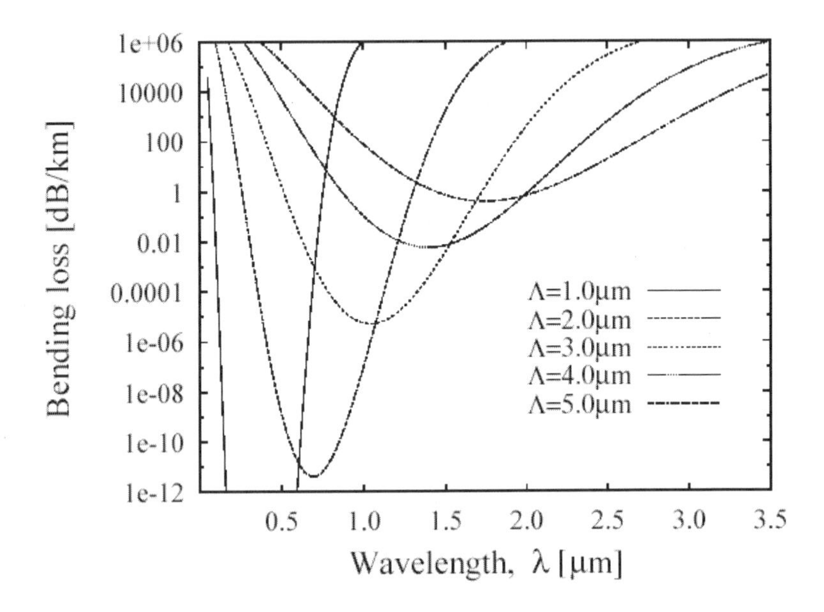

Figure 5-10. Bending properties of triangular, high-index core Photonic crystal fibres with Λ ranging from 1.0 μm to 5.0 μm for a fixed normalised hole diameter of $d/\Lambda = 0.25$.

5.3.4 Dispersion properties of index-guiding PCFs

To analyse the dispersion properties of index-guiding PCFs, point of reference will again be taken in the design shown in Figure 5-2, i.e., a triangular, high-index core PCF with a pitch of 2.3 μm. For fibres with relatively low air-filling fractions (ensuring single-mode operation), the dispersion properties calculated using the full vectorial, plane-wave method is shown in Figure 5-11. The dispersion is determined using the assumption that the refractive index of silica is wavelength independent (with a value of 1.45), and that the waveguide dispersion may be added to the material dispersion to provide the total dispersion of the PCFs. From Figure 5-11, it is first noted that for very small air-filling fractions, e.g., when the influence of the air holes is limited, the dispersion curve is expectedly very close to the material dispersion of pure silica (zero-dispersion wavelength around 1.28 μm). As the diameter of the air holes is increased, the waveguide dispersion becomes increasingly stronger. In this manner, it is found that the waveguide dispersion of PCFs may become positive at wavelengths below 1.28 μm, while - at the same time - the fibres may remain single-mode. Such a feature is typically not available from standard optical fibres due to limited refractive index contrasts. The first experimental measurements of dispersion in index-guiding PCFs that was published by Gander *et al.* [5.17] are in agreement with the information provided in Figure 5-11. Gander *et al.* measured the dispersion of a PCF with Λ=2.3 μm, and holes of (relative) size d/Λ=0.27, to be -77 ps/nm/km at a free-space wavelength of 813 nm. The silica material dispersion is -100 ps/nm/km at this wavelength, thereby, confirming that single-mode PCFs with anomalous waveguide dispersion at wavelengths below 1.3 μm.

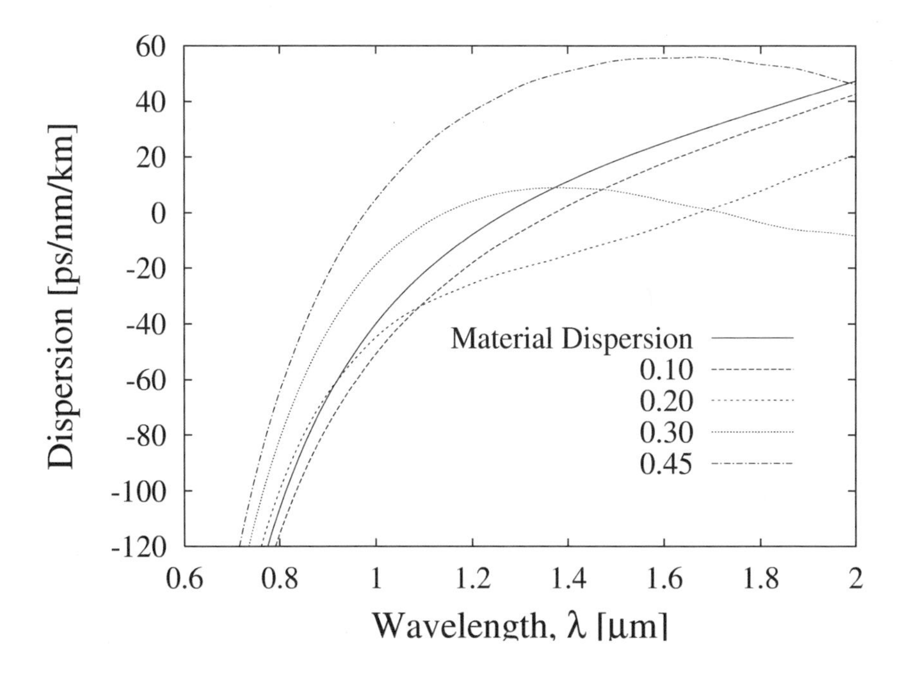

Figure 5-11. Dispersion properties of triangular, high-index core PCFs with Λ fixed at 2.3 μm and various (relatively small) air-hole sizes as predicted by the full-vectorial method.

Figure 5-11 points out a further interesting feature of PCFs, namely their ability to exhibit broadband, near-zero dispersion-flattened behavior. Here, for a fibre with d/Λ= 030. Due to the exhibition of anomalous waveguide dispersion at short wavelengths, the dispersion-flattened range is, in fact, extended to wavelengths below 1.28 μm down to approximately 1.1 μm. We will return to the issue of dispersion-flattened fibres in more detail in Section 7.8.By utilizing larger air holes, even stronger anomalous waveguide dispersion may be obtained than shown in Figure 5-11. This is illustrated in Figure 5-12 for PCFs with similar triangular structure as for Figure 5-11, but for air holes sizes of d/Λ = 0.60, 0.75, and 0.90. The results demonstrate a strong shifting of the zero-dispersion wavelengths all the way towards visible wavelengths. The first experimental demonstration PCFs with zero-dispersion wavelength shifted to around 800 nm was performed by Ranka *et al.*[5.21]. Ranka *et al.* used such PCFs for continuum light generation spanning the whole visible range. As a result of very small core size (a diameter of around 1.7 μm), strong mode confinement, and near-zero total dispersion at an available femto-second pump wavelength, continuum generation was feasible using relatively low-power laser pulses (peak powers in the kW range). We will return to the issue of continuum generation in more detail in Section 7.2.

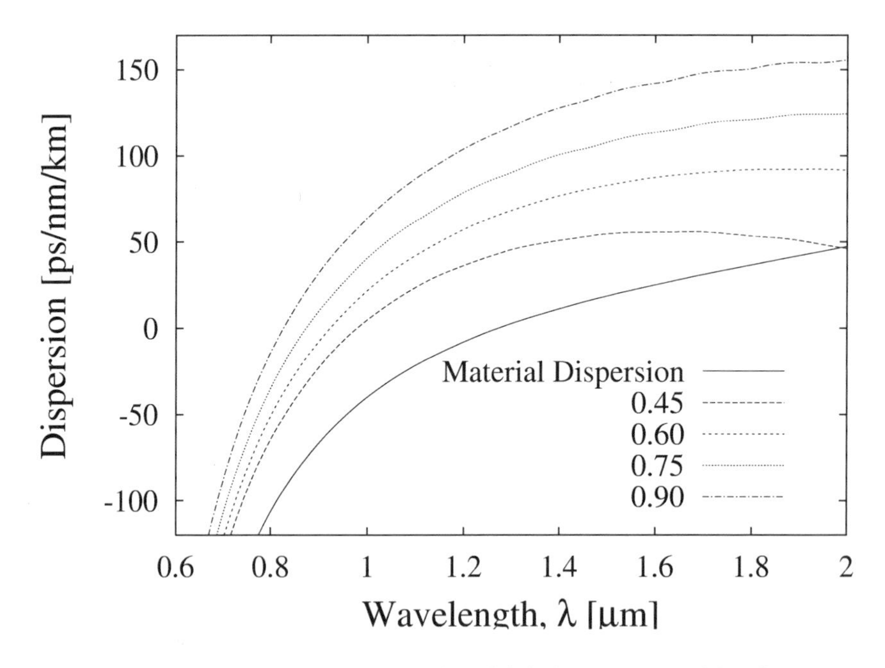

Figure 5-12. Dispersion properties of triangular, high-index core PCFs with Λ fixed at 2.3 μm and various (relatively large) air-hole sizes as predicted by the full-vectorial method. The dispersion is calculated for the fundamental mode of the fibres.

While a fixed PCF design and pitch, Λ, were chosen for Figure 5-11 and 5-12 to illustrate the various freedoms and possibilities of tuning the dispersion of PCFs, it should be emphasized that the dispersion is determined by several other parameters apart from the hole size. Most importantly, a larger pitch results in reduced waveguide dispersion for fixed wavelength and hole size. This will also be discussed in more details in Section 7.8.

Of numerical interest, it should be mentioned that the dispersion results obtained by the full-vectorial method is in agreement with results presented by Monro *et al.* in [5.18] (based on the localized function method as proposed by Mogilevtsev *et al.* in [5.19]), and with results presented by Ferrando *et al.* [5.20] (see Chapter 3 for further details on these methods). However, the scalar SIF-analogy approach and early dispersion results presented in [5.16] display a discrepancy at increased air-hole sizes compared to the more accurate above-mentioned method.

Finally, an important aspect for the development of future PCFs concerns the requirements on the actual periodicity of the microstructure of the cladding. This section has focused only on full periodic cladding structures, but whereas a high degree of periodicity is required for the fibres operating by PBG-effects, the requirements for index-guiding PCFs are much lower. In fact, the unusual properties addressed in this chapter, regarding cut-off, macro-bending, and dispersion only rely on cladding refractive indices that are highly wavelength dependent. Hence, other types of cladding morphologies than the triangular arrangement of air holes – even completely non-periodic morphologies – may prove advantageous, and we will discuss this further in the following section.

5.4 FIBRES WITH NON-PERIODIC OR NON-CIRCULAR CLADDING STRUCTURES

In this chapter, we have discussed, how very interesting waveguiding properties are obtainable in fibres, where a solid core is surrounded by a cladding material formed by a solid material with numerous air holes. The cladding may be considered as an artificial material having an effective refractive index with unique wavelength dependency. Another issue of this effective-index interpretation is that a lower refractive index than in the core may actually be obtained also for fibres, where the holes are not distributed in a periodic pattern. It should, however, also be stressed that in order to obtain well-defined spectral properties, the periodicity or at least a high degree of ordering in the localization of the holes is required. It is on the other hand interesting to look a little closer at the work by Monro *et al.* [5.25] from 2000, in which fibres with random cladding-hole distributions are presented.

The numerical investigations performed by Monro *et al.* [5.25] was done using the localized-basis-function (LBF) method as described in Section 3.3, which allows for the random distribution of air holes. Monro *et al.* analysed several distributions of air holes, and generated these by random hole distribution with the additional conditions that no air holes can be located within a distance Λ from the centre of the structure, and that two holes have to be placed with a given minimum separation. An example of a fibre structure having a random hole distribution and hole size in the cladding is shown in Figure 5-13. The scalar modal analysis made by Monro *et al.* [5.25] on such structures demonstrate that the holes in a microstructured fibre do not need to be arranged in the more conventional periodic lattice to guide light. It was also found numerically that features such as single-mode operation over a wide wavelength range are present in microstructured fibres with random cladding-hole structure.

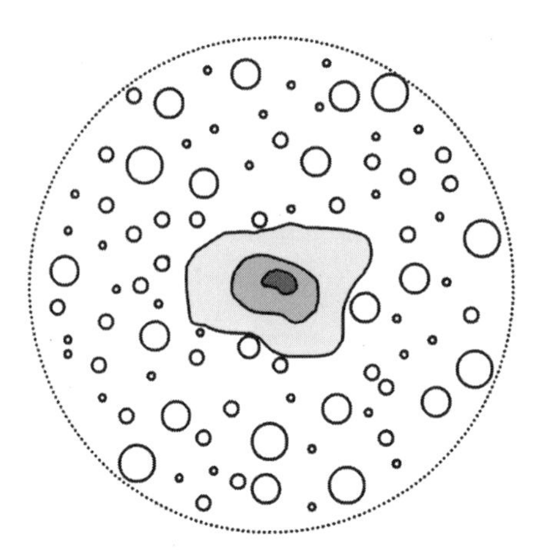

Figure 5-13. Schematic representation of cross section of a microstructured optical fibre with random hole positions and hole sizes. The fundamental mode is sketched in the core region of the fibre.

It should be pointed out that the fact that waveguiding is possible in a high-index core region placed in a cladding with random hole distribution has been one of the advantages in the early days of fabrication of microstructured optical fibres, because also non-perfect structures may be used to obtain some information about waveguiding properties, base material quality etc. However, when it comes to fibres with prescribed properties concerning spotsize, dispersion, birefringence etc., a significant process control is required, and a much higher degree of structural control is necessary.

The issue of symmetry and mode degeneracy in microstructured optical fibres has been studied by Steel *et al.* [5.26] in 2001. The key questions in this work are directed towards the existence of birefringence in PCFs, and Steel *et al.* choose to perform an analysis based on group representation theory. Steel *et al.* [5.26] find that a fibre with rotational symmetry of order higher than 2 has modes that either are non-degenerate and support the complete fibre symmetry or are twofold degenerate pairs of lower symmetry. The latter case applies to the fundamental modes of perfect PCFs, guaranteeing that such fibres are not birefringent. The findings in [5.26] based on group theory are also shown to be in very good agreement with numerical modelling – of course provided that sufficient care is taken in the numerical representation, e.g., if a plane-wave expansion method is used, sufficient basis states must be applied to correctly describe the degeneracies. It is in particular noteworthy that the theory of Steel *et al.* has established that microstructured optical fibres with sixfold symmetry are not birefringent.

Concerning the symmetry of photonic crystal fibres, it is also relevant to note that the holes used to form the cladding (or core) structure may not always be represented by circles. This may be a fact resulting from the specific fibre fabrication conditions, where the manufacturer has utilised non-circular hole shapes to obtain improved waveguiding properties such as increased air-filling fraction, and higher effective-index contrast. Examples of a fibre structure having non-circular shapes of the near-centre holes are shown in Figure 5-14. In this case, a very high degree of symmetry is still maintained, but the fibre designs make use of more "drop-shaped" holes to better confine the light to the core. We will return to more specific examples of practical fibre structures in Chapter 7.

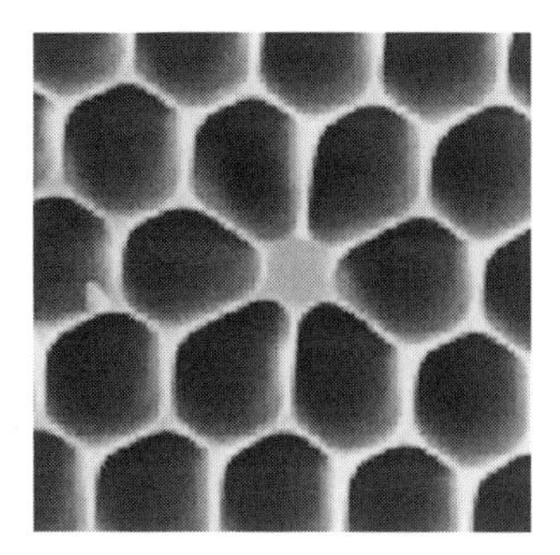

Figure 5-14. Example of a photonic crystal fibre structure using non-circular inner air-holes to obtain a very high index contrast between core and cladding. The photograph is kindly provided by Crystal Fibre A/S.

5.5 HOLE-ASSISTED LIGHTGUIDING FIBRES

In this chapter, we have discussed the important class of photonic crystal fibres known as either high-index core PCFs or index-guiding-PCF. We have seen that although these fibres have some properties that resemble those of standard optical fibres, they also allow for the realisation of waveguides having unique spectral properties.

In chapter 6, we will take the step of the PCF technology even further, but before doing so, we will here look closer at another kind of index-guiding fibre, which may be seen as a hybrid between a standard optical fibre and a microstructured fibre. This fibre was first demonstrated in 2001 by Hasegawa *et al.* [5.27], and it was given the name hole-assisted lightguide fibre (HALF).

The HALF is an optical fibre, which is realised with a high-index core (of GeO_2-doped silica), a low-index cladding (of pure silica), and several holes surrounding the core, as shown in Figure 5-15. In an effective-index interpretation, the holes may be seen as a depressed index ring surrounding the core and hereby assisting the mode confinement.

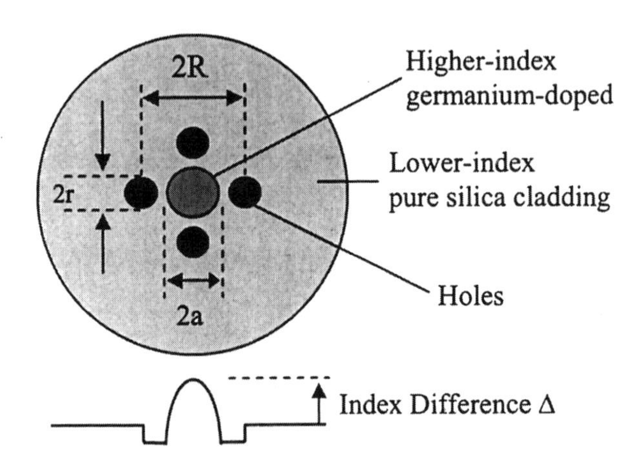

Figure 5-15. Schematic illustration of the cross section of a hole-assisted ligthguide fibre (HALF). The effective index profile of the fibre is shown at the bottom of the figure. The illustration is drawn from the reports [5.28-5.29].

The motivation behind the realization of the HALF was to a large extent the desire of making an improved optical fibre with lower transmission losses than those demonstrated during 2001, where the best loss values was 3.2 dB/km as shown by Kubota *et al.* [5.30] and 2.6 dB/km as demonstrated by West *et al.* [5.31].

A further resolution behind the work of Hasegawa *et al.* [5.27-5-29], was the wish of obtaining novel dispersion characteristics without severe degradation in transmission loss. Because of the structural proximity to the conventional fibres, and the low power fraction in the holes, the HALF has been shown to exhibit low loss below 1 dB/km at the 1550 nm wavelength (a loss of 0.82 dB/km was reported in [5.27]), and for the specific fibre reported in [5.27], the dispersion was measured to be +34 ps/(km·nm) at the wavelength of 1550 nm.

Further improvement of the HALF fibres was published by Hasewava *et al.* [5.29] in late 2001, and fibres with a loss as low as 0.41 dB/km and a dispersion of +35 ps/(km·nm) at 1550 was demonstrated. Modelling of the loss in the HALF structures indicated that the wavelength independent loss is dominant, and comparisons among result of the loss modelling for several HALF structures suggests that the contribution of the holes and the drawing tention are the most probable courses of the wavelength independent loss [5.29].

It is also relevant to note that the anomalous dispersion of the HALF can be further enlarged by increasing the air-filling fraction and the index difference between core and cladding [5.27]. The increased air-filling fraction may be easily obtained by introducing more air holes than the 4 holes schematically shown in Figure 5-15. Hasewava *et al.* [5.29] investigated fibres with core-cladding index differences of 0.6% to 1.4%, and the fibres had hole radii and distances from the core of $r = 0.4a$ and $R = 1.5a$, respectively, where a is the core radius, and the parameters r and R are shown in Figure 5-15.

Hasewava *et al.* [5.29] shows that larger index difference is preferable for lowering the relative dispersion slope (RDS) (dispersion slope normalized by dispersion), and RDS values of +0.0021 nm^{-1} was realised for the HALF having a loss value of 0.41 dB/km.

5.6 SUMMARY

This chapter has documented, how PCFs may support index-guided modes in a solid, silica core region surrounded by a microstructured cladding having a strongly wavelength dependent refractive index, which is lower than the core index. Regarding the operational principle, the high-index core PCFs, therefore, are comparable to conventional fibres. These PCFs do, however, display a number of properties, that separate them significantly from conventional fibres. These concern (among others) cut-off, macrobending, and dispersion properties, which have all been addressed from a fundamental viewpoint in this chapter. With respect to cut-off properties, PCFs with air holes below a normalized size, d/Λ, of 0.45 were found to exhibit so-called endlessly singlemode behaviour. This may potentially be utilised for design of large-mode area PCFs capable of transmitting high powers, while maintaining single-mode operation. However, as further analysis presented in this chapter pointed out, macrobending losses for the PCFs will in practice put limitations to frequency ranges in which they may be operated. As a unique feature of index-guiding PCFs, they were found to exhibit both a steep bend loss edge for long as well as for short wavelengths. The latter (and for standard fibres completely unusual) bend loss edge may be attributed to the vanishing core–cladding refractive index contrast in the high-frequency regime. The analysis of the bending losses was made possible through the employment of an approximate scalar method based upon an analogy between triangular, high-index core PCFs and SIFs. Owing to the operation based on index-guidance, such an analogy could be established and this chapter has outlined some of the basic core parameter considerations that must be made. Using the plane-wave method several attractive dispersion features were found, including PCFs with a broadband, near-zero dispersion flattened behaviour at near-infrared wavelengths, and PCFs with zero-dispersion wavelengths shifted significantly below 1.28 μm. The chapter has also included a short review of some of the possibilities of forming waveguides without a strict need for periodically distributed air holes. Finally, the idea of using non-circular index-reducing voids in the cladding is described as well as the possibility of applying micron-sized air holes to modify a doped fibre structure.

REFERENCES

[5.1] K.P. Hansen, J.R. Simonsen, J. Broeng, P.M.W. Skovgaard, A. Peterson, and A. Bjarklev,
"Highly nonlinear photonic crystal fiber with zero-dispersion at 1.55μm",
Optical Fiber Communication Conference, OFC 2002 Postdeadline Paper, FA9-1

[5.2] B.J. Eggleton, C.Kerbage, P.S. Westbrook, R.S. Windeler, and A. Hale,
"Microstructured optical fiber devices",
Optics Express, Vol.9, No.13, Dec.2001, pp.698-713.

[5.3] T. M. Monro, Y. D. West, D. W. Hewak, N .G. R. Broderick, and D. J. Richardson,
"Chalcogenide holey fibres",
IEE Electronics Letters, Vol.36, No.24, Nov.2000, pp.1998-2000.

[5.4] A. Argyros, I. M. Bassett, M. A .van Eijkelenborg, M. C. J. Large, J. Zagari, N. A. P. Nicorovoci, R. C. McPhedran, and C. M. de Sterke,
"Ring structures in microstructured polymer optical fibres",
Optics Express, Vol.9, No.13, Dec.2001, pp.813-820.

[5.5] P. Kaiser, E. A. J. Marcatili, and S. E. Miller,
"A new optical fiber",
The Bell System Technical Journal, Vol.52, No.2, pp. 265-269, Febr. 1973.

[5.6] E. A. J. Marcatili,
"Slab-coupled waveguides",
The Bell System Technical Journal, Vol.53, No.4, pp. 645-674, April 1974.

[5.7] P. Kaiser, and H. W. Astle,
"Low-loss single-material fibers made from pure fused silica",
The Bell System Technical Journal, Vol.53, No.6, pp. 1021-1039, July-August 1974.

[5.8] J. Knight, T. Birks, P. Russell, and J. Sandro,
"Properties of photonic crystal fiber and the effective index model",
Journal of the Optical Society of America A, vol. 15, pp. 748–52, March 1998.

[5.9] A. Snyder and J. Love,
"Optical waveguide theory",
Kluwer Academic Publishers, 2000, ISBN: 0-412-09950-0.

[5.10] J. Knight, T. Birks, P. Russell, and D. Atkin,
"All-silica single-mode optical fiber with photonic crystal cladding",
Optics Letters, vol. 21, pp. 1547–9, Oct. 1996.

[5.11] D. H. Smithgall, T. J. Miller, and R. E. Frazee Jr.,
"A novel MCVD process control technique",
IEEE Journal of Lightwave Technology, Vol.4, No.9, pp.1360-1366, 1986.

[5.12] D. B. Keck, P. C. Schultz, and F. W. Zimar,
US Patent 3,737,393.

[5.13] Y. Ohmori, F. Hanawa, and M. Nakahara,
"Fabrication of low-loss Al_2O_3-doped silica fibres",
IEE Electronics Letters, VOL.19, PP.261-262, 1983.

[5.14] T. Birks, J. Knight, and P. Russell,
"Endlessly single-mode photonic crystal fiber",
Optics Letters, vol. 22, pp. 961–963, July 1997.

[5.15] J. Sakai, and T. Kimura,
"Bending loss of propagation modes in arbitrary-index profile optical fibers",
Applied Optics, vol. 17, pp. 1499–1506, 1978.

[5.16] A. Bjarklev, J. Broeng, S. Barkou, and K. Dridi,
"Dispersion properties of photonic crystal fibers",
European Conference on Optical Communications, pp. 135-136, Madrid, Sept. 20-24, 1998.

[5.17] M. Gander, R. McBride, J. Jones, D. Mogilevtsev, T. Birks, J. Knight, and P. Russell, "Experimental measurement of group velocity dispersion in photonic crystal fibre", *IEE Electronics Letters,* vol. 35, pp. 63-43, Jan. 1999.

[5.18] T. Monro, D. Richardson, and N. Broderick, "Holey fibres: an efficient modal model", *Journal of Lightwave Technology,* vol. 17, pp. 1093-1102, June 1999.

[5.19] D. Mogilevtsev, T. A. Birks, and P. St. J. Russell, "group-velocity dispersion in photonic crystal fibres", *Optics Letters*, Vol.23, pp.1662-1664, Nov.1998.

[5.20] A. Ferrando, E. Silvestre, J. Miret, J. Monsoriu, M. Andres, and P. Russell, "Designing a photonic crystal fibre with flattened chromatic dispersion", *IEE Electronics Letters,* Vol. 24, pp. 276-278, March 1999.

[5.21] J. Ranka, R. Windeler, and A. Stentz, "Efficient visible continuum generation in air-silica microstructure optical fibers with anomolous dispersion at 800 nm", *Conference on Laser and Electro-Optics, CLEO'99, Baltimore,* May 1999. CPD8.

[5.22] B. T. Kuhlmey, R. C. McPhedran, and C. M. de Sterke, "Modal cutoff in microstructured optical fibres", *Optics Letters*, Vol.27, No.19, Oct.2002, pp.1684-1686.

[5.23] N. A. Mortensen, "Effective area of photonic crystal fibres", *Optics Express*, Vol.10, pp.341-348, 2002.

[5.24] T. P. White, R. C. McPhedran, L. C. Botten, G. H. Smith, and C. M. de Sterke, "Calculations of air-guided modes in photonic crystal fibers using the multipole method", *Optics Express*, Vol. 11, pp. 721- 732, 2001.

[5.25] T. M. Monro, P. J. Bennett, N. G. R. Broderick, and D. J. Richardson, "Holey fibers with random cladding distribution", *Optics Letters*, Vol.25, No.4, Febr. 15, pp.206-208, 2000.

[5.26] M. J. Steel, T. P. White, C. M. de Sterke, R. C. McPhedran, and L. C. Botten, "Symmetry and degeneracy in microstructured optical fibers", *Optics Letters*, Vol.26, No.8, April 15, pp.488-490, 2001.

[5.27] T. Hasagawa, E. Sasaoka, M. Onishi, M. Nishimura, Y. Tsuji, and M. Koshiba, "Novel hole-assisted lightguide fiber exhibiting large anomalous dispersion and low loss below 1 dB/km", *Optical Fiber Communication Conference*, OFC'2001, Anaheim, California, USA, March 17-22, Postdeadline paper PD5, 2001

[5.28] T. Hasagawa, E. Sasaoka, M. Onishi, M. Nishimura, Y. Tsuji, and M. Koshiba, "Modelling and design optimization of hole-assisted lightguide fiber by full-vector finite element method", *European Conference on Optical Communications*, ECOC'2001, Amsterdam, The Netherlands, Sept.30 – Oct.4, Paper We.L.2.5, 2001.

[5.29] T. Hasagawa, E. Sasaoka, M. Onishi, M. Nishimura, Y. Tsuji, and M. Koshiba, "Hole-assited lightguide fiber for large anomalous dispersion and low optical loss", *Optics Express*, Vol.9, No.13, Dec.17, pp.681-686, 2001

[5.30] H. Kubota, K. Suzuki, S. Kawanishi, M. Nakazawa, M. Tanaka, and M. Fujita, "Low-loss, 2 km-long photonic crystal fiber with zero GVD in the near IR suitable for picosecond pulse propagation at the 800 nm band", *CLEO'2001*, Paper CPD3, 2001.

[5.31] J.A. West, N. Venkataraman, C.M. Smith, and T. Gallagher, "Photonic crystal fibers", *European Conference on Optical Communications*, ECOC'2001, Amsterdam, The Netherlands, Sept.30 – Oct.4, Paper Th.A.2.2, 2001.

[5.32] J. W. Hicks, "Hollow tube method for forming an optical fiber" US patent 4551162, 1985.

[5.33] V. Vali, and D. B. Chang "Low index of refraction optical fiber with tubular core and/or cladding", US Patent 5155792, 1992.

Chapter 6

LOW-INDEX CORE FIBRES – THE TRUE PHOTONIC BANDGAP APPROACH

6.1 INTRODUCTION

The most fascinating and challenging element in the recent development of photonic crystal fibres is without doubt the potential of fabricating fibres in which light is guided in fibre cores having a lower effective refractive index than the surrounding cladding material. The attraction is that this was in contrast to any former fibre, and that such a discovery, therefore, inevitably would lead to new design and application possibilities. The basic motivating element behind this is also the fundamental curiosity of human beings to try to open up new areas, and in this case simply to "guide light in a manner nobody else had done before". This fundamental driving force may be seen as a very academic task, but at the same time it has the highly practical aspect of containing the potential of creating new fibres and components with radically different properties compared to the existing possibilities.

Whereas the high-index core fibres with microstructures, as described in the previous chapter, had their origin in quite early work on air-suspended waveguides [6.1-6.2], the fundamental idea of the photonic bandgap structure was formed in 1987 by Yablonovitch [6.3] and John [6.4], and followed by the work of Knight, Russell, Birks and co-workers [6.5] in 1996 with specific focus on the fibre aspect.

In this chapter, we will first present a fundamental description of the creation of photonic bandgaps in silica-air structures, presenting simple triangular structures as well as hexagonal (or honeycomb) structures of the cladding. This basis leads to a description of general design considerations for photonic crystal fibres.

After this more fundamental discussion, the focus will be moved to the first demonstration of the photonic bandgap fibre operating in the optical regime. This first experimental demonstration, which was first presented in 1998 by Knight *et al.* [6.6], was based on the fibre designs outlined by Broeng *et al.* [6.7]. After the discussion of the first experimental results and some of their implications, a further discussion of the properties of PBG-fibres is included. In this context, issues such as dispersion properties and birefringence is presented, and the option of having a truly single-mode (single-polarization) fibre using the bandgap effect is also described.

One of the most important application-aspect of PBG-fibres is the potential of providing low-loss waveguidance in air. This chapter shall, therefore, further address the requirements for obtaining leakage-free air-guidance in PBG-fibres, which will include a discussion of the optimum basis cladding-structure and a numerical analysis of the influence of the core-design on the number of air-guided modes. The analysis points out important issues to be addressed for experimental studies of air-guiding PBG-fibres and further discusses specific applications in the field of optical communications and high-power deliverance in the mid-infrared. The chapter also refers recent progress concerning the air-guiding fibre performance, as reported by West *et al.* [6.8] in 2001 and by Venkataraman *et al.* [6.9] in 2002.

6.2 SILICA-AIR PHOTONIC CRYSTALS

As we already have been discussing, the main part of this book will focus on silica-air based 2D photonic crystals, i.e., photonic crystal fibres. However, the presented results and ideas may to a large extent also be applied to fluoride, chalcogenide, or polymer-based photonic crystals and applications of these (in the latter cases of course with the proper replacement of refractive index values etc.).

6.2.1 Simple triangular structures

In 1995, Birks *et al.* demonstrated that, although the low index contrast between silica and air does not allow for complete PBGs for the in-plane case, complete 2D PBGs may open up in the case of out-of-plane wave propagation [6.10]. This was found in a simple triangular structure with a relatively large air filling-fraction of $f = 45\%$ and circular air holes. An illustration of the out-of-plane PBGs for such a photonic crystal is presented in Figure 6-1. As seen, the silica-air photonic crystal does not display in-plane PBGs, but several out-of-plane PBGs are exhibited. For a calculation of the 15 lowest frequency bands, four complete 2D PBGs were found for the $k_z\Lambda$-range from 0 to 20. The four PBGs were found to appear between bands 4 and 5 (the lowest frequency PBG at a fixed β-value), 6 and 7, 8 and 9, and 12 and 13 (the highest frequency PBG). Note that β is the propagation constant along the fibre axis (the invariant direction of the PCF). As seen from Figure 6-1, the PBGs are relatively narrow, but in order to obtain a more qualitative idea about this, their relative sizes are calculated and illustrated in Figure 6-2. Note that the relative size of a photonic bandgap here is defined as the difference between the frequencies defining the upper and lower bandgap edges divided by the centre frequency (the frequency found in the middle of the bandgap). Figure 6-2 further includes a calculation of the same PBGs for two other triangular photonic crystals with filling fractions of 30% and 70%. The largest PBG is seen to appear for the triangular photonic crystal with the highest filling fraction, and it is the PBG between bands 4 and 5. This bandgap has a relative size of up to 4.5% and occurs around a $\beta\Lambda$-value of 17.5, where the normalized centre frequency of the gap is $\Lambda/\lambda_{T,max} \approx 2.25$. For the triangular photonic crystal with $f = 45\%$, the largest PBG is seen to appear for the PBG between band 12 and 13 - having a relative size of up to 2.5%. Figure 6-2 also illustrates, how all PBGs are strongly suppressed, when decreasing the air hole sizes. For f-values below 30%, no PBGs with a relative size above 1% were found.

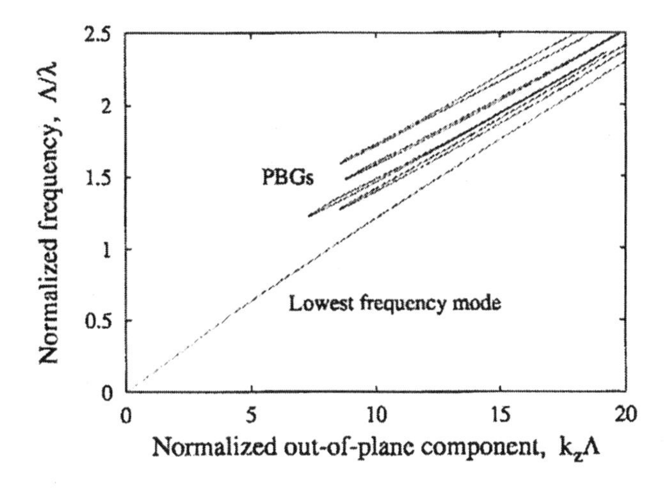

Figure 6-1. Illustration of the out-of-plane PBGs exhibited by a triangular silica-air photonic crystal with air-filling fraction f = 45%. The figure also illustrates the lowest-frequency allowed mode of the photonic crystal.

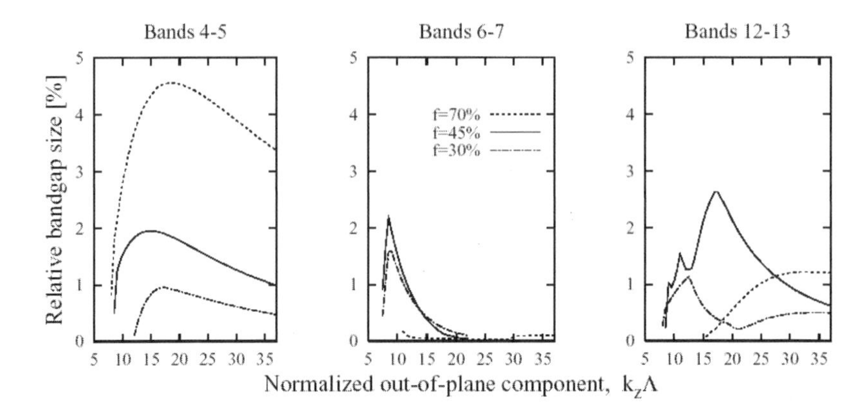

Figure 6-2. Illustration of the relative size of out-of-plane PBGs appearing between bands 4 and 5, 6 and 7, and 12 and 13 for three triangular photonic crystals with f = 30%, 45% and 70%. These filling fractions corresponds to hole sizes of d/Λ = 0.58, 0.70, and 0.88, respectively. The relative size of a PBG is defined as the frequency difference between the PBG edges divided by the centre frequency at a given k_z-value.

6.2.2 Effective-index considerations

It is at this point useful to consider wave propagation in a homogeneous medium. For such a medium, the dispersion relations are given by:

$$\sqrt{\bar{k}_{\|}^2 + k_z^2} = \frac{\omega n_h}{c} \tag{6.1}$$

where ω is the angular frequency and n_h is the refractive index of the homogeneous medium. Although describing the wave propagation in a homogeneous medium, the decomposition of the wave vector into $\bar{k}_{\|}$ and k_z is chosen to be consistent with the decomposition used for 2D photonic crystals. Considering next, the propagation of a plane wave in such a homogeneous medium, we may choose to adjust the coordinate system so that the plane wave is propagating fully in the z-direction. Hence, it is found that the refractive index of the homogeneous medium may be determined as:

$$n_h = \left.\frac{k_z c}{\omega}\right|_{\bar{k}_{\|}=\bar{0}} \tag{6.2}$$

Therefore, it appears intuitively correct to define an effective refractive index to photonic crystals by use of the lowest possible frequency, ω_{fsm}, of an allowed mode in a given photonic crystal (this lowest frequency mode is also known as the fundamental space-filling mode [6.11]). Hence, n_{eff} for a photonic crystal may be defined as:

$$n_{eff} = \frac{k_z c}{\omega_{fsm}} \tag{6.3}$$

since the lowest frequency mode is characterized by $\bar{k}_{\|} = \bar{0}$ (as may e.g., be seen from the band diagrams in Figure 2-4, where the lowest frequency mode is seen to be at the Γ-point).

With the aim of utilizing photonic crystals for optical fibres, it is at this stage advantageous to rename k_z to β in order to be consistent with the terminology used for optical fibres. For the same reason, the out-of-plane wave-vector component will also be referred to as the propagation constant. Using this terminology, it is directly seen that the above expression (6.3) for the effective index is identical to the - from optical fibre technology - well-known expression describing effective indices:

$$n_{eff} = \frac{\beta}{k} \tag{6.4}$$

where k is the free-space wave number (= $2\pi/\lambda$). Hence, by plotting β/k, the effective index of the photonic crystal may be illustrated. This is done in Fig. 6-3. As a check of the determination of the effective index using the plane-wave method (see Section 3.4), the results it provides may be compared to analytic and intuitive results in the two extremes of low and high frequency. In the low-frequency limit, we have $k = 0$ and $\beta = 0$, and the plane wave method results in an effective index value of 1.23, as may be seen from Figure 6-3. Analytic expressions of various types of structured dielectrics may be found in [6.12]. In the low-frequency limit, the effective dielectric constant of a triangular periodic arrangement of circular holes/rods (with a ralative permittivity ε_1) in a background material (with a relative permittivity ε_2) is:

$$\varepsilon_t = \varepsilon_2 \frac{\left[1 - f\dfrac{(\varepsilon_2 - \varepsilon_1)}{(\varepsilon_2 + \varepsilon_1)}\right]}{\left[1 + f\dfrac{(\varepsilon_2 - \varepsilon_1)}{(\varepsilon_2 + \varepsilon_1)}\right]} \tag{6.5}$$

This expression gives an effective index of $n_{eff} = \sqrt{\varepsilon_t} = 1.23$ for $f = 0.45$, $\varepsilon_1 = 1.0$, $\varepsilon_2 = 2.1$, and is, therefore, in agreement with the result obtained from the plane-wave method. In the other extreme, the high-frequency limit, the wavelength approaches zero and the electromagnetic fields will be able to completely avoid the low-index regions (in the above-example: the air holes) and an effective index equal to the background index should be expected. Although not observable from Figure 6-3, β/k was found to approach asymptotically a value of 1.45 for large β-values - i.e., $\beta/k \rightarrow \sqrt{\varepsilon_2}$ in the high-frequency limit.

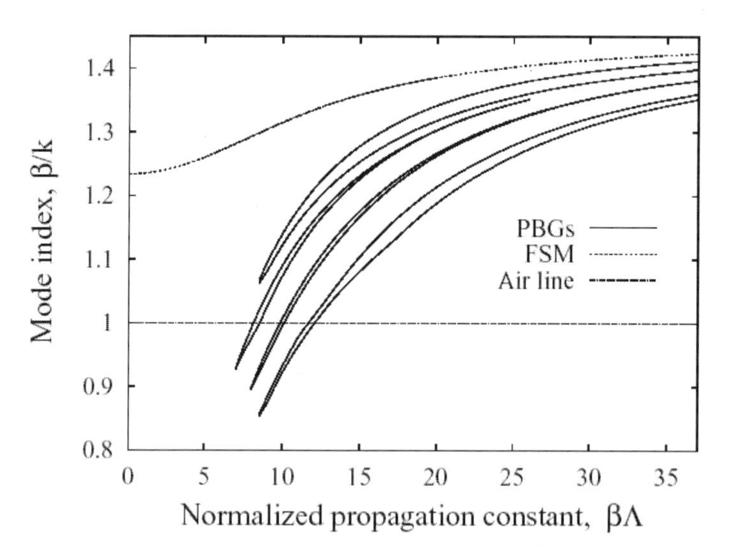

Figure 6-3. Modal-index illustration of the PBGs exhibited by a triangular silica-air photonic crystal with $f = 45\%$. The curve labelled 'FSM' indicates the effective index of the photonic crystal determined using the lowest-frequency allowed mode (the fundamental space-filling mode). The line labelled "Air-line" indicate the $\beta/k = 1.0$ case that corresponds to the dispersion relation for vacuum.

Using effective index considerations, it is realized that the forbidden region below the lowest-frequency band may be utilized to reflect light incident upon the photonic crystal from an adjacent medium, provided the medium has a *higher* effective index than the photonic crystal, and that the light is incident at a grazing angle (below a certain critical value). This is identical to total internal reflection (TIR) utilized for years in optical waveguides. 2D photonic crystals may, therefore, be used as cladding structures in optical fibres that operate in a traditional manner. The effective index of photonic crystal is, however, strongly wavelength dependent and - as documented in Chapter 5 - gives rise to significantly different features compared to conventional optical fibres.

Apart from the traditional way of reflecting light as described above, the attractive potential of photonic crystals is that they also allow reflectance in a new manner. Figure 6-3 further provides the effective index corresponding to the PBG edges and as seen, the 2D photonic crystals has the unique potential of providing total reflection of light incident upon the crystal from an adjacent material that has a *lower* refractive index than the effective index of the photonic crystal itself. This allows us to design radically new optical fibres with novel waveguiding properties. A particularly interesting feature is seen from Figure 6-3, namely that some of the PBG regions cross the air-line (defined by $\beta/k = 1.0$), which indicates that the triangular structure is able to reflect electromagnetic waves, which are incident from air. This allows us to utilize 2D photonic crystals for providing leakage-free waveguidance in air regions. It should be noticed that for light at a fixed frequency, the reflectance caused by the 2D photonic crystals only occurs within certain limited β-intervals. Hence, light at an arbitrarily small grazing angle of incidence may not be reflected from an adjacent low index material, but this can only occur within a limited oblique angular range. Therefore, the 2D photonic crystals may only provide the desired reflectance within certain fixed intervals, and this has the important consequence for PBG-fibres that their operation becomes limited within certain spectral windows.

6.2.3 Hexagonal or honeycomb structures

The initial attempts to realize triangular silica-air photonic crystals that would exhibit complete 2D PBG effect failed to succeed due to the technological difficulties in realizing high air-filling fractions and sufficient uniformity along the invariant direction [6.6]. Therefore, in the early research in photonic bandgap fibres, focus was put on a search for novel photonic crystal structures that might be more readily fabricated. In this section, investigations of novel types of hexagonal photonic crystals will be presented. Apart from the application-oriented interest directed towards optical fibres, such a search is, naturally, also of large fundamental interest in the field of photonic crystals.

The investigations will start out by considering silica-air photonic crystals with circular air holes arranged in a so-called honeycomb or 2D graphite lattice structure [6.13]. The properties of the honeycomb structure have been intensively studied in the case of in-plane wave propagation in high-index contrast photonic crystals (such as GaAs), mainly by Cassagne *et al.* [6.13-6.14]. Therefore, the area of interest for use in optical fibres, namely out-of-plane properties of the low-index contrast material system of silica/air needs special attention.

The honeycomb structure is indicated in Figure 6-4. Similar to the triangular structure, the honeycomb structure is also characterized by a hexagonal symmetry. The honeycomb structure, however, has a larger unit cell containing two holes/rods compared to a single hole/rod for the triangular structure, and it has in total one third less air holes/rods in a given volume. Hence, the filling fraction of a honeycomb structure is:

$$f = \frac{\pi}{3\sqrt{3}} \frac{d^2}{\Lambda^2} \tag{6.6}$$

where Λ is the centre-to-centre spacing between two nearest holes/rods. As for triangular structures, Λ will be used for normalization of frequencies and the propagation constant. The reason for choosing to normalize with respect to hole spacing, and not e.g., the period of the structure, is that the hole spacing is the most representative figure related to the demands on fabrication of the structure.

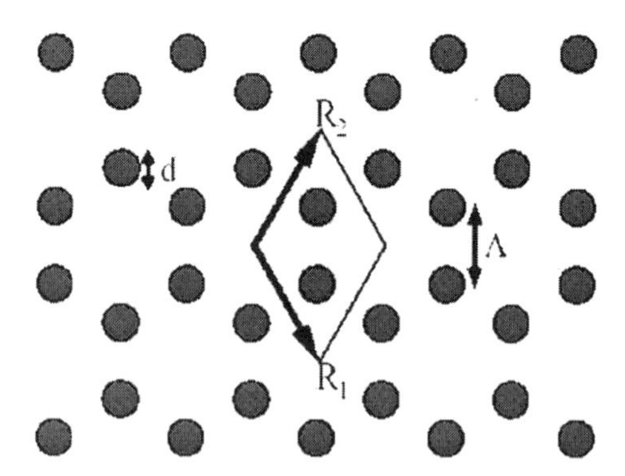

Figure 6-4. Illustration of the real-space primitive lattice vectors R_1 and R_2 and unit cell of a honeycomb photonic crystal. Λ is the spacing between centres of two nearest holes/rods, and *d* is the diameter of a hole/rod.

To compare the PBGs exhibited by honeycomb and triangular structures, the size of the three largest out-of-plane PBGs of a honeycomb structure with filling fractions f = 20%, 30%, and 47% is illustrated in Figure 6-5. These filling fractions correspond to holes sizes identical to those presented for the triangular structure (i.e., d/Λ = 0.58, 0.70, and 0.88). Figure 6-5 reveals that two PBGs with very high maxima in PBG size are exhibited by the honeycomb structures between bands 2 and 3 and between bands 6 and 7. The largest PBG exhibited by the honeycomb structure is even found to have a maximum that is higher than for a triangular photonic crystal with similar sized air holes (see Figure 6-2). A second difference between the honeycomb and triangular photonic crystals is that the PBGs in the honeycomb case is seen to open up at lower $\beta\Lambda$-values, but the size of the PBGs are more rapidly decreased for large $\beta\Lambda$-values. The maxima for the largest PBGs (between bands 2 and 3) are seen to occur around a $\beta\Lambda$-value of 0.6, where the normalized centre frequency of the gap is $\Lambda/\lambda_{H,max} \sim 0.75$. The reason for the above-described behaviour may to some degree be understood from considerations on the high- and low-index regions of the photonic crystals and their separation. Section 6.3, later in this chapter, will present a more elaborate discussion of the behaviour of the PBGs and illustrate how a simple design route is developed to provide photonic crystals with larger PBGs.

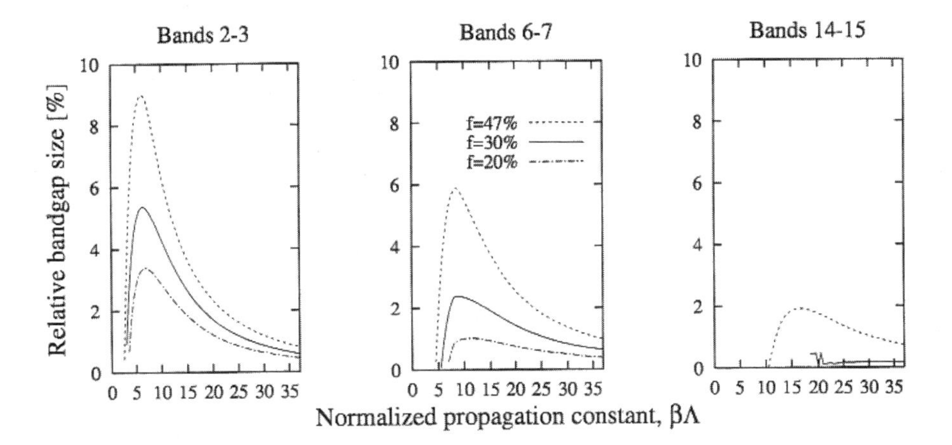

Figure 6-5. Illustration of the relative size out-of-plane PBGs appearing between bands 2 and 3, 6 and 7, and 14 and 15 for three honeycomb photonic crystals with f = 20%, 30% and 47%. The relative size is defined as the frequency difference between the PBG edges divided by the centre frequency at a fixed β-value.

To further investigate the properties of triangular and honeycomb photonic crystals, the influence on the size of the PBGs from modifying the structures using an increased number of holes/rods shall be analysed.

6.2.4 Modified triangular and honeycomb photonic crystals

These investigations were initiated by the interest in the influence of small interstitial holes, which were found to remain in the triangular structures after fabrication [6.15]. The fabrication process is described in Chapter 4, but is, in short, typically a process, where circular silica tubes/canes are stacked in a close-packed manner and drawn into fibre-form. The interstitial holes are located at mid-position between centres of three silica tubes/canes. The interstitial holes and their location are indicated Figure 6-6, whicht also illustrates the unit cells used for analysing the modified photonic crystal structures. Larger in-plane PBGs for high-index contrast 2D photonic crystals through the use of symmetry-breaking have been reported for square and triangular lattices by Anderson *et al.* [6.16]. However, the types of modifications, which are analysed in this book, was until recently not addressed for in-plane PBG-studies.

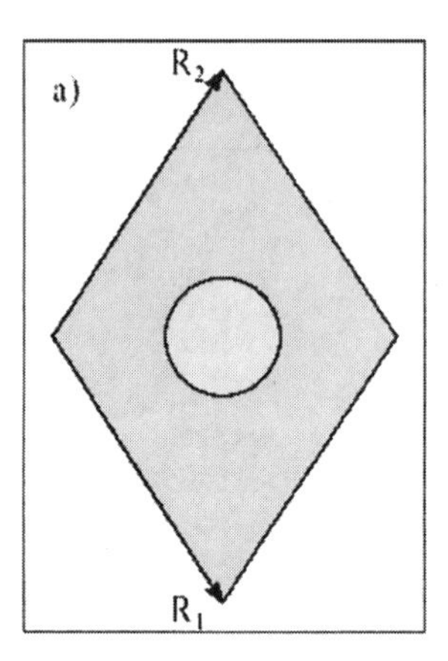

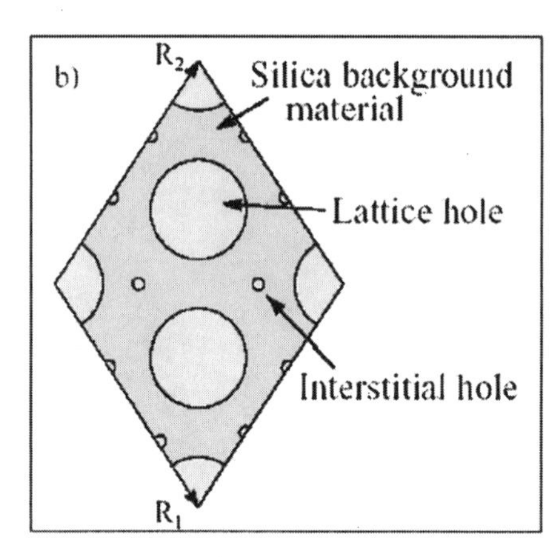

Figure 6-6. (a) Unit cell of the basic hexagonal (triangular) lattice. (b) Unit cell employed for the general study of hexagonal photonic crystals. Setting the radius of the corner holes equal to the radius of the two middle holes results in a triangular lattice with interstitial holes. Setting the radius of the corner air holes to zero results in a honeycomb lattice with interstitial holes.

In Figure. 6-7, the relative size of the PBG, which opens up at the lowest $\beta\Lambda$–value, is illustrated for the two cases of a triangular lattice with and without interstitial holes. The triangular structure without interstitial holes has an air-filling fraction of $f = 45\%$, and the addition of the interstitial holes raises the total air-filling fraction of the structure to $f = 50\%$. This PBG is labelled "primary", and it is the PBG that occurs between band 6 and 7. Λ is in both cases equal to the distance between two non-interstitial holes. For both structures, the primary PBG opens up at $\beta\Lambda \approx 7$. The maximum size and the behaviour of the PBG are seen to be almost identical in the two cases (with a maximum of about 2.1% at $\beta\Lambda = 8.5$). A secondary gap (the PBG between band 12 and 13) opens up at $\beta \approx 8.5/\Lambda$ for the structure without interstitial holes and reaches a maximum of 2.5% at $\beta = 17.5/\Lambda$. Considering, on the other hand, the structure with interstitial holes, a very different behaviour of this PBG is observed. Despite the very small extent of the interstitial holes ($f_{int} = 5\%$), the presence of these causes, in fact, a strong suppression of the secondary PBG. Not only is it reduced, but also a shifting of the gap towards higher values of β takes place. An important aspect of the interstitial holes is, however, that for high βΛ-values (which corresponds to high normalized frequencies), the secondary PBG of the triangular structure with interstitial holes is seen to become larger than the PBG of the structure without interstitial holes. The secondary PBG of the structure with interstitial holes is seen to reach a maximum of approximately 1.6% at a relatively large βΛ-value of 26. The normalized centre frequency, $\Lambda/\lambda_{T1,max}$, is here 3.5. The reason for different behaviour of the two triangular structures will be addressed in the proceeding section. The other smaller PBGs, which exist for the triangular lattice (see Figure 6-3), have been left out in Figure 6-7 for reasons of clarity.

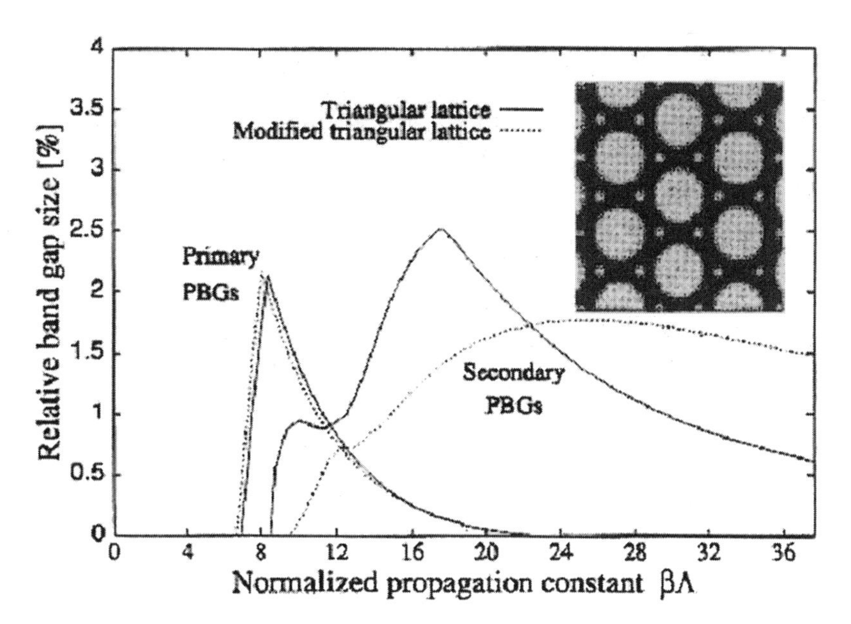

Figure 6-7. Relative size of the photonic band gaps for a triangular lattice with air-filling fraction $f = 45\%$. For the lattice with interstitial holes (dotted lines), a suppression of the secondary band gap is observed. The insert shows the geometry of the triangular lattice with interstitial holes ($f_{int} = 5\%$)

The results for the triangular photonic crystal could lead to the idea that modification of the photonic crystals through the use of interstitial holes is not a fruitful way of achieving larger PBGs for silica/air structures. However, when looking at the honeycomb structure, very interesting results do, in fact, occur by modifying the structure using interstitial holes.

Figure 6-8 illustrates the PBGs between band 2 and 3 (primary PBG) and between band 6 and 7 (secondary PBG) for a photonic crystals with air holes arranged in a honeycomb lattice, both with and without interstitial holes (the modified honeycomb structure is seen in the inset in Figure 6-8). Again, the specific location of the interstitial holes is chosen due to the currently used fabrication technique. The air-filling fraction, f, for the holes in the honeycomb lattice (the non-interstitial holes) is 30%. For comparison, the primary PBG of the triangular lattice without interstitial holes (f = 45%) has been included in Figure 6-8. Apart from providing larger PBGs compared to the triangular structure with similar-sized air holes, it is seen that the introduction of the interstitial holes is solely advantageous with respect to the size of the PBGs. In particular, the interstitial holes are seen also to increase the secondary PBG, which is in contrast to the partly degrading influence in the case of a triangular lattice.

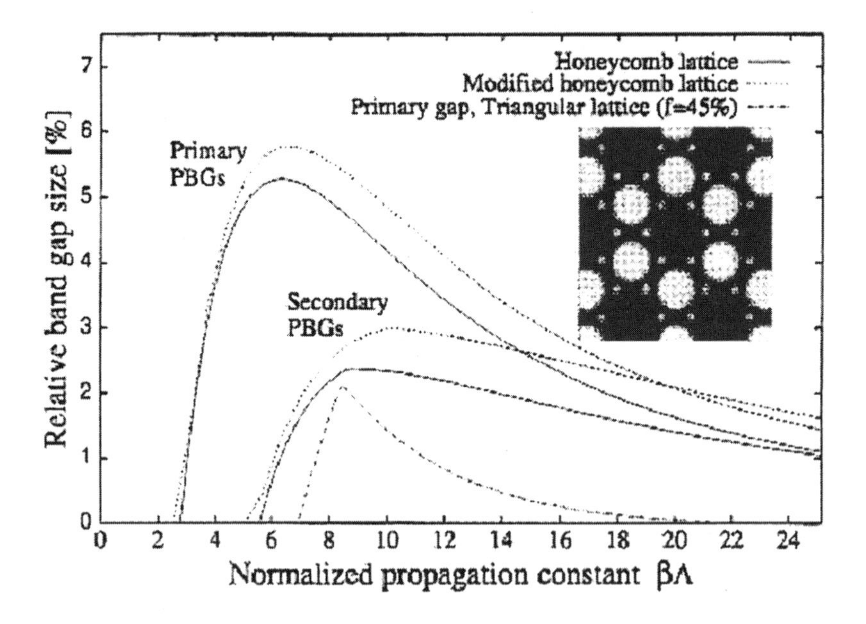

Figure 6-8. Relative size of the photonic band gaps for a honeycomb silica-air photonic crystal with filling fraction f = 30%. Interstitial holes are seen to have the effect of increasing the PBGs. The insert shows the geometry of the honeycomb lattice with interstitial holes (f_{int} = 5%).

Before presenting additional results on modified silica-air photonic crystals, it is useful to consider the already-obtained results using a concept, where the photonic crystals are described using high-index regions isolated by low-index voids. This will serve to gain further knowledge on how to optimise the photonic crystals with respect to large PBGs.

6.3 DESIGNING LARGE-BANDGAP PHOTONIC CRYSTALS

In the following discussion, the simple triangular structure (without interstitial holes) will be regarded as consisting of nodes that are the high-index regions between three adjacent air holes, and veins that are regions bridging two nearest nodes. In accordance with this, the honeycomb structure will be regarded as a structure, where the nodes are the high-index regions surrounded by six air holes and the veins are the connecting regions. The node-vein concept applied to the triangular and honeycomb structures is illustrated in Figure 6-9.

The previously presented results unambiguously indicate that increasing the size of the air holes provides larger PBGs. Hence, by using the above-described concept, a favourable first design step towards large-bandgap photonic crystals appears to be to isolate the nodes the most possible. In agreement with this, it is seen that the honeycomb structures, which exhibit the larger PBGs, intrinsically persists more isolated nodes and relatively narrower veins compared to the triangular structures.

While this design step appears correct with respect to the maximum of the size of the PBGs, the PBGs of the honeycomb structure were, however, found to decrease more strongly at higher frequencies compared to the triangular structure. For the honeycomb structure, the maximum PBGs were found to be exhibited at a normalized frequency of approximately $\Lambda/\lambda_{H,max}$ = 0.75 and for the triangular structure without interstitial holes $\Lambda/\lambda_{T,max}$ = 2.25 was found. Hence, the maximum PBG of the triangular structure occurs at a three times shorter wavelength than for the honeycomb structure. However, by considering the structures using the node-vein concept, a similar factor is, in fact, found between the separations of the high-index node of the two structures (for the honeycomb structure, the separation between the nodes are $\Lambda_{H,node} = \sqrt{3}\Lambda$ and for the triangular structure $\Lambda_{T,node} = 1/\sqrt{3}\Lambda$, where $\Lambda_{H,node}$ and $\Lambda_{T,node}$ is the spacing between two nearest nodes in honeycomb and triangular structures, respectively). Hence, these considerations reveal that an almost identical relation between the optimum operational frequency and the separation of the nodes occurs for the two different photonic crystal structures. The relation that is found is: $n\lambda_{\max} \approx \Lambda_{node}$, where the factor n is found to be identical to the modal index of the PBG at the optimum frequency (for the two studied cases $n \sim 1.3$). Therefore, a very simple second design step may be postulated, which simply states that for optimum design, the PBG structure should have a separation between the high-index nodes, which is equal to the optical wavelength in the photonic crystal.

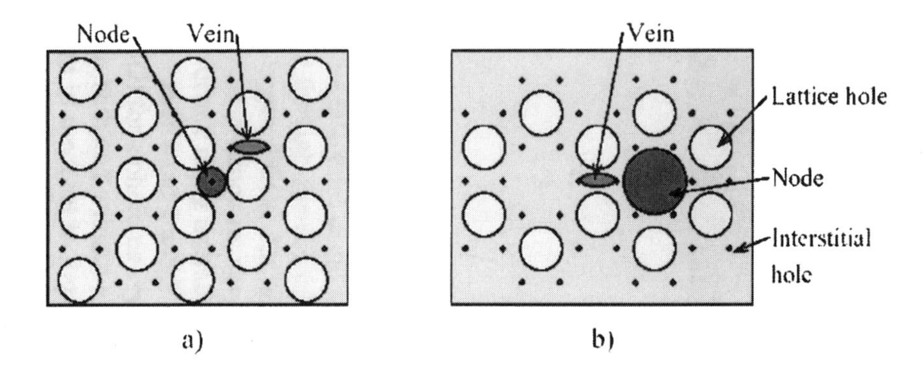

Figure 6-9. Schematic illustration of the concept of nodes and veins for (a) triangular structures and (b) honeycomb structures. The concept may aid in providing an increased understanding of the difference between the behaviour of the PBGs for the two structures and of the influence of interstitial holes.

As a further support of the above-described design-route, the influence of the interstitial holes is considered for both types of structures. In the case of the honeycomb structures, it is realized that the interstitial holes act to further narrow the veins, while leaving the nodes largely undisturbed. For the triangular structures, on the other hand, the interstitial holes are seen to fall right in the centre of the nodes, and thereby damage their ability to act as high-index regions. However, the node-vein concept is indeed also able to predict that a structure, which in not optimum in one wavelength-range, may become advantageous at a different. This corresponds e.g., to the properties that are found for the triangular photonic crystals with interstitial holes. For this structure, the introduction of the interstitial holes were seen to become advantageous at high frequencies, where the optical wavelength was comparable to the separation, $\Lambda_{TI,node}$, of those nodes that are isolated by two large holes arranged in the regular triangular lattice and by two interstitial holes. This structure has $\Lambda_{TI,node} = 1/3\Lambda$ and indeed the maximum size of the secondary PBG was found at a wavelength of approximately

$$\lambda = \frac{1}{\sqrt{3}} \frac{n_T}{n_{T1}} \lambda_{T,\max} \tag{6.7}$$

where n_T and n_{T1} are the modal indices of the corresponding PBGs, which were found to be 1.30 and 1.25, respectively.

From the above-described considerations, it follows that to optimise photonic crystal structures with respect to obtaining large PBGs, the structures should have the most isolated nodes (high-index regions), and the nodes should contain the material parts with the highest refractive index.

In order to increase the PBGs of silica-air photonic crystals, a strongly modified honeycomb-based structure is, finally, in focus, where the effects of introducing several differently-doped silica glasses are studied. Current doping technologies allows us to modify the dielectric constant of silica up to 3% and down to 1% of its nominal un-doped value. A schematic of the structure is shown in Figure 6-10, where the rhomboidal solid line represents the unit cell. The structure has a background material, which comprises interstitial holes and up to three different dopant levels of silica. In Figure 6-11, the relative size of the primary PBG is illustrated for the basic honeycomb structure with an air-filling fraction of 30%. In the figure is also included three cases of modified honeycomb structures: (a) interstitial holes introduced (f_{int} = 8%), (b) interstitial holes and three differently-doped silica glasses (n_1 = 1.44, n_2 = 1.45, n_3 = 1.47), (c) three differently-doped silica glasses (n_{s1} = 1.47, n_{s2} = 1.45, n_{s3} = 1.44). The indices refer to the three silica types indicated in Figure 6-10, and for the structures comprising three silica-dopants, the outer diameter of the tube and cane (silica type 1 and 3, respectively) have been set as large as possible (i.e., equal to Λ). For the basic structure (un-doped silica and air), a maximum PBG size of 5.3% is observed, and this size may be increased to 6.4% by adding interstitial holes to the structure, and further increased to 7.1 % by lowering the refractive index of the silica surrounding the air holes (silica type 1), while increasing the index of the silica forming the canes (silica type 3). By regarding the honeycomb photonic crystals using the node-vein concept, it is found that the results are in agreement with the belief that increasing the node index and narrowing the connecting veins is a fruitful route for designing large-bandgap photonic crystals. To further support this, it is observed how the size of the primary PBG is, indeed, decreased for the structure (c) modified by lowering the index of the nodes, while increasing the index of the silica surrounding the air holes. Although not illustrated here, the same conclusions were found also to be valid for the size of the secondary PBGs. The maximum of the PBGs is seen to shift slightly towards higher βΛ-values (corresponding to shorter wavelengths). Although, the separation of the high-index nodes are fixed for the four studied honeycomb-based structures, this shifting is counteracted by the shifting of the modal index of the corresponding PBGs, and a similar relation between the optimum wavelength and the separation, in agreement with the previously-postulated relation, was found.

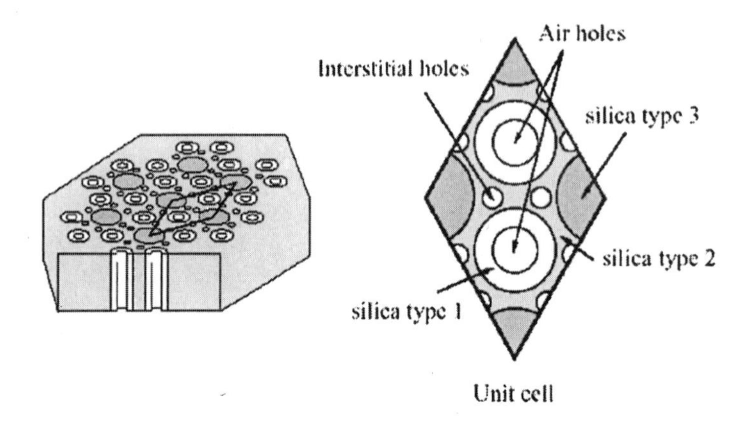

Figure 6-10. Unit cell employed for the analysis of optimised silica-air photonic crystals with up to three different dopants introduced.

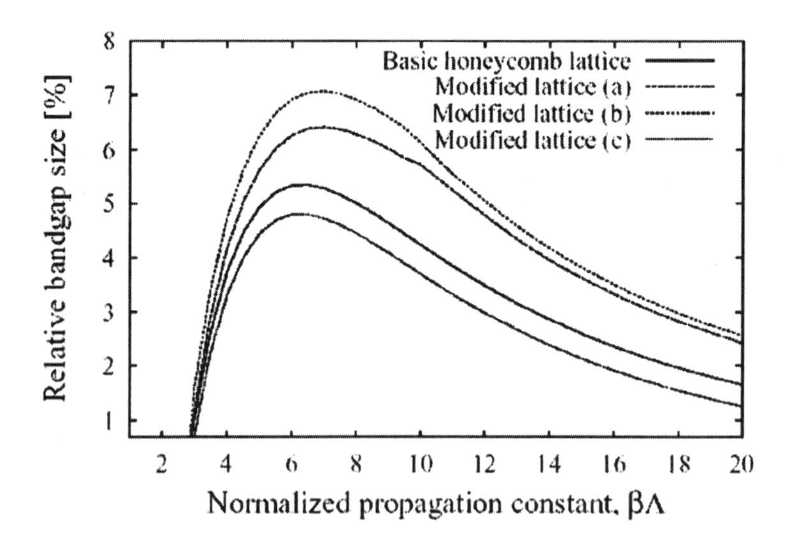

Figure 6-11. Relative bandgap size as a function of normalized propagation for a basic honeycomb lattice ($f = 30\%$), and three modified versions of this. The maximum PBG size of honeycomb-based photonic crystals may be either increased or decreased dependent on the location and degree of doping.

6.4 THE FIRST EXPERIMENTAL DEMONSTRATION OF WAVEGUIDANCE BY PHOTONIC BANDGAP EFFECT AT OPTICAL WAVELENGTHS

This section will present the background of the work, which led to the first experimental demonstration of PBG-guidance in optical fibres. The first PBG-fibres were fabricated using the stacking method as described in Chapter 4, and in the early years of the technological development, this approach placed a considerable limitation on the obtainable hole size of the fabricated fibres. As already indicated in Sections 6.2 and 6.3, one of the most feasible ways to obtain large bandgap sizes using a limited air-filling fraction is through the application of honeycomb structured fibres. We will, therefore, first point to some of the specific issues of the realization of the first bandgap fibre of this type. Thereafter, we will present some of the basic properties of the honeycomb-structured fibres. This will form the necessary basis for the understanding of the thoughts leading to the first experimental results on these fibres. Finally, the section will contain a short comparison between experimental and numerical results on accurately modelled fibre structures.

6.4.1 Considerations on fabrication of honeycomb fibres

The most important motivation to the work towards the photonic bandgap fibre was the research result from 1996, when Knight *et al.* presented an index guiding fibre with periodically spaced holes [6.5]. This initiated an intensive search for the first photonic bandgap at optical wavelengths. However, at that time, the possible air-filling fractions that could be manufactured were significantly smaller than the ones recently demonstrated, and simple triangular or close-packed structures (with small holes) could not provide guiding by the photonic bandgap effect. However, as described in Section 6.2, the relative size of the photonic bandgaps is enlarged in a honeycomb fibre with a relative low air-filling fraction compared to a fibre with triangular structured hole distribution. Broeng *et al.* [6.7], therefore, suggested that the aim should be on honeycomb-structured fibres. By doing so, the desired hexagonal arrangement of the tubes/rods still resulted from a close packing of the circular tubes/rods, and the honeycomb design was, therefore, directly feasible (see Figure 4-3). Note that the method, also introduce additional air gaps in the fibre preform (the interstitial holes analysed in Section 6.2). With respect to honeycomb photonic crystals, the interstitial holes were previously found to have a predominantly positive effect, and the modified preform fabrication method, therefore, is further advantageous by favouring these interstitial holes.

As indicated in Figure 4-3, the periodicity-breaking core region was introduced in a very simple manner by replacing a single rod by a tube of similar size as the remainder of tubes forming the fibre preform. For the first honeycomb fibres, the periodic cladding structure was terminated after about 4 honeycomb cells with a single surrounding layer of solid silica rods. Other types of termination of the periodic cladding structure may be thought of, e.g., utilization of large silica over-cladding tubes as known from conventional fibre fabrication [6.17].

After stacking, the capillaries and rods were held together by thin tantalum wires and fused together during an intermediate drawing process, where the preform was drawn into several approximately 1 meter long preform-canes with a diameter of 1.5 mm. This intermediate step was introduced to provide a large number of preform-canes for the development and optimization of the later drawing of the PBG-fibres to their final dimensions. During this step, the outer lying tubes/rods experienced some distortion, but the core region and its nearest surroundings (the about three honeycomb periods surrounding it) retained - to a large degree - the desired morphology.

The drawing was performed in a conventional drawing tower operating at a temperature around 1900°C. The honeycomb air-hole arrangement was well preserved after the drawing process, although surface-tension forces had reduced the initial air-filling fraction and distorted the holes to a triangular shape. These forces were also responsible for the collapse of the interstitial holes that, owing to their smaller size compared to the honeycomb-arranged holes, would collapse at an earlier stage in the drawing process. Also the core region was well preserved in the final fibre. In contrast to the cladding region, however, the interstitial holes were seen to remain in the core region.

Before we go on with the description of the first PBG-fibre, it will be useful to study some of the basic properties of honeycomb-based fibres.

6.4.2 Basic properties of honeycomb-based fibres

To illustrate the operation of PBG-fibres, and to serve as a basis for the investigations of their more advanced properties (in section 6.5), this section will only focus on a specific fibre based on the previously analysed photonic crystal structure.

6.4.2.1 The waveguiding principle of honeycomb fibres

The PBG-fibres is characterized by a simple honeycomb photonic crystal cladding having air holes with a diameter of $d_{cl} = 0.4\,\Lambda$ (corresponding to a filling fraction of $f = 10\%$) and a central periodicity-breaking air hole with $d_{co} = 0.4\,\Lambda$.

Figure 6-12 shows the PBG boundaries of the two lowest-frequency PBGs, which are occurring between bands 2 and 3, and between bands 6 and 7 (named primary and secondary PBG, respectively) for a corresponding full periodic honeycomb photonic crystal. Also illustrated in Figure 6-12 is the β / k *-curve* for the lowest-frequency allowed cladding mode (the effective index of the fibre cladding). Above this curve is a semi-infinite "PBG", where no modes are found. This is the region in which all Total Internal Reflection (TIR)-based fibres operate, including PCF's with a high-index core [6.18], since a high-index defect causes at least one mode to appear between this curve and the refractive index of the core (≈ 1.45 in the case of a silica core at near-infrared to visible wavelengths). Such index-guided modes are seen not to be a feature of the low-index core honeycomb PBG-fibre. What is, on the other hand, observed is a single doubly-degenerate defect mode that is traversing the primary PBG from $\beta\Lambda \approx 5$ to $\beta\Lambda \approx 17$. As previously mentioned, this defect mode is caused solely by the introduction of the extra air hole in the honeycomb structure, and - for a full periodic structure - the exact same PBG boundaries (with no modes inside) are found as for the crystal including the defect.

It should be expected that the defect-mode is strongly localized to the region comprising the extra air hole (albeit this is a low-index region), and the amplitude of the H-field for the mode is illustrated in Figure 6-13. The mode was calculated for a $\beta\Lambda$-value of 7.0. For this value, the defect mode is approximately in the middle of the primary PBG, and a strong localization of the defect mode is apparent.

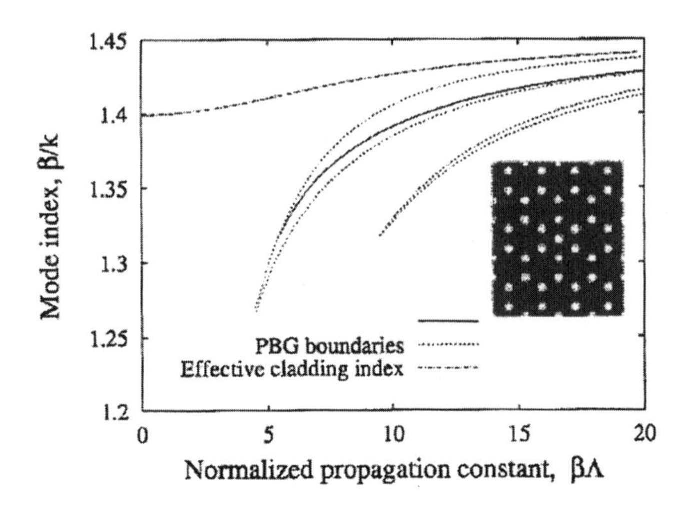

Figure 6-12. Illustration of the two lowest-frequency PBGs of a honeycomb PB-fibre with a cladding air-filling fraction of 10% and a defect hole with same size as the cladding holes. Within the primary PBG, a single degenerate mode is found. This defect mode may not propagate in the cladding structure (due to the photonic bandgap effect), and the mode is expected to be strongly localised to the region that breaks the periodicity of the photonic crystal, i.e., the region containing the extra air hole forming the core of the PBG-fibre. The inset shows the refractive index distribution of the core region, and the inner honeycomb photonic crystal cladding structure of the fibre. A super-cell of size 5 x 5 simple honeycomb cells and 16384 plane waves were used for the calculation.

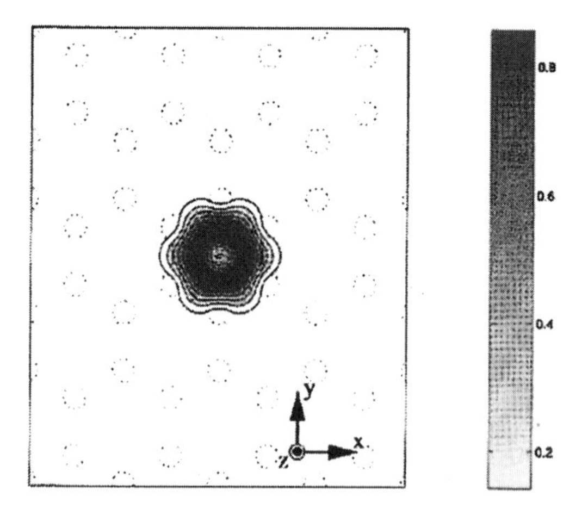

Figure 6-13. H-field distribution in the cross-section of a honeycomb PBG-fibre with $d_{cl} = d_{co} = 0.4\Lambda$. The mode is strongly confined to the central defect and has a non-zero propagation constant along the fibre axis ($\beta\Lambda = 7.0$) and will, therefore, be guided without leakage by the PBG-fibre. The air holes in the fibre cross-section are indicated by the dotted lines. The defect air hole is placed in the centre of the figure (see inset in Figure 6-12).

Since the defect mode is confined in the transverse plane of the fibre, but has a non-zero propagation constant in the invariant direction of the fibre, it may ideally be guided without leakage along the fibre. In this manner, low-loss PBG-guidance may, in principle, be achieved over very long lengths - and this type of defect modes will hereafter be referred to as core modes or PBG-guided modes.

An important difference between TIR-based fibres and PBG-fibres is that the fundamental PBG-guided mode (or any other PBG-guided modes) may not be guided over an infinite frequency interval as a fundamental TIR-guided mode, in principle, may be. As an example of the spectral transmission window of the honeycomb PBG-fibre, it could e.g., be designed for operation around 1.55 μm. Using the previously outlined design-route, an optimum absolute centre-to-centre hole spacing around 1.25 μm is predicted (a β/k-*value* of 1.37 was found at the maximum PBG size). For such dimensions, the core-mode will fall within the primary PBG in a spectral range from approximately 0.6. μm to 2.0. μm. Therefore, although single mode waveguidance may not be obtained over an infinitely broad frequency range (as for TIR-based fibres), the PBG-fibres may, in theory, exhibit broad single-mode spectral ranges. However, realization of PBG-fibres with centre-to-centre hole spacing as small as indicated above is a severe technological challenge, and even fibres with twice the spacing are critical to realize, while maintaining an air-filling fraction of 10% or more.

A second significant difference compared to conventional fibres, is that no doping of the silica material is required to obtain waveguidance in the PBG-fibres. Hence, pure fused silica may be used for the fabrication of PBG-fibres. This may be of importance, e.g., for reducing the propagation losses of optical fibres. A further issue in this respect is the potential of localizing the mode-field partly in the air holes. To provide a better illustration of the fraction of field distribution within the air holes, Figure 6-14 shows the field distribution along two orthogonal directions through the centre of the PBG-fibre. As seen, the field of the PBG-guided mode for the particular honeycomb PBG-fibre is mainly distributed in silica. For the illustrated mode, 6% of the H-field was found to be within the air holes at $\beta\Lambda = 7.0$. At larger $\beta\Lambda$-values (corresponding to higher normalized frequencies), this fraction was found to decrease rapidly, and a negligible fraction of the field (below 1%) was found to be within the air holes at $\beta\Lambda$-values above 15.

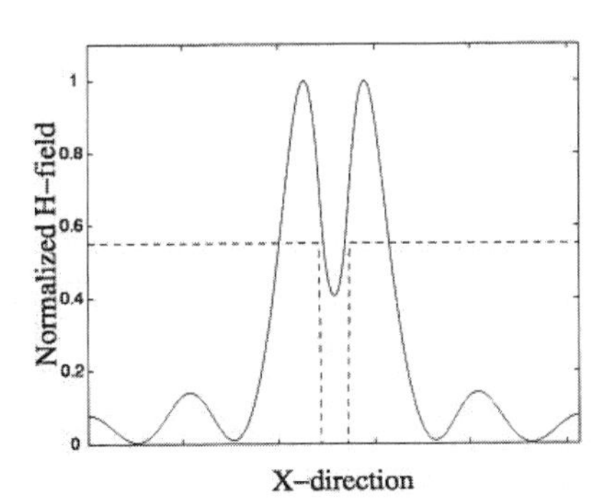

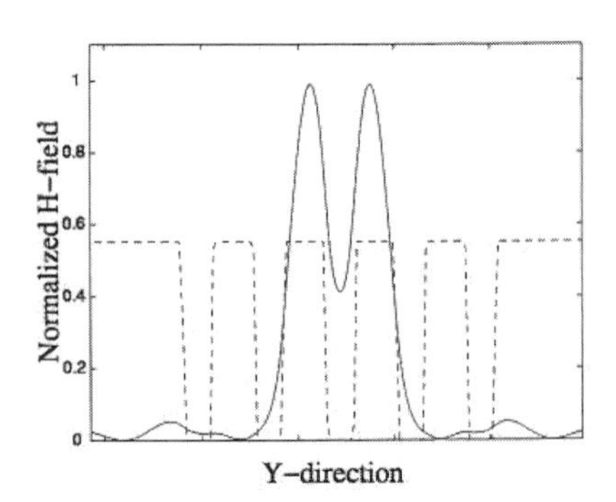

Figure 6-14. *H*-field distribution along two orthogonal directions through the centre of a honeycomb PBG-fibre with $d_{cl} = d_{cl} = 0.4$. The two directions, *x* and *y*, are shown in Figure 6-13. The dotted curves indicate the refractive index distribution along the two directions.

Although most of the field of the core mode for the analysed honeycomb PBG-fibre is distributed in the silica, full confinement of light in air is, in principle, possible for PBG-fibres. Using an illustration as in Fig. 6-12, this requires a defect mode falling inside a PBG, where both of these extend to $\beta/k < 1$ (i.e., below the air-line). Such fibres should have an exiting potential in both the telecommunications and sensor areas. More elaborate discussion of air-guiding PBG-fibres and their potential applications will be provided in Section 6.6

An important property of the PBG-fibres that has so far not been addressed is the absolute spot-size of the guided mode. As evident, the honeycomb PBG-fibres are characterized by a much different mode-profile compared to conventional fibres. Apart from the lower field intensity in the centre of the mode, also a 60° symmetry is noticeable in the transversal field distribution. Therefore, a simple spot-size definition equal to the core radius may not be applied. However, defining the spot-size as the radius from the centre of the PBG-fibre, where the *H*-field has decreased to $1/e$ of its maximum value, a radius of approximately $0.8\,\Lambda$ is found for the specific mode in Figure 6-13 and 6-14. Hence, a mode with an extremely small spot-size radius around 1 μm may be achieved, if the fibre is drawn to $\Lambda = 1.25$ μm. Such a small spot-size may be very attractive for exploration of strong non-linear effects.

6.4.2.2 Simple core design considerations of PBG fibres

For comparison with the PBG-fibre having f = 10%, Figure 6-15 (a) illustrates the mode index of the core-mode of a honeycomb PBG-fibre with $d_{cl} = d_{co} = 0.7$ (f = 30%). It is remarkable that the defect mode for this PBG-fibre does not fall inside the primary PBG, but in the secondary. The illustrated mode solution does, however, retain a similar mode-field distribution as the mode for the previously studied PBG-fibre (f = 10%). No modes occurring at higher β/k-values were found, and the indicated solution may, therefore, still be classified as the fundamental mode of the PBG-fibre. An additional feature is illustrated in Figure 6-15, namely the effective refractive index of a photonic crystal that equivalents the core region structure. For the investigated honeycomb PBG-fibres, where a single extra air hole has been introduced to form a defect region, the core may be considered as having a structure identical to a triangular photonic crystal. The curve labelled, "effective core index" has, therefore, been determined from the fundamental space-filling mode of a triangular photonic crystal, which in this case has f = 45%.

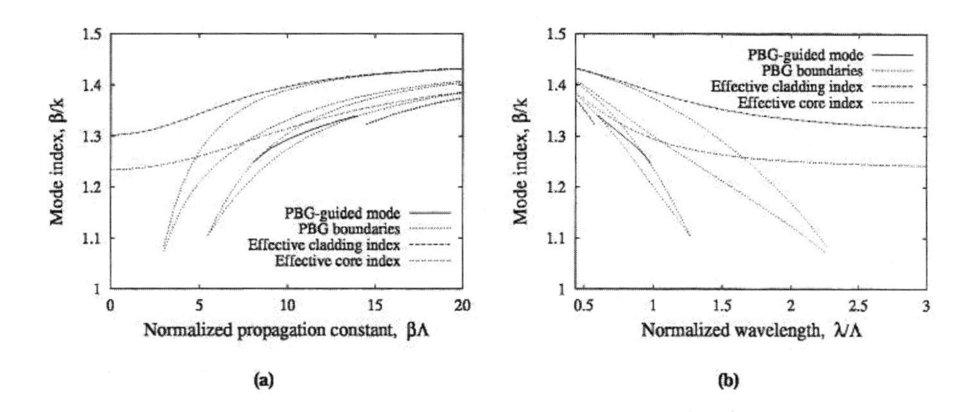

Figure 6-15. Mode-index illustration of a honeycomb PBG-fibre with $d_{cl} = d_{co} = 0.7\Lambda$ shown versus the normalized propagation constant (a), or the normalized wavelength (b), respectively. A PBG-guided mode appearing within the secondary PBG is found. The figure further illustrates the envelope of the fundamental space-filling mode of a photonic crystal equivalent to the core region.

Note that the effective index of the guided mode is lower than the effective index of the equivalent core region. This behaviour may be compared to conventional fibres for which $n_{co} > \beta_{TIR}/k > n_{cl}$, where β_{TIR} is the propagation constant of a TIR-guided mode at a given frequency, n_{co} and n_{cl} are the refractive indices of the core and the cladding, respectively. A frequently used illustration of this relation is illustrated in Figure 6-16 (a). Using this type of illustration, it may be understood that for PBG-fibres, the corresponding type of illustration is presented in Figure 6-16 (b), and the equivalent relation to that for conventional fibres is:

$$\frac{\beta_H}{k} > \frac{\beta_{PBG}}{k} > \frac{\beta_L}{k} \quad \text{for} \quad n_{eff,co} > \frac{\beta_H}{k} \quad \text{or} \tag{6.8}$$

$$n_{eff,co} > \frac{\beta_{PBG}}{k} > \frac{\beta_L}{k} \quad \text{for} \quad n_{eff,co} < \frac{\beta_H}{k} \tag{6.9}$$

where $n_{eff,co}$ is the effective index of the core region, β_{PBG} is the propagation constant of a PBG-guided mode, β_H and β_L are the upper and lower propagation constant boundaries at a fixed frequency/wavelength for a given PBG. An illustration of the mode-index for the honeycomb PBG-fibre with cladding-filling fraction $f = 30\%$ is presented in Figure 6-15 (b) as a function of normalized wavelength λ / Λ. Eqs. (6.8-6.9), therefore, partly explains the reason, why no PBG-guided modes are found within the primary PBG.

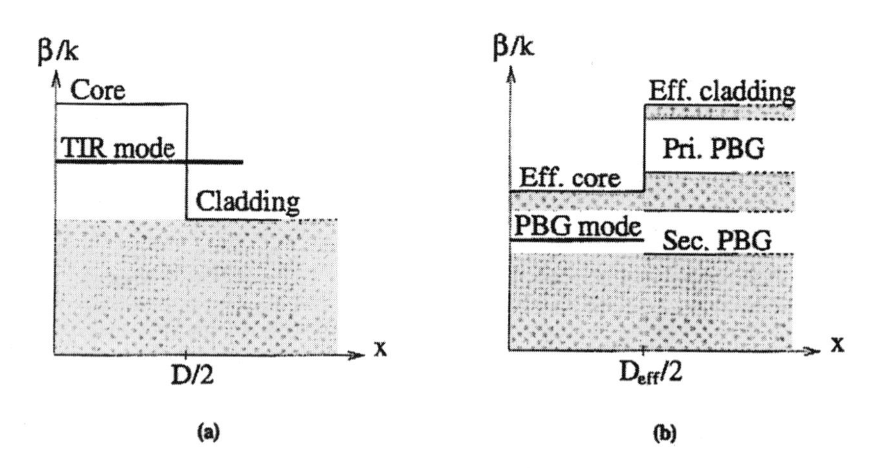

Figure 6-16. Simplified schematic illustration of the operation of (a) a conventional, circular symmetric, step-index fibre, and (b) a PBG-fibre having a PBG-guided mode within the secondary PBG. The shaded areas indicate inaccessible regions for operation of the two types of fibres.

It is worth noticing that no modes are found even for the relatively large region of the primary PBG for which $\beta_L < kn_{eff,co}$. Using a simple state-space argument it is, however, possible to estimate the possible existence of guided modes occurring for a given PBG. For a step-index fibre, the state-space argument reads [6.20]:

$$N_{conv} \approx \frac{k^2\left(n_{co}^2 - n_{cl}^2\right)\left(D/2\right)^2}{4}, \tag{6.10}$$

where N_{conv} is the number of TIR-guided modes. Applied to PBG-fibres, this reads:

$$N_{PBG} \approx \frac{\beta_H^2\left(n_{co}^2 - \beta_L^2\right)\left(D_{eff}/2\right)^2}{4} \approx \frac{\left(k^2 n_{eff,co}^2 - \beta_L^2\right)\left(D_{eff}/2\right)^2}{4} \tag{6.11}$$

where N_{PBG} is the number of PBG-guided modes, D is the core diameter for the step-index fibre, D_{eff} is the effective core diameter of the PBG-fibre, and the second expression applies in the case $k^2 n_{co}^2 < \beta_H^2$. For the PBG-fibre considered here, a estimation of N_{PBG} for the secondary PBG at $\lambda/\Lambda = 0.75$ gives $N_{PBG} \approx 1.3$, which agrees well with the observed single mode. The effective diameter for the PBG-fibre has not been accurately determined, but an approximate value of $D/2 = \Lambda$ was used for the calculation. For the primary PBG, Eq. (6.11) could be evaluated at $\lambda/\Lambda = 1.75$. At this wavelength no modes are found within the primary PBG and, in fact, a N_{PBG}-value of approximately 0.4 is found.

These results indicate that although accurate determination of possible PBG-guided modes and their properties requires utilization of advanced full-vectorial numerical methods, it is possible by (surprisingly) simple considerations to provide an important, first estimation of the possibility of supporting PBG-guided modes for a given PBG-fibre. Hence, the design process may be aided significantly by determining the PBG exhibited by the cladding structure (a much faster computational task than a full simulation of the fibre including the core), and, thereafter, determining the required core geometry for operation within a given wavelength range.

6.4.3 Basic characterisation of the first honeycomb fibres

The properties of the honeycomb fibre described in the previous section clearly underline its potential as a feasible structure for the realization of a photonic bandgap fibre. For this reason, it became the honeycomb fibre, which was chosen in 1997, as the test object for the collaboration between researches at Research Center COM, Technical University of Denmark and researchers at University of Bath, UK (first results reported in [6.5]). We will now shortly outline some of the key elements of this first demonstration.

The honeycomb fibres, which were fabricated at University of Bath, had a cladding structure with 4 periodic layers corresponding to a fibre containing 180 holes in the cladding. The first characterisation of the PBG-fibres was performed using a white light source. Fibres with a length of approximately 5 cm were mounted in a microscope setup, and white light was launched at the fibre end-face (see [6.5]). For all examined fibres, the white light was found to excite a multitude of cladding modes covering the whole visible spectrum, and for the majority of the fibres, no modes at all were found to exist within the core region. For fibres within a very narrow dimension-parameter interval, however, light within a spectrally narrow interval appeared to be guided within the core region from a visual inspection by eye. The PBG-fibres that supported waveguidance within the core region had a centre-to-centre distance around 2 μm and a total air-filling fraction in the cladding around 6%.

To examine the properties of the fibre, laser light was focused onto one end of the PBG-fibre (of length about 5 cm) using a high-power objective lens. This length was chosen, since the structure of the PBG-fibres was not sufficiently constant over longer lengths to ensure that the waveguiding properties remained uniform. Figure 6-17 shows schematically a contour map of the observed near-field distribution superimposed on a portion of the fibre surface at a wavelength of 458 nm (see [6.5]).

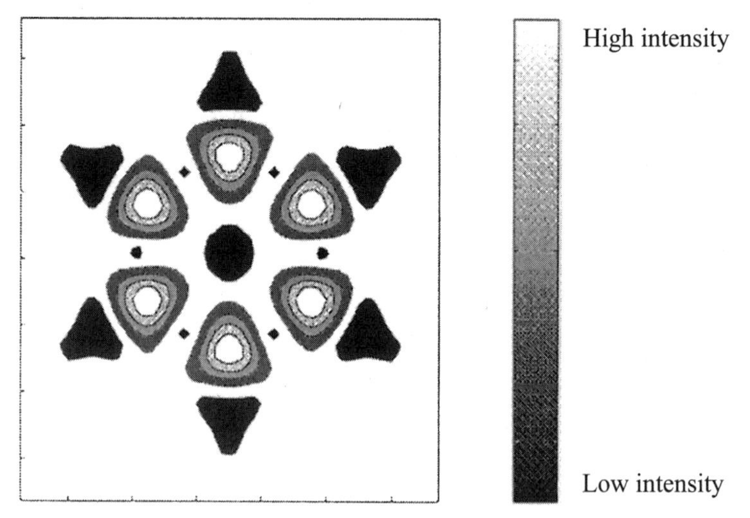

Figure 6-17. Near-field distribution of PBG-guiding mode. The morphology of the air-hole structure is seen as black areas on the illustration. The results correspond to the first PBG-guiding fibre reported by Knight *et al.* [6.5].

As seen, the mode is tightly confined around the low-index core region, and is strongly peaked in the silica. The light is concentrated just outside the central air hole, which defines the defect, and has deep minima corresponding to the locations of the residual interstitial air holes. The relative intensities of the six lobes remained fixed as the input coupling was varied, and are nearly equal, with the small difference attributed to variations in the structure between the different symmetry directions. No other mode field distributions were observed confined to the defect region. The six-lobed distribution seen on Figure 6-17 remained fixed and insensitive to input coupling conditions, except that the overall intensity of the distribution was found to be extremely sensitive to the precise input coupling. This should not be surprising, as the individual lobes of the guided mode profile were less than 1 μm in diameter.

The above-described features indicate that the observed guided mode might correspond to a higher-order PBG-mode, which was guided without the presence of a fundamental mode. In the proceeding section, a numerical analysis of the realistic honeycomb PBG-fibre will be carried out in order to provide additional insight into the specific features of the observed mode.

6.4.4 Modelling of realistic photonic bandgap fibres

The morphology of the real, fabricated honeycomb PBG-fibres (see Figure 6-17 concerning the core region) differs from that of the previously discussed theoretically studied fibres, which had circular air holes. In order to accurately model the realistic PBG-fibres, the plane-wave method (see Chapter 3.4) may be applied, as the discretization of the refractive index distribution equally well handles arbitrary index profiles as it does profiles with perfect circular features. The number of grid-points needed to accurately represent very small features, e.g., as the six interstitial holes in the core region, however, requires a significantly higher number of grid-points (and hence also expansion terms for the electromagnetic fields). This requirement, therefore, makes modeling of the realistic honeycomb PBG-fibres a computationally more extensive task than the PBG-fibres presented in the preceding section.

As well as it experimentally was found that small variations in the structure of a PBG-fibre cause the waveguiding properties to change significantly, or even to disappear completely, the refractive index distribution used for the numerical analysis must - to a very high degree - resemble the specific, realistic PBG-fibre structure in order to find similar PBG-guided modes numerically. For numerical analysis of the honeycomb PBG-fibre, a supercell as presented in Figure 6-18 was employed.

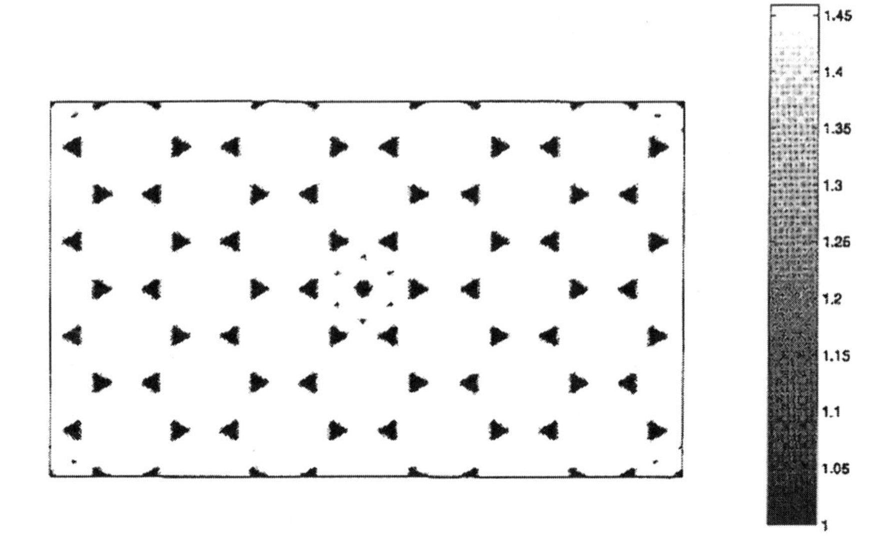

Figure 6-18. Illustration of supercell used to represent the refractive index distribution of the first honeycomb PBG-fibre (reported in [6.5]). The area included in the figure is, in fact, the area of two supercells.

As a first analysis of the realistic honeycomb PBG-fibre, the cladding structure was analysed separately and a β /k-representation of the obtained results is illustrated in Figure 6-19. The figure shows several narrow PBGs, and, as seen, a number of these do cover the visible spectrum. The figures also shows the effective core index, which was approximated by a corresponding triangular structure with f = 15% $(d/\Lambda = 0.8/2.0)$. The effective core index curve indicates that the primary PBG cannot be responsible for the observed PBG-guided mode in the relevant spectrum. The narrow higher-order PBGs, however, do potentially allow for existence of confined PBG-modes in the visible spectrum.

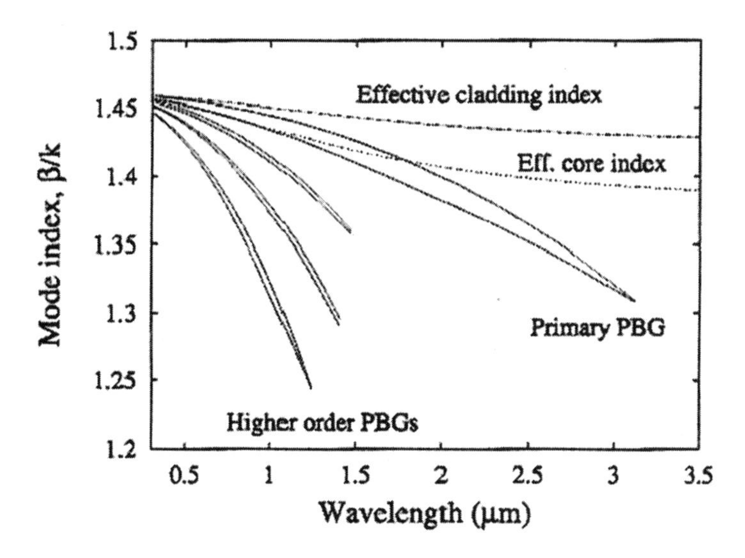

Figure 6-19. Simulation of the PBGs exhibited by the full periodic cladding structure of a realistic honeycomb PBG-fibre. The cladding air-filling fraction is 5.3%, and the air holes have a nearly triangular shape as illustrate in Figure 6-18.

As a first estimation of the possible existence of guided modes, due to higher-order PBGs, the approximate equation presented in the previous subsection (see Eq. (6.11)) may be utilized. Using this equation, $N_{PBG} > 1$ is, in fact, found for all three illustrated higher-order PBGs around $\lambda = 0.5$ μm despite their relatively narrow appearance in the figure (the highest-order PBG that is illustrated in the figures appears between bands 20 and 21 for the corresponding full periodic cladding structure, and has $\beta_H\Lambda = 34.4$ and $\beta_H\Lambda = 34.3$ at $\lambda = 0.5$ μm*)*. Hence, to search numerically for defect modes within these PBGs appears to be worthwhile. At this stage it is, however, necessary to consider the computational demands for the required super-cell calculations. Using a super-cell of size 4 x 4 simple honeycomb cells, calculation of a defect mode at a single frequency in the highest-order PBG illustrated in Figure 6-19 demands calculations of approximately 300 eigenvalues. Using a resolution of 256 x 256 = 65536 expansion terms for the numerical simulations, the corresponding processing time is several days using ordinary personal computers anno 2000. For this reason, broadband frequency-sweeps, as e.g., presented in the preceding chapters for the ideal PBG-fibres (that are computationally less extensive, since the waveguidance occurs due to the lowest-order PBG), have not been included in this description. However, supported by the experimental observations, the simulation of the realistic honeycomb PBG-fibre was refined to three different wavelengths, namely at λ = 400 nm, 500 nm, and 600 nm. These simulations both revealed a fundamental, a second-order, and a third-order mode for the specific fibre. However, only the third-order mode was found to be positioned within one of the PBGs at a wavelength of 500 nm, whereas the fundamental mode as well as the second-order mode was found to be leaky at all three wavelengths. The third order-mode appeared at Λ = 500 nm in the highest-order PBG illustrated in Figure 6-19, and its mode-field distribution is illustrated in Figure 6-20. The mode has a 180° phase reversal between neighbouring maxima in the six-lobed mode pattern, in agreement with the experimental observations.

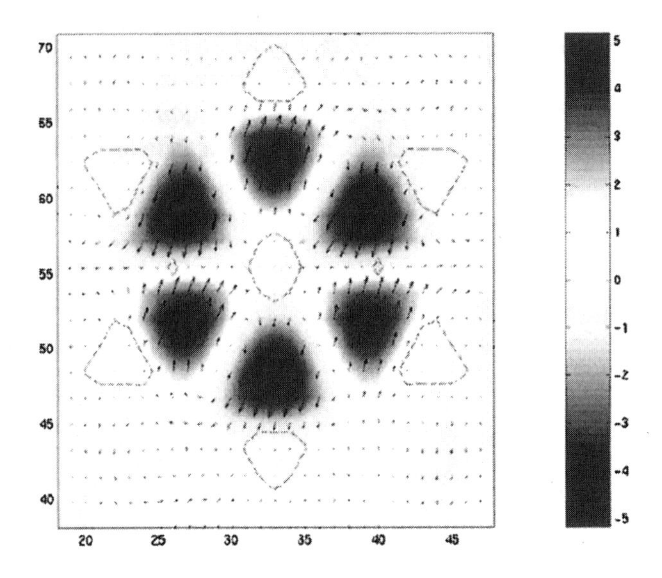

Figure 6-20. Longitudinal component of the Poynting vector for the PBG-guided mode. The illustration is calculated for a fibre with dimension Λ = 2.0 μm and an air-filling fraction in the periodic cladding structure of 5.3% at λ = 500 nm.

The above-described experimental and theoretical findings support both the demonstration of PBG-waveguidance in the fabricated honeycomb PBG-fibres, as well as certain of the unique properties of PBG-guided modes, including the existence of both an upper and lower cutoff frequency, the possibility of guiding light in a higher-order mode without the presence of a fundamental mode, and most importantly the localization of a guided mode at a low-index core region.

6.5 PROPERTIES OF PBG-GUIDING FIBRES

With an expected progress in fabrication of longer, uniform PBG-fibres, some of the immediate, most important issues to address for PBG-fibres include: determination of their propagation and bending losses, accurate experimental determination of the upper and lower limits of the spectral transmission windows, and to establish measures of the numerical aperture divergence and spot-size of the fibres as function of wavelength. A long list of other properties of the fibres are naturally of interest experimentally as e.g., the dispersion and polarization properties of the fibres, which - as demonstrated theoretically in this section - are expected to be significantly different compared to those of conventional fibres.

6.5.1 Dispersion properties of honeycomb-based fibres

The dispersion properties of conventional fibres are receiving a continuously high research interest in connection with high-capacity optical communication, soliton propagation, and control of nonlinear effects [6.19]. Accordingly, the dispersion in PBG-fibres is of high interest to investigate. With the aim of studying realistic PBG-fibres, a fibre with a relatively low cladding air-filling fraction, $f = 10\%$, and a core-defect of similar size as the cladding holes, $d_{co} = 0.4\Lambda$, is chosen for this initial analysis. A modal index illustration of this PBG-fibre has previously been presented in Figure 6-12. The mode-index curves provide an important tool for first-hand qualitative information about the dispersion properties. This qualitative information may readily be obtained by studying the curvature of the PBG-mode curve in Figure 6-12, since according to [6.20];

$$D = -\frac{\lambda}{c}\frac{d^2 n}{d\lambda^2} \tag{6.12}$$

where D is the group velocity dispersion (GVD) and n is the mode index β / k. A strong downwards bending of the mode-curve for increasing λ-value indicates a large anomalous dispersion and vice versa. Hence, from Figure 6-12, it is realized that the PBG-fibre offers the potential of a large anomalous waveguide dispersion in a wavelength range, where the fibre is single-moded. This feature is not seen in conventional fibres, and is a first indication of special dispersion properties in the PBG-fibres.

In order to obtain more detailed, quantitative information about the GVD of the fibres, Eq. (6.12) is calculated directly using the variational method presented in Chapter 3. However, while this method provides an accurate determination of the waveguide dispersion, it does, not easily allow the inclusion of the material dispersion. Therefore, as a first approximation, the

GVD is estimated by assuming the mode-fields fully distributed in silica and adding the material dispersion of silica to the waveguide dispersion. The material dispersion of silica is calculated using the Sellmeier equation [6.21]. Although, this approach may not be sufficient for high-precision determination of GVD (which e.g., may be required at a later development - stage for PBG-fibres with specifically tailored dispersion properties), it is adequate to provide answers to fundamental questions concerning the dispersion properties, and point out some of the potential applications as well as limitations of the PBG-fibres, which is the objective of this book.

6.5.1.1 Single-mode fibres with high anomalous dispersion and strongly shifted zero dispersion wavelength

The GVD and the contributions from waveguide and material dispersion for the above-discussed PBG-fibre with a pitch of 1.0 μm is illustrated in Figure 6-21. The figure reveals two important features. Firstly, the PBG-fibre is characterized by a very high anomalous waveguide dispersion, in agreement with that expected from the relatively strongly downward bending mode-curve in Figure 6-12. In particular, a GVD of up to 600 ps/(km·nm) appears feasible. It is important to notice, that in contrast to conventional fibres, this anomalous waveguide dispersion occurs, while the fibre remains single-moded. The PBG effect, hereby, provides a new type of fibre for dispersion management, and is of course of large fundamental interest. The high anomalous dispersion, along with the possibility of very small spot-sizes (see Section 6.4), may e.g., be attractive for exploration of novel nonlinear fibre devices or may be used for compensation of normal-dispersion optical transmission links. Secondly, the high waveguide dispersion allows the compensation of material dispersion of silica at short wavelengths, and provide single-mode fibres with the zero-dispersion point shifted towards wavelengths well below 1.28 μm. In particular, the specific PBG-fibre is seen to display a zero-dispersion wavelength of 0.85 μm. As the waveguide dispersion as a function of normalized wavelength may be read directly from Figure 6-21 (since the figure illustrates a PBG-fibre with Λ = 1.0 μm), it is, furthermore, realized, that by scaling the dimensions of the PBG-fibre, the waveguide dispersion may be scaled and that the zero-dispersion wavelength, thereby, may be shifted even into the visible.

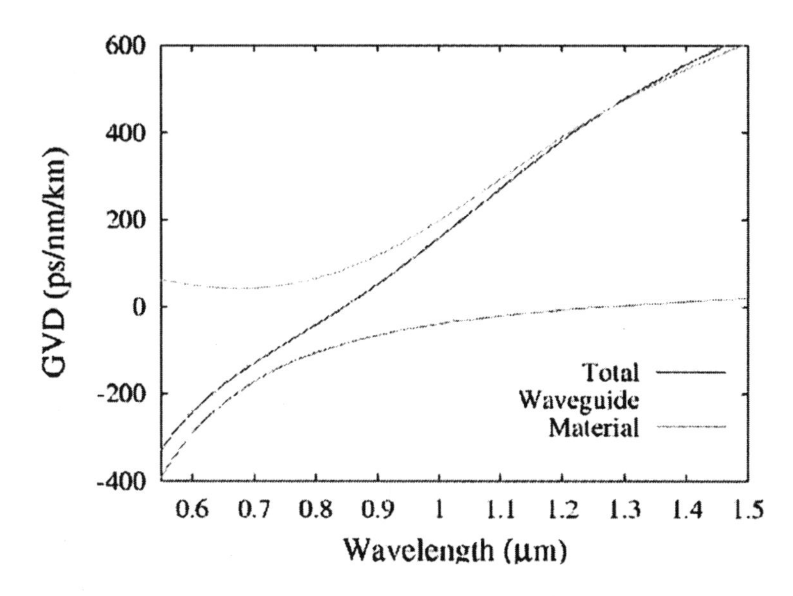

Figure 6-21. Group velocity dispersion and the contributions from waveguide and material dispersion for a honeycomb PBG-fibre with $d_{cl} = d_{co} = 0.40\Lambda$ and $\Lambda = 1.0\mu m$

This potential is further illustrated in Figure 6-22, where the GVD of the PBG-fibre is depicted for Λ-values ranging from 0.7 μm to 1.5 μm. Although the possibility of obtaining single-mode fibres with the zero-dispersion wavelength shifted below 1.28 μm is not typically exploited in conventional fibres, index-guiding microstructured fibres may also exhibit this feature as demonstrated in literature [6.22-6.25] (and discussed in Chapter 5). This aspect is, especially, highlighted by the work of Ranka *et al.* [6.23], who have demonstrated continuous light generation over the entire visible spectrum in a single-mode TIR-PCF resulting from strong nonlinear optical effects. These were obtained through a shift of the zero-dispersion wavelength to 800 nm and a very small mode-area.

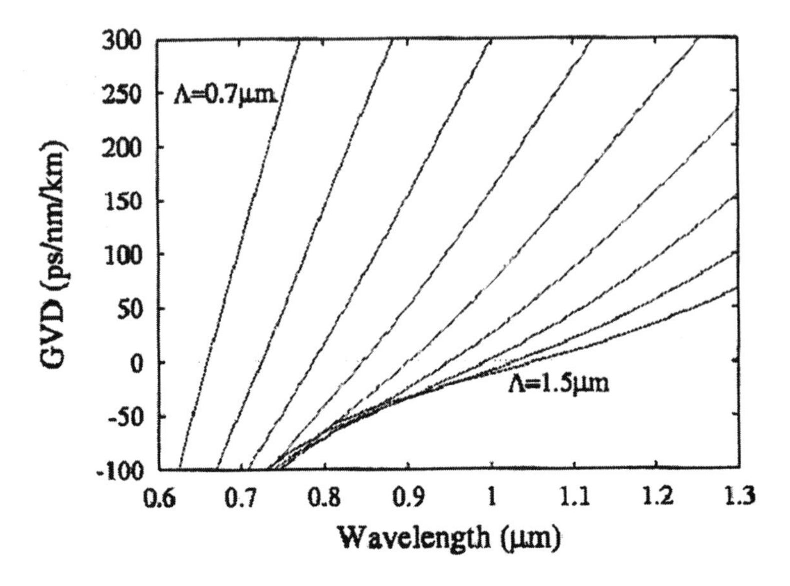

Figure 6-22. GVD of honeycomb PBG-fibres with $d_{cl} = d_{co} = 0.40\Lambda$ and Λ ranging from 0.7 μm to 1.5 μm in steps of 0.1 μm.

6.5.1.2 Single-mode fibres with broadband, near-zero, dispersion-flattened behaviour

While the above-discussed special dispersion properties appear of interest for non-linear and dispersion manipulating fibres, a further attractive property of the PBG-fibres is that they may exhibit flat, near-zero dispersion over a broad wavelength range. Such a property is desirable for various non-linear applications, and this section shall, therefore, explore PBG-fibres with broadband dispersion-flattened characteristics. The main wavelength-range of interest for modern communication systems is located around the 1.55 μm transmission window [6.26], where bulk silica displays a dispersion of approximately 20 ps/nm/km with a positive slope. Therefore, to obtain near-zero dispersion-flattened PBG-fibres, the waveguide dispersion must be tailored to compensate both the nominal value and the slope of the silica dispersion. While the above-analysed PBG-fibre does not meet these requirements, a second group of PBG-fibres with higher air-filling fraction in the cladding structure is focussed on. Keeping the central core-hole fixed at $d_{co} = 0.40\Lambda$, Figure 6-23 illustrates the waveguide dispersion for a series of PBG-fibres with cladding hole sizes ranging from $d_{cl} = 0.40\ \Lambda$ to $d_{cl} = 0.70\ \Lambda$.

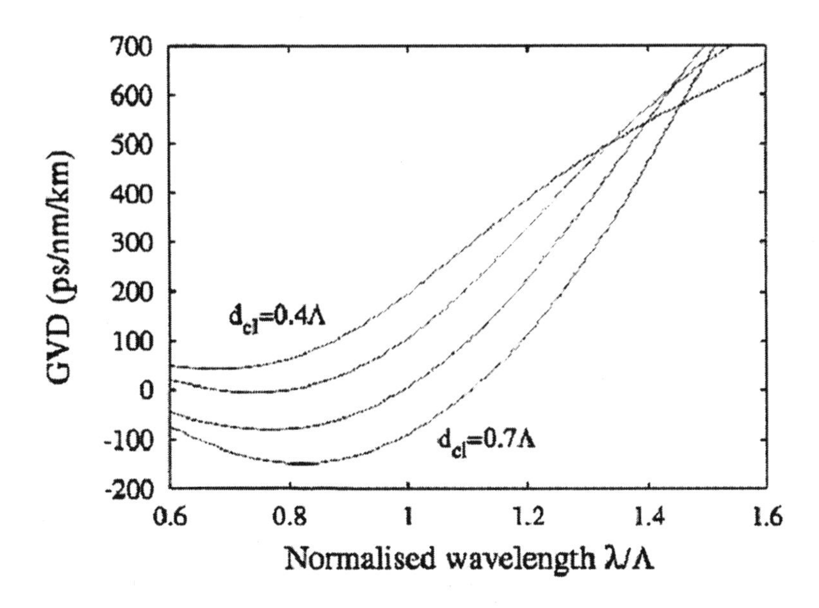

Figure 6-23. Waveguide dispersion of honeycomb PBG-fibres with d_{co} = 0.40Λ and d_{cl} ranging from 0.4Λ to 0.7Λ in steps of 0.1Λ.

As seen from Figure 6-23, the minimum of the dispersion-curve may be pushed towards lower values by increasing the cladding-hole sizes, and for d_{cl} -values below 0.50 Λ, the PBG-fibres may exhibit wavelength-regions of anomalous waveguide dispersion. In particular, Figure 6-24 indicate that for d_{cl} -values in the range from 0.50 Λ to 0.60 Λ, the desired waveguide-dispersion values may be obtained. With respect to the second requirement that must be met, namely a negative dispersion slope, it is further found that the PBG-fibres are displaying this feature around a normalized wavelength of 0.7. Hence, for operation around a free-space wavelength of 1.55 μm, this indicates an optimum pitch of Λ = 1.6 μm/0.7 = 2.3 μm. The GVD for a series of PBG-fibres within the above-specified parameter range is illustrated in Figure 6-24. As seen, the material dispersion of silica may be compensated to shift the zero-dispersion wavelength within the desired wavelength-range around 1.55μm. However, as also seen, none of the PBG-fibres display sufficiently dispersion-flattened characteristics. Therefore, as a final illustration of the potentially fruitful exploration of the high degree of flexibility of the PBG- fibres, the dispersion curve shall be tune by adjusting the size of the core hole. This is done in Figure 6-25, for a series of PBG-fibres with fixed cladding hole sizes of d_{cl} = 0.60 μm and core holes ranging from d_{co} = 0.37 μm to 0.44 μm.

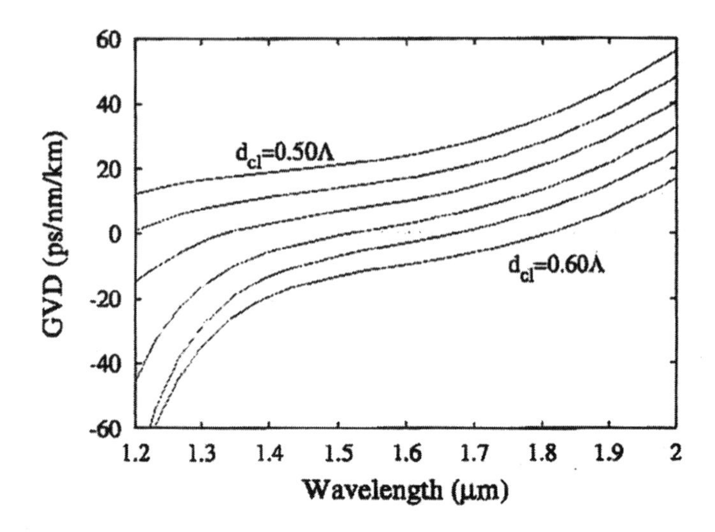

Figure 6-24. Total GVD of honeycomb PBG-fibres with d_{co} = 0.40Λ and d_{cl} ranging from 0.5Λ to 0.6Λ in steps of 0.02Λ. The pitch of the fibres is fixed at Λ = 2.3 μm.

Figure 6-25 shows, how the dispersion curve may be precisely tailored, and for d_{co} = 0.39, a very flat dispersion of around -2 ps/nm/km is obtained over a more than 100 nm broad wavelength range centered at 1.55 μm. While the above-presented results documents how the high flexibility of the PBG-fibres may be utilized to design fibres with specific dispersion properties, they do, however, also illustrate how even small diameter-changes of the air holes (a few percent) may change the dispersion by several tenths of ps/nm/km.

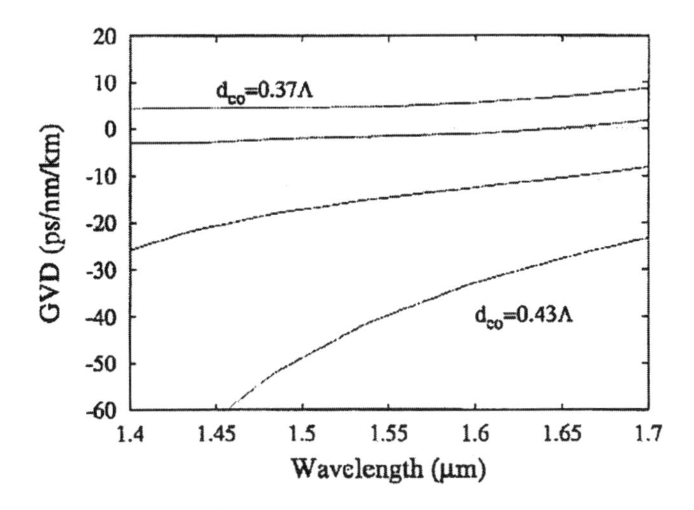

Figure 6-25. Total GVD of honeycomb PBG-fibres with d_{cl} = 0.60Λ and d_{co} ranging from 0.37Λ to 0.43Λ in steps of 0.02Λ. The pitch of the fibres is fixed at Λ = 2.3 μm.

To further investigate the potential of PBG-fibres, the following section shall present an analysis of their polarization properties. This analysis will also touch upon the possible influence of unintentional structural non-uniformities in the fibre cross-sections.

6.5.2 Polarization properties of honeycomb-based fibres

Until this point, the theoretical study of PBG-fibres has only concerned fibres with a perfect 60-degree symmetry - having no unintentional imperfections or asymmetries. Such "ideal" fibres are characterized by a fundamental guided mode, which is doubly-degenerate and, therefore, has two orthogonal polarization states with identical propagation properties.

This is comparable to the fundamental doubly-degenerate mode of conventional fibres having a perfect circular symmetry. However, as well known from conventional fibres, deviations from the ideal structure cause the two polarization states to experience different β/k-values, while propagating along the fibre. This is commonly known as birefringence, and it results in different waveguiding properties for the two polarization states [6.20]. For certain applications, it is desirable to enhanced such deviations and create high-birefringent fibres as it is e.g. the case for polarization-maintaining fibres [6.19]. For other applications, however, even small deviations may be strongly degrading to the system performance. In particular, for high-speed, long-distance optical communication systems (operating at bit-rates of 10 GHz and beyond), an increased attention is now being paid to polarization effects. For such fast systems even very low degrees of birefringence (< 10^{-7}) occurring in conventional fibres may cause a crucial dispersion differences between the two polarization states, thereby setting upper limits for transmission speed. To explore the potential of PBG-fibres, it is, therefore, important to predict and understand the polarization effects, which may result from non-uniformities and asymmetries in and around the core region of a PBG-fibre.

6.5.2.1 Polarization effects from non-uniformities

To analyse the polarization effects resulting from structural non-uniformities in the fibre cross-section, a honeycomb-based PBG-fibre design with relatively small air holes was chosen in order to perform a simulation for realistic fibres. The design is illustrated in Figure 6-26, where non-uniformities were simulated by breaking the 60-degree symmetry of the structure in the immediate proximity to the centre-hole defect. For the specific fibres analysed here, a 10% diameter difference is introduced for the two air holes *P* compared to the remainder of holes in the fibre structure.

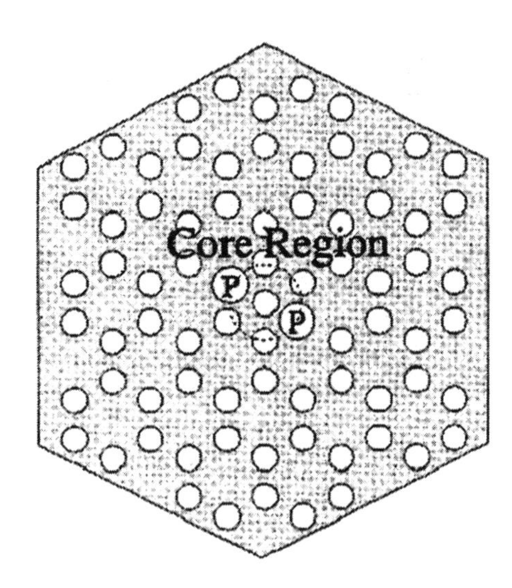

Figure 6-26. Schematic illustration of the basis structure used for the analysis of polarization properties of honeycomb photonic bandgap fibres with non-uniformities in the core region.

A modal index analysis of three different fibres is presented in Figure 6-27. For all three fibres, the holes in the periodic cladding structure have a diameter, d_{cl}, of 0.40Λ and a similar sized core hole. For the "ideal" fibre, the holes *P* are identical to the cladding holes, while the type 1 fibre has *P* holes with a diameter d_p = 0.44Λ, and the type 2 fibre has d_p = 0.36Λ. For the fibre type 1 (with increased hole sizes), Figure 6-27 indicates a slight shift in the mode index towards the lower bandgap edge, i.e., towards lower index values.

This may readily be understood from the fact that the imperfection causes a small increase in the air-filling fraction of the core region. In accordance with this, the mode index for the fibre type 2 is shifted towards higher index values. As is apparent, these shifts do not cause significant alterations in the operational window of the fibres, which remains approximately from Λ/λ = 0.7 to 1.4. However, the non-uniformities do result in important changes in the polarization properties of the fibres. While the two polarization states of the ideal fibre - as expected - coincide, the degeneracy of the guided mode is strongly lifted for both fibre type 1 and fibre type 2. The birefringence, introduced solely from the 10% diameter-change of the holes, *P,* is for fibre type 1 illustrated in Figure 6-28. The figure reveals a birefringence around 10^{-4} over the operational window of the fibre.

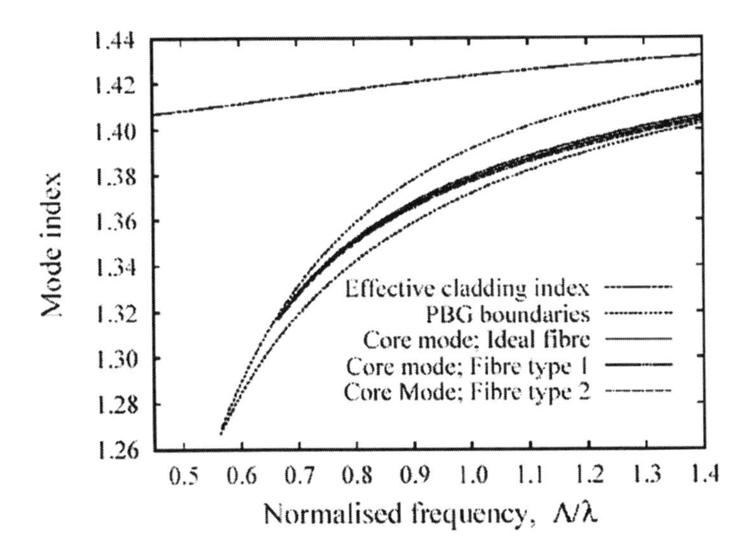

Figure 6-27. Modal index analysis of three types of fibres. The ideal fibre has identical air holes with a size $d_{co} = d_{cl} = d_p$ = 0.40Λ, Fibre type 1 has $d_{co} = d_{cl}$ = 0.40Λ, d_p = 0.44Λ, and Fibre type 2 has $d_{co} = d_{cl}$ = 0.40Λ, d_p = 0.36Λ.

The inaccuracy on the determination of the birefringence was estimated from the birefringence erroneously introduced by the numerical simulation for the ideal fibre. The fibre type 2 showed a slightly smaller degree of birefringence. Since non-uniformities are almost unavoidable with the present technology for fabricating PBG-fibres, the relatively large birefringence occurring from the modest diameter changes for the two holes

P indicates a potential problem for the utilization of PBG-fibres as the basic transmission medium in a long-distance, high capacity communication links. On the other hand, the high degree of birefringence points strongly towards applications of PBG-fibres as polarization maintaining devices. This will be further elaborated on in the following section, where fibres designed for exhibiting a high birefringence are being investigated.

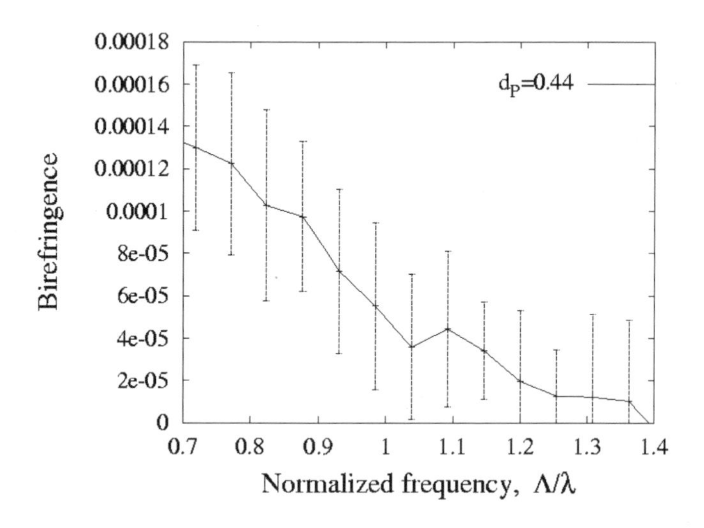

Figure 6-28. Birefringence in fibre of type 1 with $d_{co} = d_{cl} = 0.40\Lambda$, $d_p = 0.44\Lambda$ (see Figure 6-26 and Figure 6-27). The vertical bars indicate numerical error.

6.5.2.2 High-birefringent fibres and single-polarization state fibres

Based on the discussion in the previous section, it may at present seem most relevant to explore the high-birefringence potential of PBG-fibres. With this aim, a fourth fibre with a low air-filling fraction is analysed. This fibre is set to have $d_{co} = d_{cl} = 0.40\Lambda$, $d_p = 0.44\Lambda$, and the positions of the holes P shifted a distance $d_p/2$ towards the centre-defect. As the properties of the guided mode is mainly determined by the core geometry - while the cladding structure primarily determines the operational window of the fibre - a change in position towards the centre was expected to have the largest effect on the polarization properties. The birefringence for this fibre was found to be as large as $5 \cdot 10^{-4}$ at a normalized frequency, Λ/λ, of 0.9. The field distributions of the two orthogonal polarization states are illustrated in Figure 6-29. Commercially available polarization-maintaining fibres, as e.g., PANDA-fibres [6.19], have a birefringence of up to $2 \cdot 10^{-4}$, and the high-degree of birefringence found for the low-filling fraction honeycomb fibres, therefore, supports utilization of PBG-fibres for polarization maintaining and manipulating purposes. Fibres, which have higher air-filling fractions, may exhibit wider PBGs (i.e., PBGs covering a wider index range), and, therefore, allow for an even larger splitting of the core modes. Hence, it should be emphasised that the above-presented results only document the first analysis of polarization effects in PBG-fibres, and the found birefringence-values may not be representative for PBG-fibres optimized for exhibiting high birefringence.

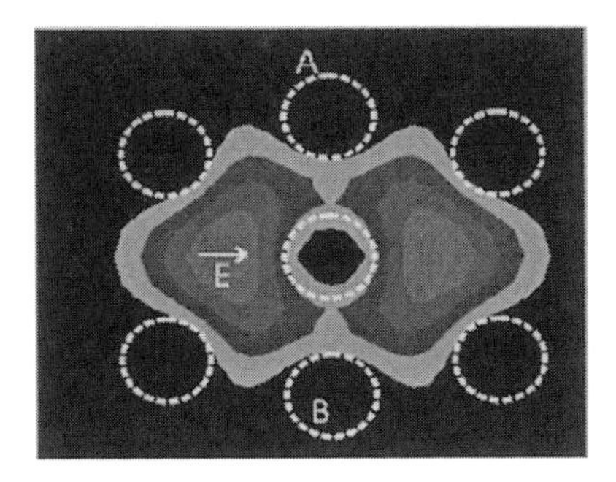

Polarization state 1:
Vertical ***E***-field orientation

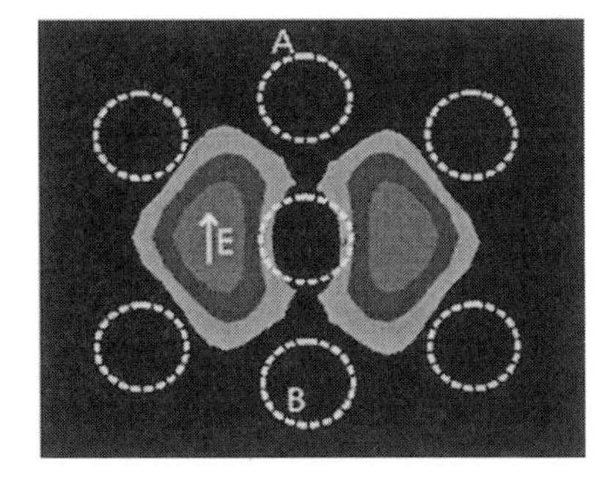

Polarization state 2:
Horizontal ***E***-field orientation

Figure 6-29. Illustration of the field distributions of the two orthogonal polarization states (dark areas indicate low intensity). The fibre geometry is indicated by the dashed lines, and the arrows illustrate the transverse vector components.

It was previously found that the cladding structure mainly determines the operational window of the PBG-fibres, and the asymmetry of the core region primarily results in a splitting of the two polarization states. Therefore, optimising the cladding region to exhibit the widest PBGs, and designing an asymmetric core region seem the natural design-step towards high-birefringent PBG-fibre. However, the cladding and core-region may not be designed completely independently, since this may result in the core mode being pushed out of the PBG-region. On the other hand, this rises the intriguing question, if it would be possible to design the core and cladding of a PBG-fibre in a manner to exactly split the two polarization states to a level, where only one of them would remain within the PBG-region, while the other falls outside? To answer this question, a fibre with a relatively high air-filling fraction is considered. The reason for this choice is that the mode falling inside the PBG region should remain well positioned inside the bandgap in order for the fibre to be robust. Basis for investigating the potential of such a single-polarization state fibre will be taken in a honeycomb PBG-fibre with an air-filling fraction of 30% in the cladding (d_{cl} = 0.70Λ). Fibres with such cladding structures were previously presented in this chapter, and by varying the size of the central air hole - but keeping it perfectly symmetric - it was found possible to shift the core mode from the primary to the secondary bandgap. Elaborating upon this fibre, a strong asymmetry is now applied to the central part of the core region. A modal index illustration for a fibre with an elliptic centre-hole having first and second main axis equal to Λ and 0.35Λ, respectively, and a total area of $0.3\Lambda^2$, is illustrated in Figure 6-30.

From Figure 6-30 is seen, how the two polarization states are highly separated, and a birefringence as large as $2 \cdot 10^{-2}$ at Λ/λ = 0.7 is found. In the normalized frequency range from approximately Λ/λ = 0.5 to 0.9 both modes are within the PBG-region, hence, in this range both modes will be guided. Moving towards higher frequencies, it is observed, how the transversely elongated core hole causes one of the polarization states (the one having transverse vector components parallel to the long axis of the elongated core hole) to be pushed out of the PBG-region. Thus, from approximately Λ/λ = 0.9 to 1.2 only the state being polarized perpendicular to the elongated core hole falls within the PBG-region, and the fibres may, hence, be classified as single-polarization state (or truly single-mode) in this frequency interval. The field distribution of the guided polarization state at a normalized frequency of 0.9 is illustrated in Figure.6-31.

Also more realistic PBG-fibres having less strict two-fold symmetry should display frequency interval regions, where they are single-polarization stated. Naturally, these intervals will be narrowed with decreasing core ellipticity and PBG sizes. Experimental realization of PBG-fibres with well-controlled single-polarization state behaviour seems very attractive for a range of non-linear, sensor, and telecommunication applications.

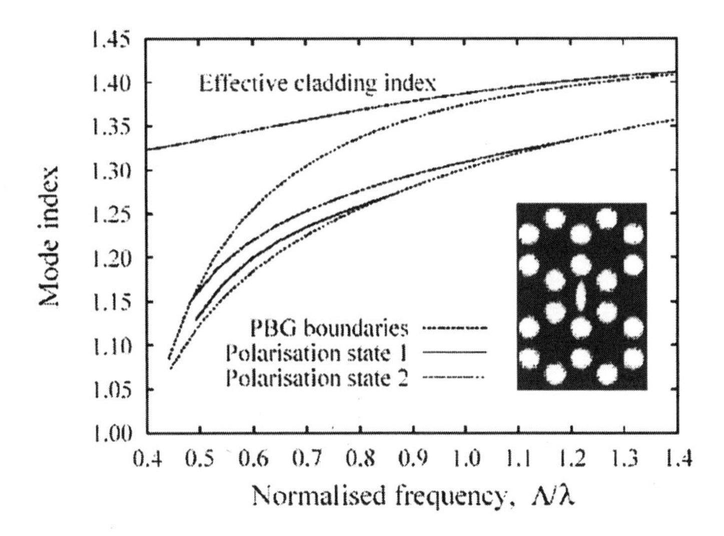

Figure 6-30. Modal index illustration for a honeycomb PBG-fibre having a highly asymmetric central core hole.

Focusing on telecommunications, a truly-single moded PBG-fibre would naturally eliminate the problems of polarization dispersion discussed previously in this section, and may, therefore, well be an interesting candidate as the basic transmission fibre for future high capacity optical communication systems. A second aspect of interest is that the information of a well-preserved polarization over an optical fibre link, may be utilized for achieving lower bit-error rates (perhaps causing a novel interest in coherent communication systems [6.27]). Therefore, although rare-earth doped optical amplifiers have almost eliminated the distance-barriers for optical communications [6.26], truly single-mode PBG-fibres may provide longer repeater-less links - or alternatively a reduction in the number of repeater stations. This requires, naturally, a long maturity process for PBG-fibres, and that several demands may be met. These demands concerns among others - tailoring of the dispersion properties of the fibres, providing relatively large mode areas (or relatively low light-intensity in silica), and bringing down the propagation losses in PBG-fibres to a level comparable to that of conventional fibres.

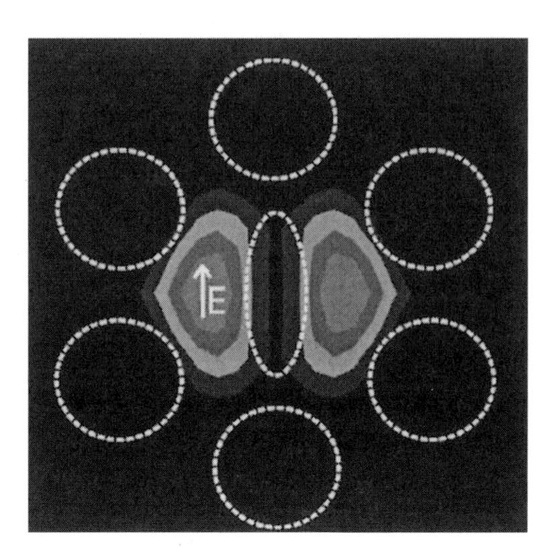

Figure 6-31. Illustration of the field distribution for a single-polarization-state optical fibre.

Concerning the issue of single-polarization-state fibres, it is relevant to mention recent results by Ferrando *et al.* [6.31]. In this work, single-polarization single-mode intra-band guidance (meaning guidance in a photonic bandgap) is theoretically predicted in so-called supersquare photonic crystal fibres. The supersquare PCF structure suggested by Ferrando *et al.* [6.31] is illustrated in Figure 6-32. Ferrando *et al.* performs a full-vectorial analysis of the structure (for an example having a lattice pitch $\Lambda = 2.0$ μm and a hole radius of d = 0.7 μm) using the biorthonormal-basis method (see Section 3.5). The calculations of the band structure predict the existence of a large bandgap, and by the introduction of the off-lattice air hole (the core hole) of radius b = 0.5 μm, intra-band guidance is produced. The supersquare PCF is highly anisotropic, and it may be considered as a highly-birefringent-polarization fibre, because the mode-index gap between two polarization modes is determined to be $\Delta n = 10^{-4}$, comparable to standard values found in conventional strongly birefringent fibres. This inherent birefringence of the supersquare PCF may be further enhanced by a slight alteration of the basic structure, i.e., by a modification of the ratio between the horizontal and vertical distances between consecutive holes [6.31] (a ratio of 1:1.15 leads to a birefringence around $\Delta n = 4 \cdot 10^{-3}$). However, the most interesting feature of the supersquare PCF is the existence of a threshold frequency at which one of the polarization modes is cutoff. For the examples outlined by Ferrando *et al.*, a wavelength window between 1350 nm and 1700 nm (see Figure 6-33) appears, within which only one of the two polarizations can be propagated as a guided mode within the bandgap. With the other polarization mode located outside the bandgap, it naturally becomes a highly interesting task to investigate how large a difference in actual attenuation that may be found in a practical fibre. This is one of the key questions that need to be answered in the years to come.

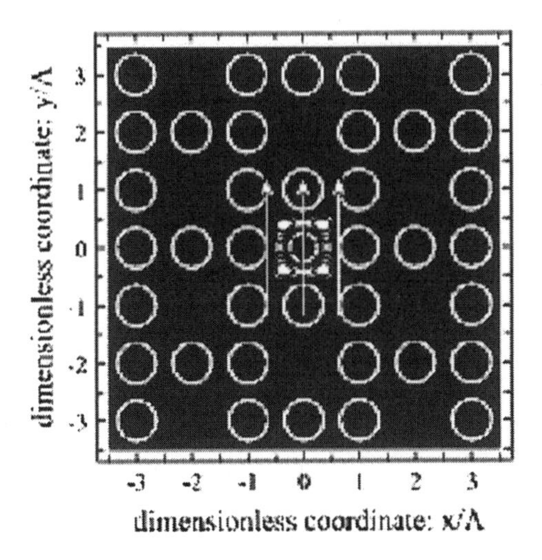

Figure 6-32. Illustration of supersquare PCF structure as suggested by Ferrando *et al.* [6.31]. Also shown is the modal irradiance distribution for the fundamental mode of the supersquare PCF with center-hole radius b = 0.5 μm, cladding-hole radius d = 0.7 μm, and pitch Λ = 2.0 μm. Arrows indicate the polarization state of the magnetic field. The figure is kindly provided by Professor Miguel Andrés, University of Valencia, Spain.

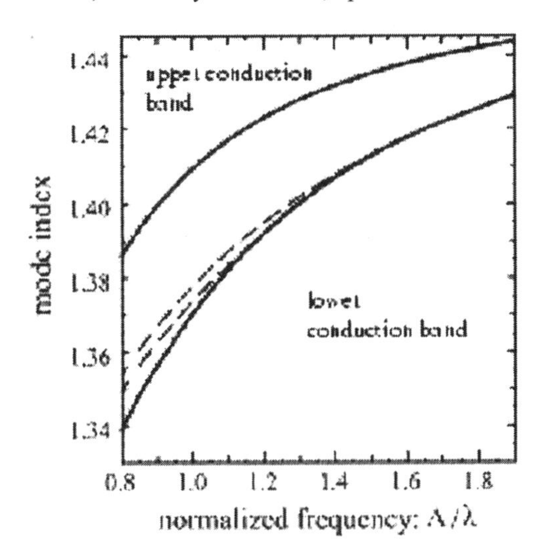

Figure 6-33. Modal intra-band dispersion curves for a slightly deformed supersqueare PBG-PCF (deformation ratio 1.0:1.15), with d = 0.7 μm and b = 0.5 μm (from [6.31]) . The figure is kindly provided by Professor Miguel Andrés, University of Valencia, Spain.

As previously mentioned, the loss-mechanisms in the PBG-fibres are not yet well understood. However, one potential aspect of PBG-fibres allows them to have zero absorption losses. This is the case for PBG- fibres guiding light in air or rather vacuum. Apart from the elimination of absorption losses, such PBG-fibres have a series of other interesting features, and the air-guiding PBG-fibres shall be addressed in the proceeding section.

6.6 AIR-GUIDING FIBRES

Whereas the honeycomb PBG-fibres may be attractive for fibres with special dispersion or polarization properties, a common characteristic is that the guided modes are substantially distributed in silica (or whatever base material is chosen). It is, however, indeed a possibility with PBG-fibres to guide light inside a hollow core (containing air or a vacuum). Cregan *et al.* [6.28] presented in 1999 the first experimental demonstration of air-guiding PBG-fibres, and in this section a numerical analysis of these fibres shall be presented. Apart from documenting the operational principle of the fibres, the analysis reveals unique properties and points out important issues to be addressed for their further development and characterization.

6.6.1 Cladding requirements for obtaining leakage-free waveguidance in air

One of the requirements for obtaining waveguidance in air is that the cladding structure is able to exhibit at least one PBG, which covers β/k-values equal or less than one, as shall be discussed later in this section. From a modal index illustration, this may be observed as PBGs extending below the air-line, as discussed in Section 6.2 (see Figure 6-3). It is not a simple task to determine how to design periodic structures exhibiting PBGs at low β/k-values. However, using a heuristic argument, the ability of a periodic structure to exhibit PBGs extending below the air-line is related to the total air-filling fraction of the structure. From this, it may be understood that triangular structures with respect to air-guidance are intrinsically superior to honeycomb structures (a honeycomb structure has one third less air holes compared to a triangular structure). Close-packing of elongated, circular elements results in a triangular structure, and, therefore, makes a triangular arrangement of air holes appear the optimum basis-structure for high filling fraction 2D photonic crystals.

Using the variational method, a triangular photonic crystal cladding structure with a relatively high air-filling fraction of 70% has been analysed (such a high air-filling fraction is beyond the maximum obtainable for a regular honeycomb structure). For a fixed $\beta\Lambda$-value of 9.0, the photonic band structure diagram of the triangular photonic crystal is illustrated in Figure 6-34. Of particular importance, the triangular structure is seen to exhibit a PBG overlapping the special in case of $k\Lambda = \beta\Lambda$. This case corresponds to plane waves propagating in a homogeneous medium with a refractive index equal to one, such as air (or vacuum). Hence, light propagating in a large air region surrounded by the particular triangular photonic crystal is not allowed to escape this air region, but will be totally reflected due to PBG-effects. To realize a photonic-crystal-cladding structure, which exhibits PBGs covering the $\beta / k = 1$ case is, therefore, a basic requirement to obtain air-guiding PBG-fibres. Although the triangular structure is seen to exhibit a PBG covering the air-line around $\beta\Lambda = 9.0$, this is not the case for arbitrary $\beta\Lambda$-values. This is illustrated in Figure 6-35, where the variation of the PBGs as a function of normalized frequency is illustrated. As seen, several PBGs appear but only within narrow intervals do they overlap the air-line.

Although only five PBGs, which cross the air-line, are illustrated in Figure 6-35, there may well be others appearing at even higher Λ/λ-values. Indeed, for the first experimental realisation of air-guiding PBG-fibres demonstrated by Cregan *et al.* [6.28], PBGs crossing the air-line around $\Lambda/\lambda = 10$ appear to be utilized. However, with the aim of analysing general properties of air-guiding PBG-fibres, we will in the following section focus on structures with lower-frequency PBGs, which are easier to model.

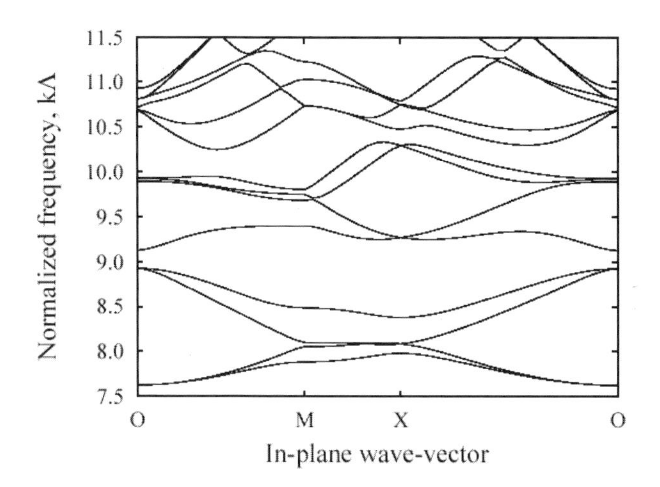

Figure 6-34. Photonic band structure diagram for a silica-air, triangular photonic crystal with an air-filling fraction of 70%.

6.6.2 Core requirements of air-guiding fibres

Having established the cladding requirements for air-guiding PBG-fibres, the requirements of the core region and the properties of the air-guided modes shall now be addressed.

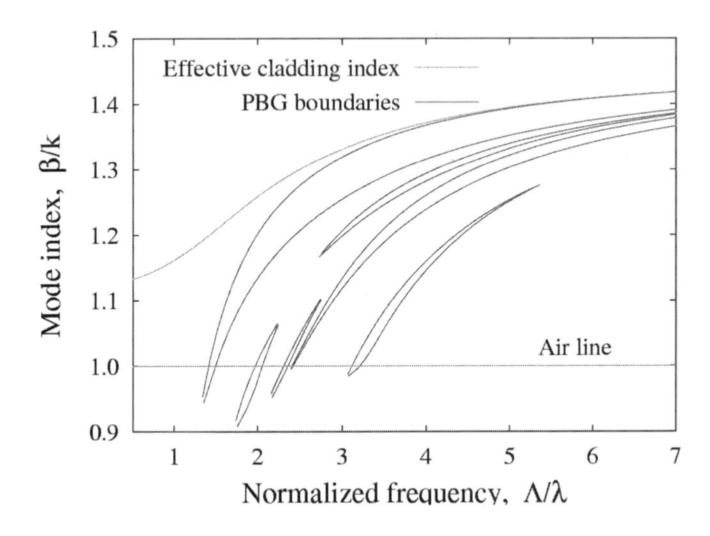

Figure 6-35. PBGs exhibited by a triangular photonic crystal structure with a relatively high air- filling fraction of 70%. The air-line β/k= 1.0 is indicated by the dashed dotted line.

To design the core region, and give a first estimate of the number of air-guided modes supported by the PBG-fibres, the approximate equation, Eq. (6.11), may be used. For the PBG-fibre considered here, having an air-core (n_{co} = 1.0) with a radius, $D_{eff}/2$, of $\sqrt{7}\Lambda/2$, Eq. (6.11) predicts the fibre to have $N_{PBG} = 1.2 \approx 1$ guided modes at $k\Lambda = 9.0$. Using the variational method, an accurate determination of the number of air-guided modes may be performed, and, indeed, only one single mode is found to be confined in the air-core at $k\Lambda = 9.0$. This fundamental mode consists of two orthogonal, quasi-linear polarization states, and its mode-field distribution at $k\Lambda = 9.0$ is illustrated in Figure 6-36. At a slightly lower normalized frequency, $k\Lambda$ = 8.8, a second mode was found to enter the PBG and to become confined within the air-core. Again this may partly be predicted by Eq. (6.11), according to which $N_{PBG} = 1.7 \approx 2$. The second-order mode is four-fold degenerate and has a two-lobed field-distribution with a 180°-phase reversal between the two maxima.

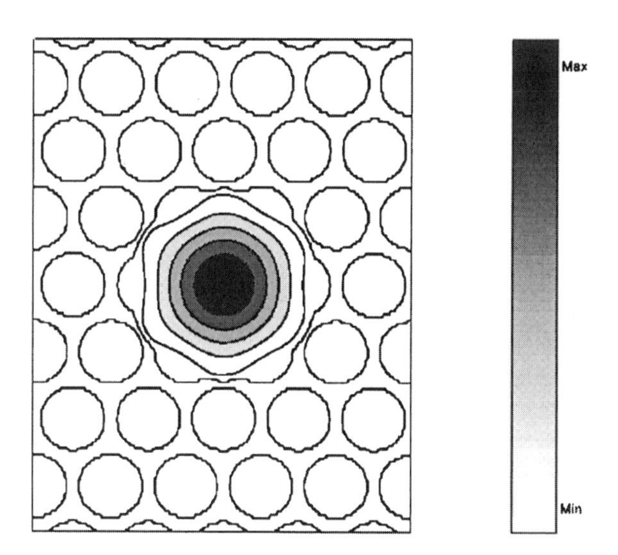

Figure 6-36. Field distribution of the fundamental, air-guided mode. The mode consists of two orthogonal, quasi-linear polarization states and is calculated at a normalized wave-number, $k\Lambda$, of 9.0. The thin black lines indicate the air holes in the fibre.

While Eq. (6.11) provides a good estimate of the number of air-guided modes, we must turn to the full-vectorial method for further detailed information about these modes. An important property of the PBG-fibres was found by observing the two modes and their field distributions at frequencies outside the PBG. The inset in Figure 6-37 illustrates the trace of the two modes for a normalized propagation range of $\beta\Lambda = 7.9$ to $\beta\Lambda = 9.4$. Within the PBG, both modes are spatially confined within the air-core. Hence, the leakage-free operational window of the PBG-fibres is determined as the spectral range for which the modes are positioned inside the PBG. The term leakage-free is used to indicate the situation for an infinitely extending photonic crystal cladding structure. In practice, it is important to take into account considerations on confinement loss due to a final number of air-hole rings surrounding the hollow core – as demonstrated by White *et al.* [6.33]. For the considered PBG-fibre, the PBG range was determined to be approximately $\Delta k\Lambda = 0.75$. Outside the PBG, both modes were found to become resonant with allowed cladding modes, and for a real fibre they will, therefore, be leaky. However, for a relatively broad spectral range (of at least two times $\Delta k\Lambda$) the leaky modes were found to retain their maxima within the air-core. The mode-field distribution of a leaky second-order mode is illustrated in Figure 6-38.

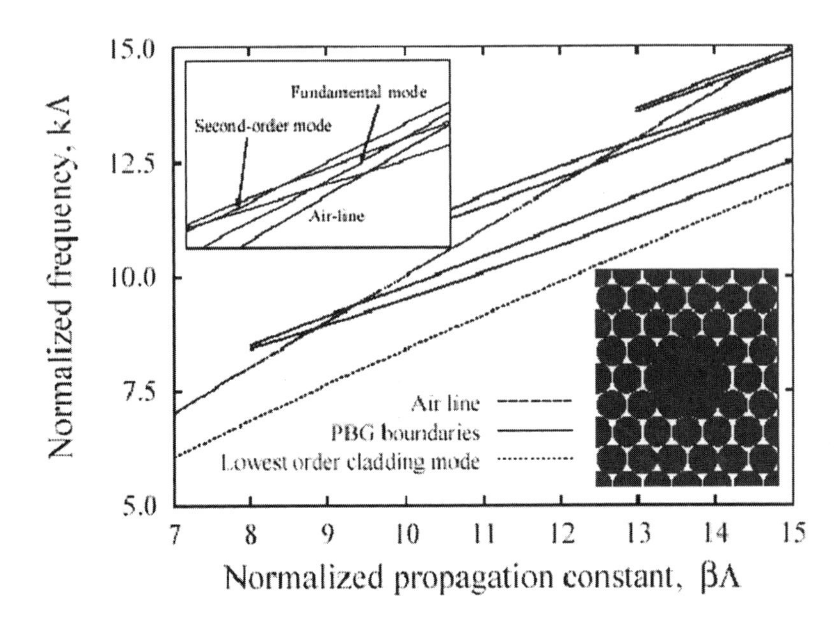

Figure 6-37. Variation of the photonic band gaps with normalized propagation constant. The photonic crystal may reflect light incident from air at the ($\beta\Lambda$,$k\Lambda$)-intervals, where the air-line falls within one of the photonic band gaps. A full analysis of the large-core photonic band gap fibre revealed the presence of two core modes, and the upper inset shows the trace of these modes for a normalized propagation constant interval from $\beta\Lambda = 7.8$ to $\beta\Lambda = 9.3$. The lower inset shows the core and inner cladding structure of the analysed fibre (white areas are silica and black areas are air).

6.6.3 Advanced properties of air-guiding photonic crystal fibres

Although the applied numerical method does not provide direct information about the fibre-lengths needed to strip the leaky air-modes, recent results indicate that the leakage-free spectral range of the PBG-fibres may be considerably more narrow than found from an experimental characterization performed on short fibre-lengths (e.g., a few centimetres such as the transmission measurements presented by Cregan *et al.* [6.28]). This was recently confirmed by Hansen *et al.* [6.34] who demonstrated air-guiding fibres of up to 345 m.

A second important property of the PBG-fibres is realized from inspection of the leakage-free spectral ranges of the fundamental and the second-order modes. These ranges do not coincide, and as a unique feature, the numerical analysis reveals a range for which only the second-order mode is confined, whereas the fundamental mode is leaky. Hence, it is, in principle, possible to strip the fundamental mode, while guiding the second-order mode over long lengths. This finding represents an important property, by which light can be guided in a higher-order mode without the presence of a fundamental mode, and opens up new ways of selecting and tailoring the mode profiles of optical fibres. It is important to notice that this property is not be restricted to the special case of air-guiding PBG-fibres, but it may also apply to other types of PBG-fibres, such as the honeycomb-based PBG-fibres.

The numerical simulations further revealed the presence of several ring-shaped resonant modes localised at the core-cladding interfaces - see Figure 6-39 (a). These modes are also positioned outside the PBG, and must be classified as leaky. However, they were found to have a small overlap with allowed modes in the photonic crystal cladding (the field distribution of a typical allowed cladding mode is illustrated in Figure 6.39 (b), and may, therefore, be difficult to strip for a real fibre. As the resonant modes are expected to play an important role as loss mechanism in real PBG-fibres, their accurate determination may prove essential for the further development of PBG-fibres. However, a detailed study of the resonant modes, the stripping of these, and their coupling coefficients with cladding modes is the subject of ongoing research, and are important issues to be addressed.

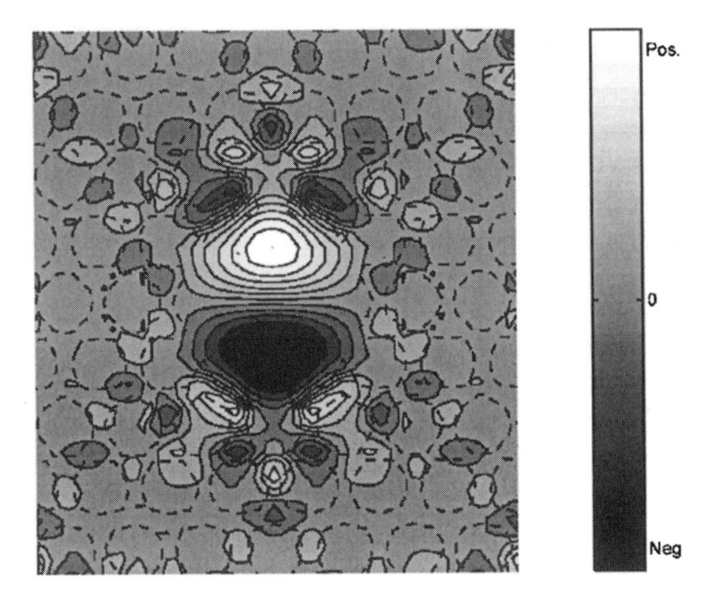

Figure 6-38. Field distribution of a leaky second-order mode in an air-guiding fibre structure as illustrate in figure 6-35. The mode has a 180° phase-shift between the two maxima, which is indicated by the different grayscaling of the field-lobes (notice that zero intensity is medium-gray). The dashed lines indicate the air holes in the fibre.

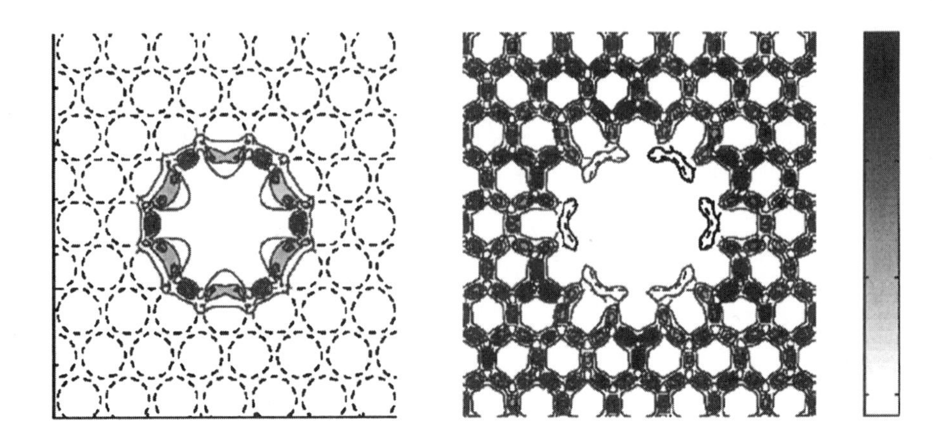

Figure 6-39. (a) Resonant defect-mode being localised to the core-cladding interface. (b) Typical field distribution of modes allowed in the cladding structure. The modes were calculated at a normalized frequency, kΛ, of 9.0.

Finally, the numerical analysis shall be directed towards specific applicational aspects of the air-guiding PBG-fibres. The fibres appear promising for short-distance delivery of high-power laser light in the mid-infrared, where e.g., the laser lines of CO_2 around 10.6 μm and of Er:YAG around 3 μm are presently used for medical purposes [6.29-6.30]. Although conventional, silica-based glass fibres have mechanical properties desirable for laser surgery, their high background losses at these wavelengths makes them unsuitable for practical use. The commonly used fibre-types in laser surgery are hollow sapphire fibres for delivery around 10.6 μm, and both fluoride-glass fibres and solid sapphire fibres for delivery around 3 μm. While fluoride-glasses are characterized by relatively low losses around 3 μm *(>0.2 dB/m),* they have the severe disadvantage of being very fragile and thus difficult to handle. In contrast to this, both types of sapphire fibres are more stable, but have higher losses. Other types of hollow fibres include fibres with inner metal coatings, with dielectric multi-layers, or with combined metal and polymer coatings [6.30]. However, all these types of hollow fibres are characterized by relatively high losses (>0.2 dB/m). It is, therefore, of relevance to consider air-guiding PBG-fibres for high-power transmission. The mechanical properties are expected to be superior to both metal-coated, dielectric multi-layer, sapphire and fluoride glass fibres as they should be comparable to those of conventional silica fibres. Since, furthermore, light may be guided in air (or vacuum), absorption losses may be considerably reduced. It is important to note that although the majority of the power may be guided in the core, a small fraction will still be located in the fibre base material. For this reason, a fibre material with reasonably low material-loss should be chosen, and for wavelengths larger than approximately 3.0 μm, alternatives to silica should be considered. Of particular interest could be the use of polymer-based PBG fibres. Polymer-materials are potentially cheap to fabricate, easy to handle, and furthermore span a significantly wider index-range than silica-glasses. Therefore, increased flexibility to PBG-fibres may be added by utilizing polymers as background material. Index-guiding polymer PCFs were first demonstration by Eijkelenborg *et al.* [6-35], and naturally, the question regarding if fabrication of low-loss air-guiding polymer fibres is feasible must be answered positively.

Finally, it must be emphasised that although only air-guided modes, which occur for the lowest-frequency PBG, were analysed, the results presented in this section point out important issues to be addressed also for experimental investigations of air-guiding PBG-fibres utilizing higher-order PBGs.

6.6.4 Loss properties of air-guiding fibres

We will end this section by looking at some of the recent experimental results concerning the realisation of air-guiding PBG fibres. The recently reported structures are from an overall view similar to the structure shown in Figure 6-36, although the silica base-structure presented in recent publications by researchers from Corning Incorporated (see [6.8-6.9]) seems to have thinner silica wall structures than the one numerically analysed in the previous text.

A very significant loss reduction has been shown over the past 1-2 years. This is illustrated by the fact that West *et al.* [6.8] as late as September 2001 reported the lowest loss values in PBG fibres to be around 1000 dB/km, but in September 2002 Venkataraman *et al.* [6.9] reported the significant improvement showing a loss value of 13 dB/km at the wavelength of 1500 nm for an air-guiding fibre with a length of 100 m. For the fibre reported in [6.9], the loss spectrum showed a loss of less than 30 dB/km over the transmission window between 1395 nm and 1520 nm, and it was stressed that the fibre structure did not show any significant variation over the length of the fibre. Venkataraman *et al.* [6.9] present the hypothesis that the uniformity of the transmission spectrum (related to the axial uniformity of the structure) over the length of the fibre is one of the principal reasons for the reduced loss compared to previous air-guiding fibres.

As further pointed out by Venkataraman *et al.* [6.9], the passing of the 20 dB/km milestone [6.32] for standard optical fibres represented the beginning of the telecommunications revolution about 30 years ago, and passing the same milestone for air-guiding fibres clearly brings these a huge step closer to practical applications. We may, therefore, expect many new and exiting research and development results to appear within this area over the next years to come.

6.7 SUMMARY

In this chapter, the properties and design of some of the most fundamental photonic bandgap fibres are described. The chosen approach has been first to examine material systems with a lower index contrast, namely silica-air 2D photonic crystals. These were analysed for regular triangular and honeycomb structures and important differences between the two structures were found. These findings included significantly higher maximum PBG sizes of the honeycomb structures. The obtained results were further discussed using considerations on the distribution of the high-index regions of the photonic crystals and a relatively simple design route to large-bandgap photonic crystals was demonstrated. The design route was used to explain the influence of modifications on the photonic crystals from introducing small additional holes at interstitial positions between the holes of the regular lattice and, further, to design novel optimised photonic crystals utilizing several levels of silica dopants.

This chapter further dealt with waveguidance by the PBG effect in optical fibres. Initial discussions on how to design PBG-fibres have been presented. The basic operation of PBG-fibres has been illustrated and important differences between conventional TIR-based optical fibres and PBG-fibres have been documented. These concern spectral extent of the transmission windows and radically different core-cladding index-difference requirements, where most importantly the leakage-free waveguidance in low-index core fibres was demonstrated to be theoretically possible. The chapter has further addressed the influence of the core and cladding hole sizes on the basic waveguiding properties of the PBG-fibre. This has documented how the guided modes of the PBG-fibres may be shifted to a higher-order PBG. Also, the effect of varying the core and cladding holes independently were studied. It was demonstrated, how this allows for a high degree of flexibility, when tailoring the PBG-fibres for specific applications.

After a presentation of some of the key properties of the first experimental demonstration of the PBG effect in optical fibres, additional aspects of PBG fibres were presented. An analysis of their dispersion properties was discussed, and it was found that the fibres may be designed to exhibit high anomalous dispersion, while remaining single mode or they may exhibit flat, near-zero dispersion over a broad wavelength range. The former property allows for example the design of single-mode fibres with their zero-dispersion point shifted towards shorter wavelengths (significantly below 1.0 μm). To further investigate the potential of the PBG-fibres, an analysis of their polarization properties was performed. Strong birefringence was seen to result from moderate non-uniformities in and around the core region. Furthermore, the potential of realizing single-polarization-state fibres was demonstrated.

Finally, a numerical analysis of large-core air-guiding PBG-fibres was presented. The analysis provided detailed information about the fibre properties, including spectral location of the band gaps, mode-field distributions, design guidelines, and an accurate determination of the single/multi-mode ranges.

REFERENCES

[6.1] P. V. Kaiser, and H. W. Astle,
"Low loss single material fibers made from pure fused silica",
The Bell System Technical Journal, Vol.53, pp.1021-1039, 1974

[6.2] V. Vali, and D. B. Chang
"Low index of refraction optical fiber with tubular core and/or cladding",
US Patent 5155792, 1992.

[6.3] E. Yablonovitch,
"Inhibited spontaneous emission in solid-state physics and electronics",
Physical Review Letters, Vol.58, pp.2059-62, May 1987.

[6.4] S. John,
"Strong localization of photons in certain disordered dielectric superlattices",
Physical Review Letters, Vol. 58, no. 23, pp. 2486-9, 1987.

[6.5] J. Knight, T. Birks, D. Atkin, and P. Russell,
"Pure silica single-mode fibre with hexagonal photonic crystal cladding,",
Optical Fiber Communication Conference, Vol. 2, p. CH35901, 1996.

[6.6] J. Knight, J. Broeng, T. Birks, and P. Russell,
"Photonic band gap guidance in optical fibers",
Science, Vol. 282, pp. 1476-1478, Nov. 20, 1998.

[6.7] J. Broeng, S. Barkou, and A. Bjarklev,
"Waveguiding by the photonic band gap effect,"
Topical meeting on Electromagnetic Optics, (Hyeres, France), pp. 67-68, EOS, September 1998.

[6.8] J. A. West, N.Venkataraman, C. M. Smith, M. T. Gallagher,
"Photonic Crystal Fibres".
ECOC'2001, Amsterdam, Netherlands, Paper Th.A.2,pp. 582-585, Sept. 30th – Oct. 4th 2001.

[6.9] N. Venkataraman, M. T. Gallagher, C. M. Smith, D. Müller, J. A. West, K. W. Koch, J. C. Fajardo,
"Low loss (13 dB/km) air core photonic band-gap fibre",
ECOC'2002, Copenhagen, Denmark, Post deadline paper PD1.1, 2002.

[6.10] T. Birks, P. Roberts, P. Russell, D. Atkin, and T. Shepherd,
"Full 2-d photonic bandgaps in silica/air structures",
IEE Electronics Letters, Vol. 31, pp. 1941-1943, Oct. 1995.

[6.11] T. Birks, J. Knight, and P. Russell,
"Endlessly single-mode photonic crystal fiber",
Optics Letters, Vol. 22, pp. 961-963, July 1997.

[6.12] B. Scaife,
"Principles of dielectrics",
Clarendon Press, 1989.

[6.13] D. Cassagne, C. Jouanin, and D. Bertho,
"Hexagonal photonic-band-gap structures",
Physical Review B, Vol. 53, pp. 7134-7142, March 1996.

[6.14] A. Barra, D. Cassagne, and C. Jouanin,
"Existence of two-dimensional absolute photonic band gaps in the visible",
Applied Physics Letters, Vol. 72, pp. 627-629, Feb. 1998.

[6.15] J. Knight, T. Birks, R. Cregan, and P. Russell,
"Photonic crystals as optical fibres -physics and applications",
Optical Materials, Vol.11, pp.143-151, Jan. 1999.

[6.16] C. Anderson and K. Giapis,
"Larger two-dimensional photonic band gaps",
Physical Review Letters, Vol. 77, pp. 2949-2952, Sept. 1996.

[6.17] J .B. Mac Chesney, D. W. Johnson Jr., S. Bhandarkar, M. P. Bohrer, J. W. Flemming, E. M. Monberg, and D. J. Trevor,
"Optical fibres using sol-gel silica overcladding tubes",
IEE Electronics Letters, Vol. 33, No. 18, p.1573, 1997.

[6.18] J. Knight, T. Birks, P. Russell, and J. Sandro,
"Properties of photonic crystal fiber and the effectice index model",
Journal of the Optical Society of America A, vol. 15, pp. 748-52, March 1998.

[6.19] G. Agrawal,
"Nonlinear fiber optics",
Academic Press, second ed., 1995.

[6.20] A. Snyder and J. Love,
"Optical waveguide theory",
Kluwer Academic Publishers, 2000, ISBN: 0-412-09950-0.

[6.21] J. Fleming,
"Material dispersion in lightguide glasses",
IEE Electronics Letters, vol.14, pp. 326-328, 1978.

[6.22] T. Monro, D. Richardson, and N. Broderick,
"Holey fibres: an efficient modal model,"
Journal of Lightwave Technology, vol. 17, pp. 1093–1102, June 1999.

[6.23] J. Ranka, R. Windeler, and A. Stentz,
"Efficient visible continuum generation in air-silica microstructure optical fibers with anomolous dispersion at 800 nm",
Conference on Laser and Electro-Optics, CLEO'99, Baltimore,May 1999. CPD8.

[6.24] A. Ferrando, E. Silvestre, J. Miret, J. Monsoriu, M. Andres, and P. Russell,
"Designing a photonic crystal fibre with flattened chromatic dispersion",
IEE Electronics Letters, vol. 24, pp. 276–278, March 1999.

[6.25] M. Gander, R. McBride, J. Jones, D. Mogilevtsev, T. Birks, J. Knight, and P. Russell,
"Experimental measurement of group velocity dispersion in photonic crystal fibre",
IEE Electronics Letters, vol. 35, pp. 63–43, Jan. 1999.

[6.26] A. Bjarklev,
"Optical Fiber Amplifiers: Design and System Applications",
Artech House, Boston-London, August 1993.

[6.27] T. Baba and T.Matsuzaki,
"Polarisation changes in spontaneous emission from GaInAsP/InP two-dimensional photonic crystals",
IEE Electronics Letters, Vol. 31, pp. 1776–1778, Sept. 1995.

[6.28] R. Cregan, B.Mangan, J. Knight, T. Birks, P. Russell, P. Roberts, and D. Allan,
"Single-mode photonic band gap guidance of light in air",
Science, Vol. 285, pp. 1537–1539, Sept. 1999.

[6.29] J. Harrington,
"Fiber optics for the delivery of 3 and 10.6 micron energy for laser surgery,"
LEOS '93 Conference Proceedings, pp. 225–226, 1993.

[6.30] Y. Matsuura and M. Miyagi,
"High power hollow fibers",
SPIE-Int. Soc. Opt. Eng., Vol. 3343, pp. 222–227, 1998.

[6.31] A. Ferrando, and J. J. Miret,
"Single-polarization single-mode intraband guidance in supersquare photonic crystal fibres",
Applied Physics Letters, Vol.78, No.21, pp.3184-3186, 2001.

[6.32] F. P. Kapron, D. B. Heck, and R. D. Maurer,
"Radiation losses in glass optical waveguides,
Applied Physics. Letters, Vol.17, p.423-425, 1970.

[6.33] T. P. White, R. C. McPhedran, L. C. Botten, G. H. Smith, and C. M. de Sterke,
"Calculations of air-guided modes in photonic crystal fibers using the multipole method",
Opt. Express, Vol. 9, p. 721-732, 2001.

[6.34] T.P. Hansen, J. Broeng, C. Jakobsen, G. Vienne, H.R. Simonsen, M.D. Nielsen, P.M.W. Skovgaard, J.R. Folkenberg, and A. Bjarklev,
"Air-guidance over 345 m of large-core photonic bandgap fiber",
Optical Fiber Communication Conference OFC'03, Post Deadline paper, 2003.

[6.35] M. van Eijkelenborg, M. Large, A. Argyros, J. Zagari, S. Manos, N. A. Issa, I. M. Bassett, S. C. Fleming, R. C. McPhedran, C. M. de Sterke, and N. A. P. Nicorovici,
"Microstructured polymer optical fibre",
Opt. Express, Vol. 9, pp. 319-327, 2001.

Chapter 7

APPLICATIONS AND FUTURE PERSPECTIVES

7.1 INTRODUCTION

In the previous chapters of this book, we have introduced the fundamental properties of photonic crystal fibres ranging from fundamental definition of photonic bandgap structures over numerical modelling to fabrication of these new fibre types. We have also discussed the fundamental differences between index-guiding PCFs and bandgap-guiding PCFs, and some examples of fibre structures and properties have been presented. However, the research field is still very young, and numerous new results and applications appear as the fibres are tested and investigated by more research groups and companies. For this reason, it is a very significant challenge to try to give an up-to-date picture of the most relevant applications of photonic crystal fibres, and just between the time of writing this chapter and to the point, when the book is printed, novel findings will be added to the field. For this reason, our ambition with the present chapter is more modest, since we have chosen to present some of the applications and ideas for the PCF technology, which primarily provides a good impression of the wide range of possibilities provided by these waveguides rather than necessarily covering all of the latest research results.

The chapter has been divided into a number of subsections, which are chosen for the following reasons; Section 7.2 describes the properties of so-called large-mode-area (LMA) photonic crystal fibres. These fibres are typically of the index-guiding type and they provide new functionality to the field of fibre technology by allowing single-mode operation at relatively low wavelengths (covering the visible part of the optical spectrum), while at the same time having so large spotsizes that more simple coupling is possible. Section 7.2 will describe some of the fundamental features of these fibres, as well as key parameters will be defined. Finally, we will also discuss some of the recent advances including novel core designs.

Section 7.3 will address the interesting class of highly-nonlinear photonic crystal fibres (HNL-PCFs). Most of these PCFs are (at least at present time) also of the index-guiding type, and they may also be seen as the complement to the large-mode-area (LMA) PCFs described in Section 7.2, because in contrast to the LMA-PCFs, the HNL-PCFs are made with the purpose of having a reduced effective mode area. The objective is that the reduced mode area, which is obtained through high mode confinement by large cladding air-filling fractions, may enhance the nonlinear coefficient of the optical fibre. In combination with the possibility of waveguide dispersion control – making anomalous fibre dispersion possible at visible wavelengths, this opens up for a completely new performance of optical fibres as tailored nonlinear elements. Section 7.3 will describe such fibres in applications for optical signal processing and as supercontinuum generating media.

In cases, where single-mode operation is not of critical concern, the photonic crystal fibre technology also opens up for a completely new class of multimode fibres as described in Section 7.4. The interesting property is that very high numerical apertures may be achieved using a large pure silica core surrounded by a ring of closely packed air holes, which forms a very high index contrast and allows for fibres that are very useful in collection of optical power.

A step even further in the exploration of unique waveguiding properties, and representing a combination of single-mode control and high-numerical-aperture structures is described in Section 7.5, where the highly relevant double-clad microstructured rare-earth-doped fibres are described.

Section 7.6 will shortly address the possibility of fabricating tuneable fibre devices through the application of microstructured fibres with filled holes. In the specific example, the approach of using a temperature sensitive polymer filling is discussed.

Section 7.7, is directed towards the application of the PCF technology in the realisation of highly-birefringent optical fibres. We here shortly address the importance of high fabrication flexibility and emphasize the existence of high index contrasts in photonic crystal fibres in connection with simple methods for modifying the typical hexagonal preform stacking to obtain non-degenerate orthogonal fibre modes.

After these interesting examples of fibres with fundamentally different – or at least enhanced – properties compared to standard optical fibres, the final parts of the chapter will address some of the more fundamental properties in connection with the practical application of PCFs. A very important parameter in this connection is the fibre dispersion, which is discussed in Section 7.8. This is a key parameter not only in connection with highly nonlinear fibres, but also in connection with low-power signal transmission, optical signal processing in general and in numerous other applications.

Section 7.9 is dealing with the issue of coupling and splicing loss, which is highly relevant, whenever PCFs are used in systems combining standard fibre and PCF technology. The issue of power loss does, however, also relate directly to the microstructured fibres themselves, and the efforts of lowering transmission loss in these components are among the most important in present day fibre development. For this reason, we have chosen to use Section 7.11 in this chapter for the description of key issues concerning loss properties. Finally, Section 7.12 includes a short summary.

7.2 LARGE-MODE-AREA PHOTONIC CRYSTAL FIBRES

Optical fibres having waveguide properties, which allows for large mode areas, while at the same time staying in single-mode operation, are highly interesting. The development of large-mode-area (LMA) fibres is important for a wide range of practical applications most notably those requiring the delivery of high power optical beams. For these applications, where large optical powers are to be transmitted, it is generally important that the transmission may be done without the influence of undesired nonlinear effects appearing due to the presence of the fibre material. For many of these applications, spatial mode quality is also a critical issue, and such fibres should preferably support just a single transverse mode.

Relatively large-moded single-mode fibres can be made using conventional fibre doping techniques such as modified chemical vapour deposition (MCVD) simply by reducing the numerical aperture (NA) of the fibre and increasing the fibre core size. However, the minimum NA that can be reliably achieved is restricted by the accuracy of the control of the refractive index difference between the core and cladding. In addition to offering large mode areas, LMA-PCFs offer other unique and valuable properties. Among the most notable are that they can be single-moded at practically all wavelengths [7.1]. This is in contrast to standard optical fibres, which exhibit a cut-off wavelength below which the standard fibre becomes multi-moded. The PCF technology also provides an alternative, and potentially more accurate route to controlling the index difference between core and cladding regions of the fibre, since the control of hole size and distribution may be used to tailor the exact effective refractive indices.

7.2.1 Characteristics of large-mode-area photonic crystal fibres

One possibility of realising LMA fibres with very high nonlinear threshold powers is of course the future application of air-guiding photonic bandgap fibres such as the ones described in Section 6.6. However, index-guiding photonic crystal fibres offer an alternative route towards large mode areas [7.1]. Index guiding LMA-PCF may be produced by designing fibres with a large hole-to-hole spacing, Λ typically around 10λ (where λ is the wavelength) or larger, and air holes of size, d/Λ, around 0.45 or smaller. Figure 7-1 shows the near field mode profile in a typical LMA-PCF superimposed on a Scanning Electron Microscope (SEM) picture of the fibre profile. This fibre is interesting, not only because it has a very large mode field diameter (MFD around 20 microns at a wavelength of 1550 nm), but also because it is single-moded for any wavelength at which the silica base material is transparent.

Figure 7-1. Cross section of a large-mode-area photonic crystal fibre having a mode field diameter of 20 microns at a wavelength of 1550 nm. The near-field picture (recorded at the output end of the fibre) of the guided mode is superimposed. The photograph is kindly provided by Crystal Fibre A/S.

Since LMA-PCFs rely on a very small effective-index contrast between core and cladding, the fibres can be sensitive to macro- and microbending, and this is further discussed later in this chapter. For this reason, the fibre properties can be highly sensitive to the precise details of the fibre structure, and, therefore, accurate fabrication techniques are required. Figure 7-2 shows cross-sectional pictures of two typical LMA-PCFs with very regularly arranged air holes.

Both of the fibres shown in Figure 7-2 have core diameters (measured as the distance between two oppositely placed holes surrounding the centre of the fibre) of 15 microns. The fibre shown on the left side has an outer fibre diameter of 125 microns, whereas the fibre having the cross section shown at the right side of the Figure 7-2 has an outer diameter of 160 μm. The thicker fibre has been manufactured in order to obtain a more rigid and, thereby, more microbending resistant fibre structure.

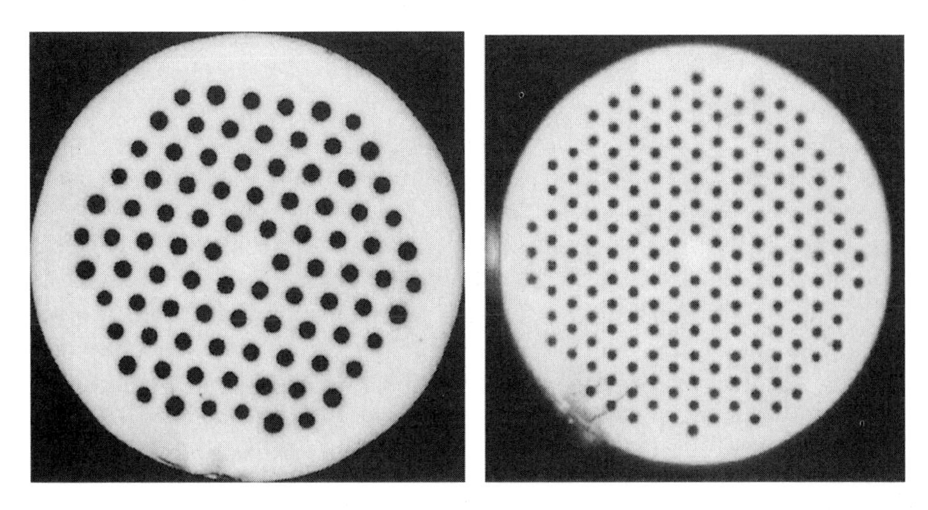

Figure 7-2. Full cross section of two typical large-mode-area photonic crystal fibres with different air-filling fractions (both have core diameters of 15 microns) and different outer fibre diameters. Outer diameter of the left fibre is 125 μm and that of the right fibre is 160 μm. Photographs kindly provided by Crystal Fibre A/S.

The numerical models described in Chapter 3 can be applied to model the optical properties of these LMA fibres, although, typically, extra care is needed due to the wide range of spatial scales present. Polarization effects are typically less important for this class of fibres, because of the very symmetric fibre structures, and it is often sufficient to use a scalar model in the analysis of basic fibre properties.

Macroscopic bend loss ultimately limits the practicality of large-mode – area photonic crystal fibres, and so understanding bend related loss is important in the design of this class of fibre. Two distinct bend loss mechanisms have been identified in conventional fibres; transition loss and pure macrobending loss [7.2]. As light travels into a curved fibre, the mode distorts, causing a transition loss (analogous to a splice loss). Pure macrobending loss occurs continually along any curved section of an optical fibre. The mechanism is as follows: At some radial distance, the tails of the mode need to travel faster than the speed of light to negotiate the bend, and are thus lost.

Using the simplified effective-index method described in Section 3.2, it is possible to derive useful formulas for the pure bend loss in these PCFs by applying standard results for the power loss coefficient of standard step-index fibres (see [7.3] and [7.4]). Utilizing this approach, the pure macro-bending loss can be derived from the coefficients of the Bessel function of the equivalent step-index fibre, as described in [7.5]. Unsurprisingly, the critical bending radius and the spectral width and location of the operational window depend strongly on hole size, d, and hole-to-hole spacing, Λ. Generally, larger holes result in broader operational windows, whereas the hole-to-hole distance roughly determines the centre position of the window (as a first approximation, the minimum bend loss occurs at a wavelength around $\Lambda/2$) [7.5]. Hence, standard telecommunications wavelengths fall on the short wavelength loss edge for large mode photonic crystals. Despite this, LMA-PCFs have been demonstrated to possess comparable bending losses to similarly sized conventional fibres at this wavelength [7.6].

The pure macrobending losses, which will be the dominant form of bend related loss in long fibre lengths. For shorter lengths of fibre, however, it can be important to investigate the impact of transition loss too. Transition and pure bend losses can be distinguished experimentally by progressively wrapping a fibre around a drum of well-defined radius [7.2] The fibre experiences a sharp change of curvature as it enters and leaves the drum surface, which results in transition losses at these points. A detailed numerical and experimental study of the relation between pure macrobending loss and transition loss is made by Baggett *et al.* [7.6]. Baggett *et al.* also show that for two different angular orientations of the same fibre, the geometry of the cladding structure has a noticeable effect on the bend loss characteristics. Hence, in order to understand and predict this type of detailed results, it is necessary to use a numerical method that accounts for the complex structure, since effective index-methods cannot account for orientationally dependent behaviour.

These loss properties becomes important with respect to the optimal design of LMA-PCFs, and we will look closer at some of the further options for realising higher performance in this class of optical fibres in the following.

7.2.2 Key parameters in describing large-mode-area photonic crystal fibres

Before we discuss at some of the most recent advances in practically implemented LMA-PCFs, it will be advantageous to look a little closer at the key parameters, which are used to describe optical fibres in general and large-mode-area PCFs in particular. Many of these parameters have been used regularly for years in connection with standard-fibre technology, and they are still highly relevant also in the case of PCFs.

One of the fundamental parameters in the description of optical fibres is the numerical aperture, which is given as

$$NA = \sin \theta_{\upsilon} \tag{7.1}$$

which may be defined [7.33] in the far-field limit ($z \rightarrow \infty$) from the half divergence angle θ_{υ} between the z-axis (see figure 7-3) and the υ-intensity point $r_{\upsilon}(z)$, i.e.,

$$\tan \theta_{\upsilon} = \lim_{z \rightarrow \infty} \frac{r_{\upsilon}(z)}{z} \tag{7.2}$$

with $r_{\upsilon}(z)$ determined from the far-field distribution ψ as

$$\frac{\left|\psi(2, r_{\perp} = r_{\upsilon})\right|^2}{\left|\psi(z, r_{\perp} = 0)\right|^2} = \upsilon \tag{7.3}$$

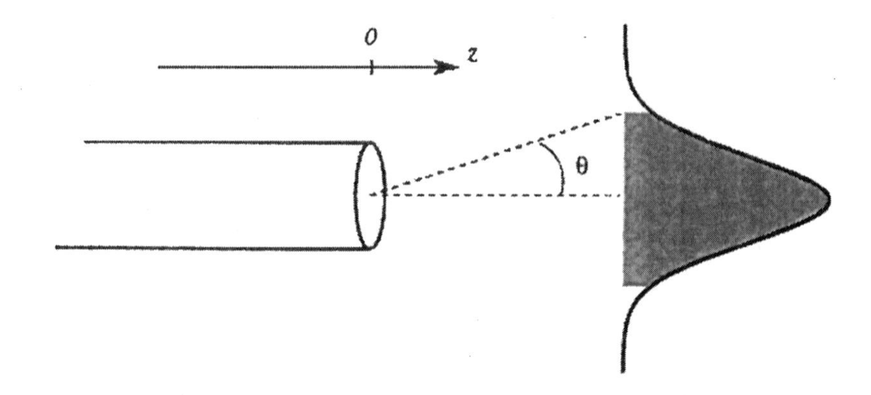

Figure 7-3. Schematic illustration of optical far-field from the end-face of an optical fibre.

For a Gaussian field ψ of width ω, the standard approximate expression for $\upsilon = 1/e^2 \cong 13.5\%$ is [7.34]

$$\tan\theta_{1/e^2} \cong \frac{2}{k\omega} = \frac{\lambda}{k\omega} \tag{7.4}$$

which is valid for $k\omega >> 1$. The spotsize value ω, is related to the effective area of the fibre as follows

$$A_{eff} = \pi\omega^2 \tag{7.5}$$

However, as pointed out by Mortensen *et al.* [7.33], the complex fibre structure of a PCF makes it necessary to consider the problem of radiation into free space from the end-facet exactly.

The problem of end-reflection at the fibre end is treated by Mortensen *et al.* [7.33] through approximate boundary conditions requiring continuity of the field ψ and its derivative $\partial\psi/\partial z$ in the length direction. At the fibre end-face, equations determining the reflection amplitude and the transmission amplitude may be formulated, and the following expression for the field radiating from the fibre end may be determined [7.33]

$$\psi_{>}(r) \propto 2\pi k^2 \int_0^\infty d\chi\chi \frac{2n_{eff}}{\sqrt{1-\chi^2}+n_{eff}} \cdot e^{-(\chi k\omega/2)^2} J_0(\chi k r_\perp) e^{i\sqrt{1-\chi^2}kz} \quad (7.6)$$

where $\chi = k_\perp / k$, J_0 is the Bessel function of first kind and order 0, and $n_{eff} = \beta / k$ is the effective mode index. Note that this field expression (7.6) represents the correct solution to the scattering problem including the small, but finite, backscattering in the fibre. Mortensen *et al.* [7.33] has found that eq. (7.6) describes a close-to-Gaussian field in the far-field limit.

We may, therefore, also with good approximation apply results from standard fibre technology relating numerical aperture and spotsize values in the case of the photonic crystal fibres, and we may also note that the effective area generally may be determined from numerical results of the magnetic field from the following expression [7.29].

$$A_{eff} = \frac{\left[\int d\bar{r} \left|\overline{H}(\bar{r},z)\right|^2\right]^2}{\int d\bar{r} \left|\overline{H}(\bar{r},z)\right|^4} \quad (7.7)$$

This definition of the effective area described by the magnetic field $\overline{H}$ is often used in connection with numerical determination of key fibre parameters, and together with eq. (7.5), which relates the effective area value to the mode-field diameter, they are central results in the description of LMA-PCFs.

7.2.3 New approaches and recent improvements of large-mode-area photonic crystal fibres

Since the publication of large-mode-area properties of photonic crystal fibres by Knight *et al.* in 1998 [7.26], a very significant number of improved designs have been developed and published. Also a number of studies of modal properties and effective areas have been presented over the past year (e.g., see the work of N. A. Mortensen *et al.* [7.35-7.36]). In this section, however, we will shortly present a few novel approaches towards the realization of large-mode-area PCFs.

In the search for alternative large-mode-area fibres, also other kinds of optical fibre structures have been proposed. One of the interesting possibilities is the so-called segmented-cladding fibre, which has been suggested by Chiang and Rastogi [7.7]. The fibre structure has been reproduced in Figure 7-4, and it may been seen as a different realisation of a microstructured fibre, which on the other side uses no air holes. The fibre structure presented by Chiang and Rastogi has - to the best of our knowledge - so far not been fabricated in long lengths, but numerical analysis indicate that a 50μm-core segmented-cladding fibre is capable of providing single-mode operation from 0.9 μm to 1.7 μm. These predictions has been made on the basis of a radial effective-index method and a finite-element method, and future research is expected to bring results on dispersion and attenuation properties of fibres having segmented-cladding designs.

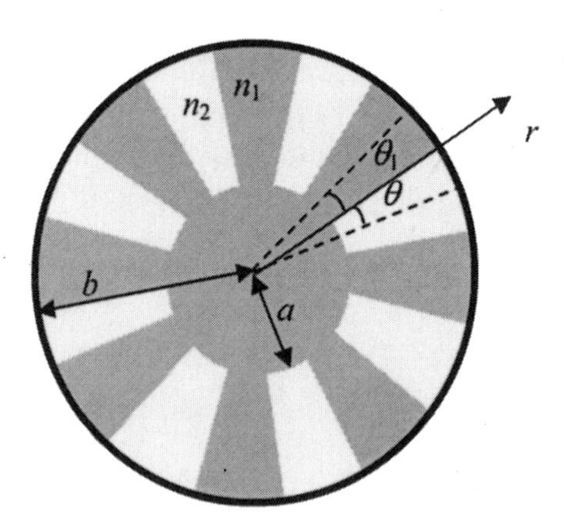

Figure 7-4. Segmented-cladding fibre with core radius a and cladding radius b. n_1 and n_2 are the refractive indices of the cladding segments, and $2\theta_1$ and $2\theta_2$ are the corresponding angular widths of these. The illustration has been drawn based on the outline published in [7.7].

Another alternative structure, which has recently been presented, is the so-called Fresnel fibre described by Canning *et al.* [7.38]. The Fresnel fibres are air-silica structured waveguides with air holes distributed along the effective Fresnel zones (see Figure 7-5), and they have been shown to support mode propagation at 1550 nm with peak intensity in the centre hole. It is stated in [7.38] that the Fresnel fibre has potential applications in reducing nonlinear and damage effects associated with high-peak-intensity light propagating in an optical waveguide, i.e., some of the characteristic properties of LMA fibres.

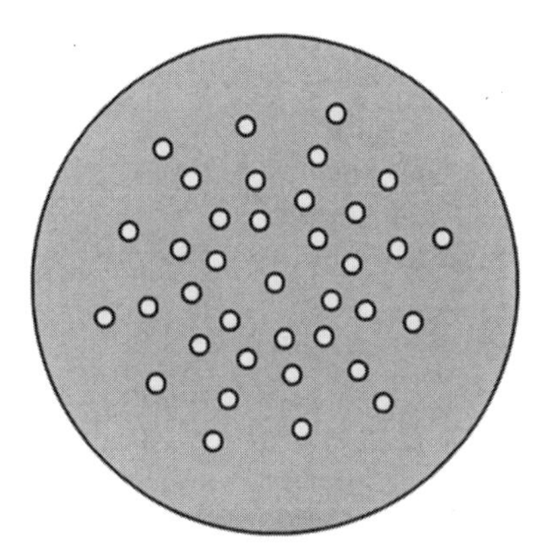

Figure 7-5. Schematic illustration of the cross section of a Fresnel fibre as presented by Canning *et al.* [7.38].

After now having looked briefly at these new design routes towards alternative implementations of LMA microstructured fibres, we will now return to some of the most recent improvements of LMA-PCFs, which have claddings consisting of a triangular array of air holes. Applying the normal terms d for hole diameter and Λ for pitch, the corresponding air-filling fraction of the triangular structured cladding is $f = \pi/\left(2\sqrt{3}\right)(d/\Lambda)^2$. The presence of the air holes results (as we have seen in previous chapters) in a strongly wavelength-dependent effective index n_{eff} of the cladding, and in the short- and long-wavelength limits we have, respectively:

$$\lim_{\lambda << \Lambda} n_{eff} = n_{si} \tag{7.8}$$

$$\lim_{\lambda >> \Lambda} n_{eff} = f n_{air} + (1-f) n_{si} \equiv \bar{n} \tag{7.9}$$

Numerical results calculated for the intermediate regime can according to Mortensen et al. [7.39] be reasonably fitted by the following expression

$$n_{eff} \approx \bar{n} + \left(n_{si} - \bar{n}\right)\cosh^{-2}\left(\frac{\alpha\lambda}{\Lambda}\right) \tag{7.10}$$

where α has order unity and is only weakly dependent on (d/Λ). The following α-values may be used with good approximation for different relative hole dimensions [7.39] (d/Λ=0.30, α=1.40; d/Λ=0.35, α=1.30;

d/Λ=0.40, α=1.27; d/Λ=0.45, α=1.25; d/Λ=0.50, α=1.24) It is these unusual dispersion properties of the cladding that facilitate the design of large-mode-area endlessly single-mode photonic crystal fibres.

As we have seen in Section 7.2.1, the normal LMA-PCF confine the light to a core region formed by a high-index defect placed in the triangular air-hole array. The normal approach is to leave out one of the air holes, and in the stack-and-pull fabrication method this means that one of the capillaries is replaced by a silica rod. We may, therefore, in short identify this fibre design by the term "single-rod" PCF. We have already in Chapter 5 seen, how such single-rod PCFs may be kept endlessly single mode no matter how large the core diameter. However, when the fibre structure is scaled up, the mode area is increased, at the cost of increased susceptibility to longitudinal modulations such as scattering loss induced by e.g., microbending and macrobending. The reason for this behaviour is that in order to increase the mode area, one must scale the pitch Λ to a large value, but this also implies that $\lambda/\Lambda \ll 1$, and in this limit, the core index approaches the cladding index. The typical approach to counterbalance the decreasing effective index step is to increase the air-hole diameter, which can be done up to $d/\Lambda = 0.45$ – the upper limit for endlessly single-mode operation. For LMA-PCFs working in the ultra-violet and visible regimes, this sets an upper limit on the mode areas that can be realised with a reasonable loss. A solution to this kind of problem is outline by Mortensen *et al.* [7.39], who demonstrate, how inclusion of more neighbouring solid rods can be used for improved endlessly single-mode LMA-PCFs. The resulting fibre described in [7.39] has a triangular core formed by three neighbouring rods, and the resulting structure is shown in Figure 7-6. It should be noted that this structure has an upper limit of $d/\Lambda = 0.25$ for endlessly single-mode operation.

Nielsen *et al.* [7.40] presented experimental results for the fibre illustrated in Figure 7-6 at the OFC′2003 conference. For the fabricated fibre, a hole size of 1.84 μm and a pitch of 6.84 μm was chosen, which yielded a relative hole dimension of $d/\Lambda = 0.27$. Although this d/Λ value is slightly above the predicted value for single-mode operation, only the fundamental mode could be excited regardless of wavelength down to a wavelength of 635 nm. Measured spectral attenuation further showed that a single-rod PCF with a mode-field diameter (MFD) of 10.5 μm showed a longer wavelength macrobending edge than a 3-rod PCF with a MFD of 11.6 μm, i.e., the 3-rod fibre is more bending insensitive despite the fact that it has a larger MFD.

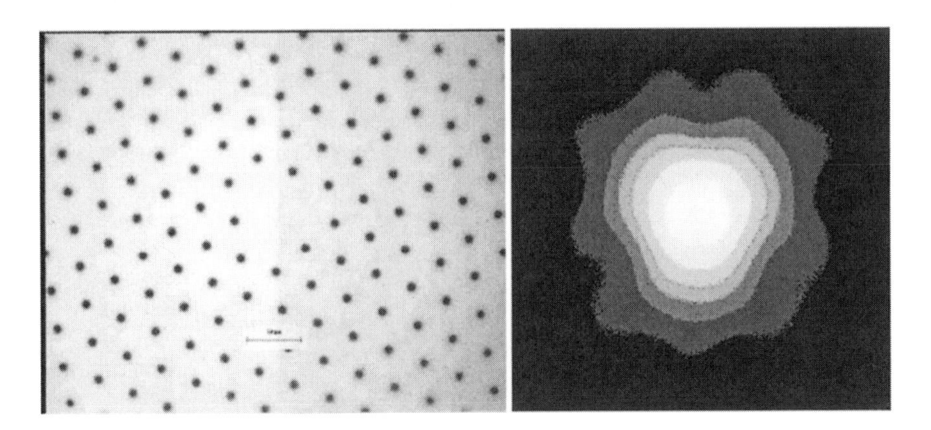

Figure 7-6. End facet of a fabricated 3-rod LMA-PCF (d = 1.8 μm and Λ = 6.8 μm) (left), and near-field of fundamental mode (right). Photograph kindly provided by Crystal Fibre A/S.

The new fibre designs, of which we have seen some examples in this section, allows for additional design flexibility, while providing enhanced effective mode areas in single-mode optical fibres. It is - as already mentioned - difficult to provide typical parameters, which will be valid for a long period of time to come. However, in order to give an impression of the present-day technological level, a few key parameters for commercially available LMA-PCFs are shown in Table 7-1.

Parameter	LMA-15 (Crystal Fibre A/S)	LMA-20 (Crystal Fibre A/S)
Effective mode area, A_{eff}	75 μm^2	150 μm^2
Core diameter	15 μm	20 μm
Cladding diameter	172 μm	230 μm
Mode field diameter	10.0 μm (@ 780 nm)	15 μm (@ 1550 nm)
Numerical aperture	0.05 (@ 780 nm)	0.08 (@ 1550 nm)
Dispersion @ 1550 nm	27 ps/(km·nm)	23 ps/(km·nm)
Dispersion slope @ 1550 nm	0.07 ps/(km·nm^2)	0.06 ps/(km·nm^2)
Attenuation	< 0.02 dB/m (@ 800 nm)	< 0.004 dB/m (@ 1550 nm)
Cutoff wavelength	None	None

Table 7-1. Key parameters of commercially available large-mode-area photonic crystal fibres.

It should also be noted that recent studies of polarization properties of single-moded LMA-PCFs by Niemi *et al.* [7.37] have demonstrated that although the polarization-mode coupling was fairly large in the tested fibres, they exhibit a greatly reduced local birefringence and polarization mode dispersion (PMD) indicating an excellent symmetry in the cladding hole structure and stress distribution. Niemi *et al.* [7.37] has reported PMD values in the order of 5-10 fs measured at wavelengths from 1455 nm to 1612 nm for a 15 micron core PCF.

7.3 HIGHLY NON-LINEAR PHOTONIC CRYSTAL FIBRES

Of the large variety of photonic crystal fibres, the highly nonlinear PCF (HNL-PCF) is today the most commonly used type. Their use are within a wide field of applications ranging from spectroscopy and sensor applications to the directly telecom oriented. The high nonlinear coefficient and designable dispersion properties makes these fibers attractive for many nonlinear applications of which supercontinuum generation has been the most intensively investigated [7.8-7.10]. The continua have been used in applications like optical coherence tomography [7.11], spectroscopy, and metrology [7.12]. Supercontinua covering several octaves as well as multi-watt output have been demonstrated [7.13]. Considerable effort has been made to develop better understanding of the complex interplay of nonlinear processes behind supercontinuum generation and many of the basic mechanisms (e.g., soliton fission [7.14-7.15], self-phase modulation [7.9], four-wave-mixing and stimulated Raman scattering [7.9]) are today understood.

Highly nonlinear fibres with zero-dispersion at the wavelength of 1.55 μm have long been pursued, as these fibres are very attractive for a range of telecom applications such as 2R Regeneration [7.17], multiple clock recovery [7.18], parametric amplifiers (OPAs) [7.19], pulse compression [7.20], wavelength conversion [7.21], all-optical switching [7.22] and supercontinuum-based WDM telecom sources [7.23]. Recently, Hansen *et al.* [7.24] demonstrated a highly nonlinear photonic crystal fibre with zero-dispersion wavelength at 1.55 μm and a nonlinear coefficient of 18 $(W{\cdot}km)^{-1}$ - two times that for standard nonlinear fibre. As we will discuss later in this section, the fibre has been utilized in an all-optical nonlinear optical loop mirror, demultiplexing a bit-stream of 160 GB/s down to 10 Gb/s. This was achieved with only 50 m of fibre, compared to the 2.5 km of dispersion shifted standard fibre normally required [7.25-7.26].

New production processes, where the fibre preforms are extruded in soft lead-glass has also been demonstrated for production of highly nonlinear fibers [7.28].

In this section, we will look closer at some of the properties, which make the nonlinear PCF one of the most interesting waveguide components in present day research.

7.3.1 Design considerations of highly nonlinear PCFs

Among the key elements in the potential of highly nonlinear PCFs is found that even modest optical powers can induce significant nonlinear effects. The interesting fact is that even though silica is not intrinsically a highly nonlinear material, its nonlinear properties can be utilized in silica optical fibres, if high light intensities are guided within the core. Such high light intensities become possible, because of the significant index contrast between silica core and air-filled cladding. This property results in the accessibility of high effective nonlinearity per unit length, and it hereby becomes feasible to reduce device lengths and the associated optical power requirements for fibre-based nonlinear devices.

One commonly used measure of fibre nonlinearity is the effective nonlinearity γ, [7.29] which is given by:

$$\gamma = \frac{2\pi n_2}{\lambda A_{eff}} \qquad (7.11)$$

where n_2 is the nonlinear coefficient of the material ($n_2 \approx 2.2\times10^{-20} m^2/W$ for pure silica), A_{eff} is the effective mode area discussed in section 7.2, and λ is the free space optical wavelength. A typical value for comparison with other fibres may be derived for a standard Corning SMF28 fibre, which has a nonlinear coefficient of $\gamma \approx 1\ W^{-1}km^{-1}$. By modifying conventional fibre designs, values of γ as large as 20 $W^{-1}km^{-1}$ have been achieved [7.30]. This is done by reducing the diameter of the fibre core, and using high germanium concentrations within the core, which both increases the numerical aperture (NA) and enhances the intrinsic nonlinearity (n_2) of the material. Both modifications act to confine light more tightly within the fibre core, and thus increase the nonlinearity γ by reducing the effective mode area A_{eff}. However, the maximum NA that can be achieved by conventional means limits the obtainable nonlinearity of conventional fibre designs.

Index-guiding photonic crystal fibres can have a significantly larger NA than conventional solid fibre types, because the cladding region can be mostly comprised of air. In particular, using PCFs with small-scale cladding features and large air-filling fractions (i.e., large values of d/Λ), light can be confined extremely tightly within the core, thus resulting in small mode areas, and large values of γ. An example of a highly nonlinear silica photonic crystal fibre is shown in Figure 7-7. Effective mode areas as small as $A_{eff} \sim 2.8\mu m^2$ have been demonstrated at 1550 nm corresponding to $\gamma \sim 35\ W^{-1}km^{-1}$ [7.31]. This is one of the best result in terms of nonlinearity reported to date in a pure silica fibre. Pure silica PCFs can be designed to have A_{eff} at least as small as 1.7μm^2 at 1550 nm [7.32]. Hence, nonlinearities as high as $\gamma \sim 52$ $W^{-1}km^{-1}$ are practical in these fibres, which is more than 50 times higher than in standard telecommunications fibre and 2 times higher than for the large-NA conventional designs described above. This emerging class of waveguides offers an attractive new route towards efficient, compact fibre-based nonlinear devices. We will, therefore, look closer at the optical properties of these small-core fibre designs, and present an overview of some of the emerging device applications of this new class of fibres.

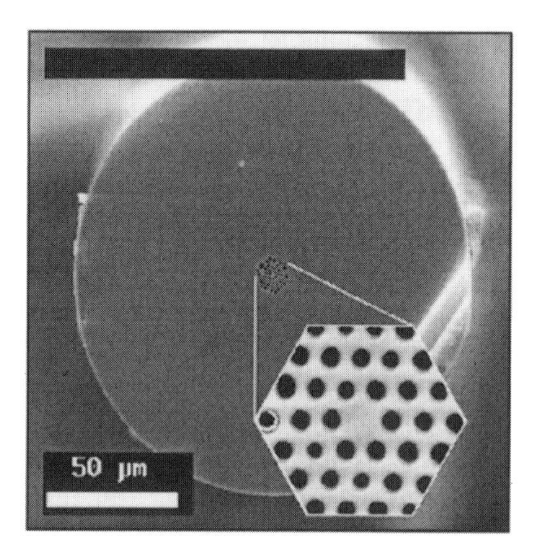

Figure 7-7. A typical highly nonlinear silica photonic crystal fibre. In this case, a small core diameter combined with a large air-filling fraction results in a fibre that confines light tightly within the solid central core region. Photograph kindly provided by Crystal Fibre A/S.

7.3.2 Dispersion management in highly-nonlinear photonic crystal fibres

Small-core PCFs can also exhibit a range of novel dispersive properties of relevance for nonlinear applications [7.41]. By modifying the fibre profile, it is possible to tailor both the magnitude and the sign of the dispersion to suit a range of device applications. They can exhibit anomalous dispersion down to a wavelength of 550 nm [7.42], which has made soliton generation in the near-IR and visible spectrum possible for the first time. An application of this regime was reported in [7.43], in which the soliton self frequency shift in an ytterbium-doped PCF amplifier was used as the basis for a femto-second pulse source tunable from 1.06 μm to 1.33 μm. Shifting the zero-dispersion wavelength to regimes, where there are convenient sources, also allows the development of efficient super-continuum sources [7.8], which are attractive for DWDM transmitters, pulse compression and the definition of precise frequency standards. It is also possible to design nonlinear PCFs with normal dispersion at 1550nm [7.45]. Fibres with low values of normal dispersion are advantageous for optical thresholding devices, since normal dispersion reduces the impact of coherence degradation [7.46] in a nonlinear fibre device.

In this connection, it is relevant to refer to the recent results of Hansen *et al.* [7.24], in which a combined low dispersion and high nonlinear coefficient was obtained through the application of an advanced PCF design. These results, which were presented in the beginning of 2002, represent a significant step forward with respect to simultaneous control of nonlinearity and near-zero dispersion at wavelengths relevant for telecom applications. The fibre developed by Hansen *et al.* [7.24] has a cross section as shown in Figure 7-8.

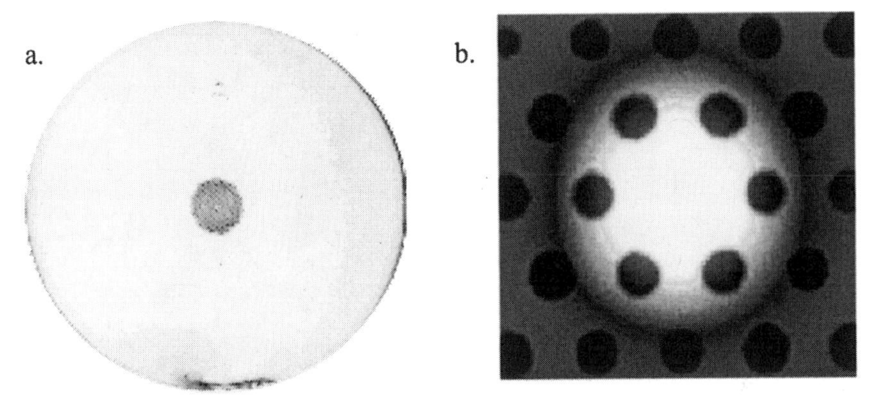

Figure 7-8. Cross section of highly nonlinear silica PCF having a zero-dispersion wavelength in the 1550 nm wavelength region. (a) The full fibre cross section. (b) Close-up on the central part of the fibre core region with near-field distribution of the optical mode superimposed. Illustrations are kindly provided by Crystal Fibre A/S.

A number of fibres similar to the one having a cross-section as the one shown in Figure 7-8 were drawn by Hansen *et al.*, and the resulting dispersion curves are shown in Figure 7-9. It is noted that zero-dispersion wavelengths in the range from 1552 nm to 1561 nm was measured, and the fabricated fibres had nonlinear coefficients around $\gamma = 10\ W^{-1}km^{-1}$.

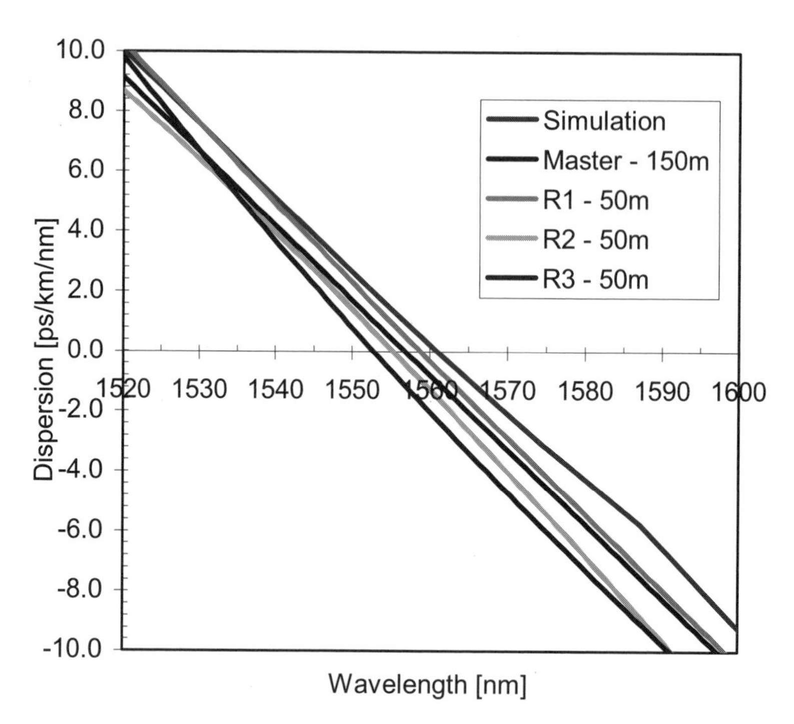

Figure 7-9. Dispersion as function of wavelength determined for different draws of highly nonlinear photonic crystal fibres having fibre cross sections as shown in Figure 7-8. The illustration is kindly provided by Crystal Fibre A/S.

Small-core PCFs pose a number of challenges for effective modelling. The high index contrast inherent in these fibres generally necessitates the use of a full-vector method. In addition, any asymmetries or imperfections in the fibre profile, when combined with this large index contrast and the small structure-scale can lead to significant geometrical (or form) birefringence. Consider, for example, the fibre shown in Figure 7-10, which is highly birefringent due to the elliptical shape of the core. In general, even small asymmetries can led to noticeable birefringence for these small-core fibres. Hence it is often necessary to use the detailed fibre profile in order to make accurate predictions of the optical properties.

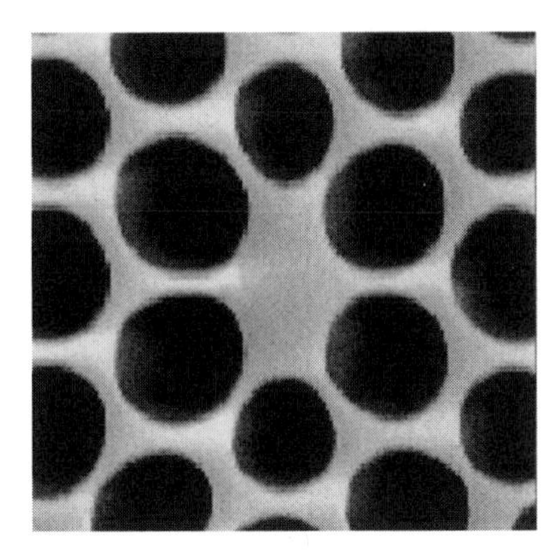

Figure 7-10. Cross section of polarization maintaining fibre. Photograph kindly provided by Crystal Fibre A/S.

Fibre losses as low as 1 dB/km have now been demonstrated in pure silica PCFs [7.48]. However, when the core diameter is reduced to scales comparable to (or less than) the wavelength of light guided within the fibre, confinement loss arising from the leaky nature of the modes can contribute significantly to the overall fibre loss [7.32]. Indeed, the small-core PCFs fabricated to date are, typically, more lossy than their larger-core counterparts. In this small-core regime, unless many (in other words more than 6) rings of holes are used, the mode can leak into the finite cladding region. Thus, it is important to be able to reliably calculate the confinement loss characteristics of PCFs, and the multipole technique [7.49] has been demonstrated to be a useful tool for exploring the impact of confinement loss in PCFs. Here, we briefly outline some general design rules for designing low-loss high nonlinearity PCFs [7.32].

To get an idea of the range of effective mode areas that can be achieved using silica glass at 1550 nm, one may consider the extreme case of a rod of diameter Λ suspended in air. As the diameter of the rod is reduced, the mode becomes more confined, and the effective mode area decreases. Once the core size becomes significantly smaller than the optical wavelength, the rod becomes too small to confine the light well and the mode broadens again. Consequently, there is a minimum effective mode area that, for a given wavelength, depends only on the refractive index of the rod. For silica, this minimum effective mode area is $1.48\mu m^2$ [7.50]. Note that PCF structures also exhibit a minimum effective mode area due to the same mechanism described. For a PCF with a triangular arrangement of cladding holes with the largest air-filling fraction (corresponding to $d/\Lambda=0.9$), a minimum effective mode area of $1.7\mu m^2$ is found, and this is only slightly larger than the value found for the air-suspended rod.

The hole-to-hole spacing (Λ) can be chosen to minimize the value of the effective mode area, and this is true regardless of the air-filling fraction. However, it is not always desirable to use the structures with the smallest effective mode area, because they typically exhibit higher confinement losses [7.32]. A relatively modest increase in the structure scale in this small-core regime can lead to dramatic improvements in the confinement of the mode without compromising the achievable effective nonlinearity significantly [7.50].

In addition, relatively lower confinement loss and tighter mode confinement can always be achieved by using fibre designs with larger air-filling fractions. Note that by adding more rings of holes to the fibre cladding, it is always possible to reduce the confinement loss. In the limit of core dimensions that are much smaller than the wavelength guided by the fibre, many rings (i.e., more than 6) are required to ensure low-loss operation, which increases the complexity of the fabrication process. With careful design, it is possible to envisage practical PCFs with small core areas ($< 2\mu m^2$) and low confinement loss (< 0.2dB/km) [7.50]. Note that although the fibre loss limits the effective length of any nonlinear device, for highly nonlinear fibres, short lengths (< 10 m) are typically required, and so loss values of the order of 1 dB/km or even much more can be readily tolerated.

7.3.3 Supercontinuum generation in silica-based index-guiding photonic crystal fibres

The high nonlinear coefficient and designable dispersion properties of index-guiding photonic crystal fibres makes them attractive for many nonlinear applications of which supercontinuum generation has been the most intensively investigated [7.8-7.10].

Before we look closer at some of the obtained results, we will for a moment take a view at the historic development concerning supercontinuum generation as outlined by Nicolov *et al.* [7.51]. If we go back to 1970, and look at the first observation of a 200-THz supercontinuum (SC) spectrum of light in bulk glass by Alfano *et al.* [7.52-7.53], it is clear that much has been done on the understanding and control of this process [7.54]. A target of numerous experimental and theoretical investigations has been the improvement of the characteristics and simplification of the technical requirements for the generation of a SC [7.54]. The first experiments in bulk materials, based on self-phase modulation (SPM), required extremely high peak powers (> 10MW). New techniques based on the use of optical fibres as a nonlinear medium for SC generation allowed lower peak powers to be used due to the long interaction length and high effective nonlinearity

[7.9,7.55-7.57]. However, the necessity to operate near the wavelength of zero group velocity dispersion, restricted the SC generation to the spectral region around and above 1.3 μm. The use of dispersion-flattened or dispersion-decreasing fibres as nonlinear media for SC generation resulted in a flat SC spanning 1400-1700 nm [7.58-7.59] and 1450-1650 nm [7.60], respectively. The spectrum was still far from the visible wavelengths and in some cases very sensitive to noise in the input [7.60]. Photonic crystal fibers (PCF) and tapered fibres may overcome these limitations. The unusual dispersion properties and enlarged effective nonlinearities make them a promising tool for effective SC generation [7.9]. PCFs and tapered fibres have similar dispersion and nonlinearity characteristics and they have the advantage that their dispersion may be significantly modified by a proper design of the cladding structure [7.24, 7.61- 762], or by changing the degree of tapering [7.8], respectively. Using kilowatt peak-power-femtosecond pulses, a SC spanning 400-1500 nm has been generated in a PCF [7.8] and in a tapered fibre [7.63]. The broad SC was later explained to be a result of SPM and direct degenerate four-wave-mixing (FWM) [7.64]. However, high- power femtosecond lasers are not necessary, since SC generation may be achieved with low-power picosecond [7.9, 7.57] and even nanosecond [7.65] pulses. Thus Coen *et al.* generated a broad SC in a PCF using picosecond pulses with sub-kilowatt peak power, and they showed that the primary mechanism was the combined effect of stimulated Raman scattering (SRS) and parametric FWM, allowing the Raman shifted components to interact efficiently with the pump [7.9]. Using 200 fs high-power pulses and a 1 cm long tapered fibre, Gusakov has shown that direct degenerate FWM can lead to ultrawide spectral broadening and pulse compression [7.64]. Nicolov *et al.* [7.51] numerically shows, how the direct degenerate FWM can significantly improve the efficiency of SC generation with sub-kilowatt picosecond pulses in PCFs, and they go one step further by optimising the effect through engineering of the dispersion properties of the PCF. It is shown that by a proper design of the dispersion profile, the direct degenerate FWM Stokes and anti-Stokes lines can be shifted closer to the pump, thereby allowing them to broaden and merge with the pump to form an ultra-broad SC. This significantly improves the efficiency of the SC generation, since the power in the Stokes and anti-Stokes lines no longer is lost.

The continua from PCFs have been used in applications like optical coherence tomography [7.11], spectroscopy, and metrology [7.12]. Supercontinua covering several octaves as well as multi-watt output have been demonstrated [7.13]. Considerable effort has been made to develop better understanding of the complex interplay of non-linear processes behind supercontinuum generation, and many of the basic mechanisms (e.g., soliton fission [7.14-7.15], self-phase modulation [7.16], four-wave-mixing and stimulated Raman scattering [7.9]) are today understood.

An example of the possibilities of supercontinuum generation in index-guiding photonic crystal fibres is presented in Figure 7-11 showing an octave-spanning spectrum broadening.

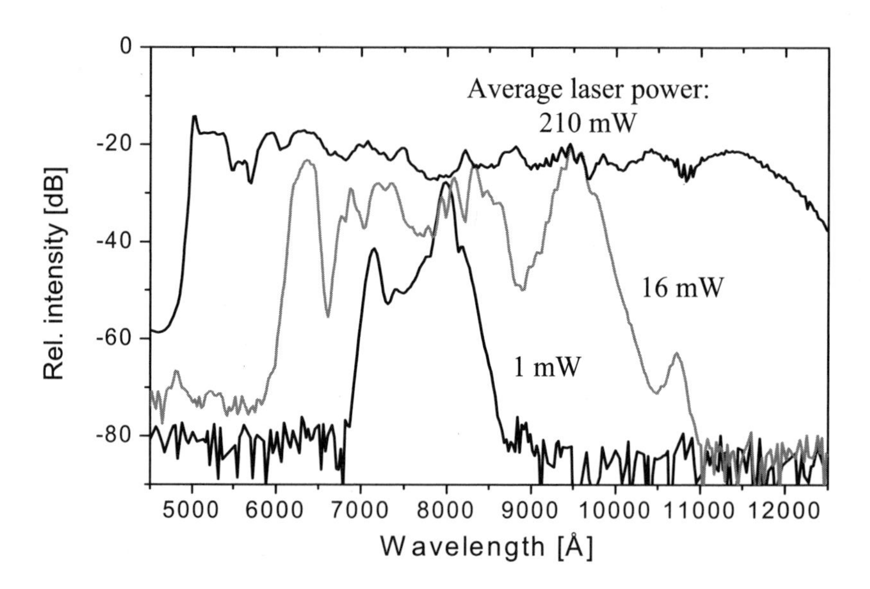

Figure 7-11. Octave-spanning spectrum broadening is made possible with highly nonlinear photonic crystal fibres. This figure illustrates the output from a 50 cm, 2 micron core fibre pumped by a 50 fs 800 nm Ti:Sapphire laser. The repetition rate of the pump laser is 76 MHz. The single mode fibre's zero-dispersion wavelength is 760 nm. The experiment was done by Jacob Juul Larsen, Aarhus University/NKT Research and the fibres were provided by Crystal Fibre A/S.

Finally, note that several groups presently are working on detailed theoretical and numerical studies of non-linear properties of index-guiding PCFs. Among the most interesting recent results is the work by Ferrando *et al.* [7.66], which demonstrates self-trapped localized modes in photonic crystal fibres. In this work, the confinement properties of nonlinearities within the fibres are taken into account.

7.3.4 Device demonstrations

A number of device demonstrations have been performed using highly non-linear silica photonic crystal fibres as reviewed by Monro *et al.* [7.50]. A highly relevant example is 2R data regeneration, which is a function that is a crucial element in any optical network, because it allows a noisy stream of data to be regenerated optically. The first demonstration of regeneration used a silica PCFs with an effective mode area A_{eff} of just 2.8 μm^2 [$\gamma = 35$ $W^{-1}km^{-1}$] at 1550 nm [7.17]. Typically, devices based on conventional fibres are ~1 km long, whereas in these early experiments, just 3.3 m of PCF were needed for an operating power of 15 W. Subsequent experiments used a 8.7 m long variant of this switch for data regeneration within an optical code division multiple access (OCDMA) system [7.21]. Significant improvements in system performance were obtained in this way.

Fibres with a high effective nonlinearity also offer length/power advantages for devices based on other processes such as Brillouin and Raman effects. The demand for increased optical bandwidth in telecommunications systems has generated enormous interest in the short wavelength (S) and long wavelength (L) bands, outside the gain band of conventional erbium-doped fibre amplifiers. Fibre amplifiers based on Raman effects offer an attractive route to extending the range of accessible amplification bands. In addition, the fast response time (< 10 fs) of the Raman effect can also be used for all-optical ultrafast signal processing applications. Despite these attractions, there is one significant drawback to Raman devices based on conventional fibres: long lengths (~10 km) are, generally, required, and so Rayleigh scattering ultimately limits their performance. Highly nonlinear fibres offer a method for obtaining sufficient Raman gain in a short fibre length, which eliminates this problem.

For example, Yusoff *et al.* [7.67] demonstrated a 70 m fibre-laser-pumped Raman amplifier. The amplifier was pumped using a pulsed fibre laser and provided gains of up to 43 dB in the L+ band for peak powers of ~7 W.

Other nonlinear device applications of PCFs that have been demonstrated include a CW Raman laser [7.68] and a WDM wavelength converter [7.21]. The CW Raman laser was pumped at 1080 nm using a high-power, cladding-pumped Yb-doped fibre laser. The laser had a CW threshold of 5 W, and slope efficiency of 70%. Note that in this experiment, the CW power density at the fibre-end facet (0.2 GW/cm^2) demonstrates that PCFs can exhibit a good resilience to damage.

7.3.5 Comparison of key parameters of highly nonlinear fibres

The nonlinear fibres have numerous properties, which makes their comparison rather complicated. We will here draw some of the most significant parameters forward and in Table 7-2, we have listed a number of values from recently published papers, which may provide some overall indications of typical fibre properties.

A very important parameter with respect to not only the nonlinearity, but also the coupling-loss properties is the effective area of the fibre as defined in Eq. (7.7). Table 7-2 shows values of A_{eff} ranging from 2.6 μm^2 for lead-silicate fibres [7.93] to 54 μm^2 for standard dispersion compensating fibres. The nonlinear coefficient (here defined according to Eq. (7.11)) is naturally of major importance, and values ranging from below 1 $W^{-1}km^{-1}$ to several hundred has been reported over the past few years. Note here that the operation wavelength should be considered, when these numbers are compared, because of the (often very strong) wavelength dependency of especially the effective mode area. A high nonlinear coefficient is naturally attractive, because it may allow the realisation of relatively short fibre devices. However, for many practical applications, it is not only the magnitude of the nonlinearity, which is of importance, but equally relevant is the attenuation of the fibre. The reason is that a given nonlinear process depends strongly on the power at a given place in the fibre, and, consequently, a large attenuation will limit the potential use of the highly nonlinear fibres. It is noted that loss values may vary several orders of magnitude depending on the specific fibre design and base material, and that present day silica-based highly-nonlinear PCFs often have attenuations from 10 dB/km or higher. We will return to the relation between nonlinearity and loss shortly.

A very important property of highly nonlinear PCFs is the potential of making fibres with anomalous dispersion below about 1.3 μm, which is the lower limit for optical fibres manufactured by standard fibre technology. Table 7-2 also shows zero-dispersion wavelengths for a number of the reported fibres, and values from below 700 nm to above 1550 nm has been realised. It should, furthermore, be noted that the present table does not provide a full picture of the potential advantages of PCFs concerning a highly controlled dispersion design, while maintaining a relatively high nonlinearity, as described by Hansen *et al.* [7.92]. For these and further details of the other reported fibres, the reader should consult the indicated references listed in Table 7-2 and in this chapter in general.

Fibre ID	A_{eff} (μm^2)	γ ($W^{-1}km^{-1}$)	α (dB/km)	λ_0 (nm)	Comment	Ref.
Silica PCFs						
2 ring structure Λ=1.8μm, d=1.5μm			80 ± 40 @850nm	740	6 modes supported @ 850nm	[7.86] [7.87]
Hexagonal CP 2a=1.7μm, d=1.3μm			50 @1μm	767	λ_c = 530 nm Bend insensitive to 1600 nm	[7.8]
Hexagonal CP	14	1.2 @ 850nm	240 @850nm		Brillouin spectra measured	[7.88]
Silica				1100	Broadband cont. spectr. 400-1850 nm	[7.89]
1 ring air holes 2a=1.7μm d=1.4μm				710-950		[7.44]
Silica		0.15 @ 647 nm	600 @647nm	677	600nm wide supercont. gen.	[7.90]
2a=1.5μm	3 @ 1060nm				Nonlin. pulse compr. and soliton prop.	[7.91]
Triangular-core		11.2	9.9 @1550nm	1550	Disp.slope 10^{-3} ps/(km nm^2) Splice loss 0.25dB to SIF	[7.92]
2a=1.6 μm	2.85	32 @1535nm	40		Raman amplifier, All-opt.modulator	[7.94]
Lead silicatePCF						
Lead silicate	2.6	640 @1550nm	9000 @1550nm		10-fold pulse compr., Raman soliton generation +80 ps/(km nm)	[7.93]
Silica - Standard fibres						
HN-fibre	12	10 @1544nm	0.7	1559	Disp.slope 0.03 ps/(km nm^2)	[7.99]
Standard DCF	54	1.75 @1550nm	0.5 @1550nm			[7.88]

Table 7-2. Key parameters of highly nonlinear photonic crystal fibres. Abbreviations: Close-packed (CP), Core diameter: 2a.

Before we leave the subject of highly-nonlinear photonic crystal fibres, it may be useful to take a further look at the relation between the already mentioned key parameters. As mentioned in connection with the discussion of Table 7-2, it is not only an issue to obtain a nonlinear coefficient as high as possible, but the attenuation also becomes important. This may be expressed through the definition of a figure-of-merit (FOM) for nonlinear fibres as follows:

$$FOM = \frac{\gamma}{\alpha} \tag{7.12}$$

where the nonlinear coefficient is defined as in Eq. (7.11) and α is the attenuation measured at the same wavelength as the nonlinear coefficient.
In order to illustrate the large span of values that may be found, if we start to look closer at these FOM values, we have in Figure 7-12 shown the calculated FOM for nonlinear fibres as function of their operation wavelength (a wavelength selected to be near or at the zero-dispersion wavelength). We first note that the dashed vertical line indicates the lower limit for the zero-dispersion wavelength for standard optical fibres, so the only results placed at the left side of this line relates to PCFs. Furthermore, we note that the loss values become a quite serious limitation on the FOM value of the lead-silicate PCF at 1550 nm compared to the other fibres. If we further look at the results in this wavelength range, it is obvious that the standard fibres (here represented by a dispersion-compensating (CD) fibre and a highly nonlinear standard fibre) have the highest nonlinear FOM as defined in Eqn. (7.12). The highly nonlinear silica PCFs optimised for this wavelength interval does, however, provide a different dispersion profile compared to that obtainable by standard fibre technology, so the preferred choice is given by the specific application, and the requirements on nonlinear coefficient, length, loss, splicing properties, power resistance, dispersion, and dispersion slope, and may not solely be evaluated from data as shown in Figure 7-12. However, we still believe that it is useful to provide comparisons as the one shown in the figure, also to be able to evaluate the progress with time as improved fibre parameters are realised. An area, where a significant development is seen these days, is within the reduction of fibre loss on the PCF products, so we must expect that the illustrated picture of Figure 7-12 will change significantly in the nearest future.

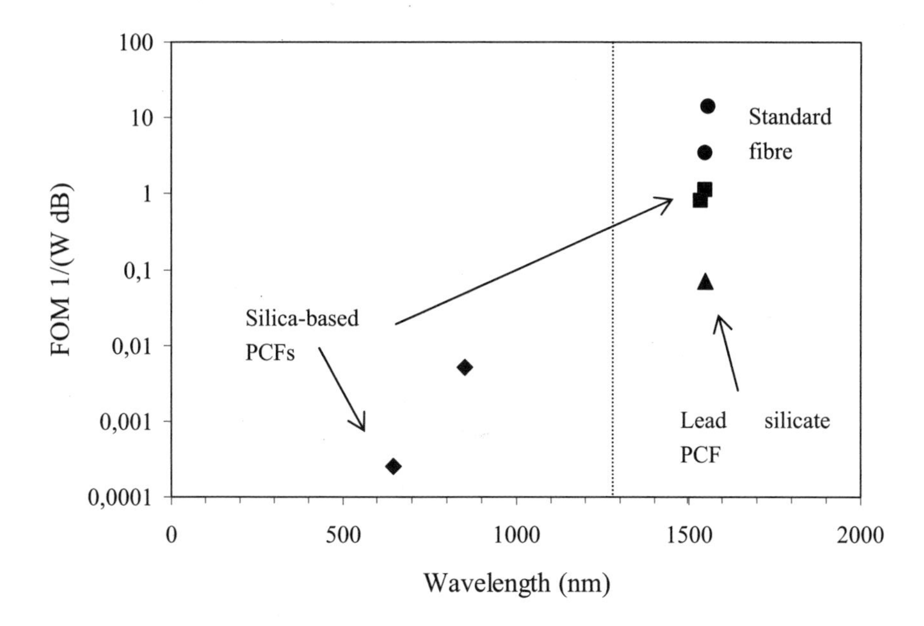

Figure 7-12. Figure-of-merit (FOM) for nonlinear fibres as defined in Eqn. (7.12) shown as function of operational (near-zero-dispersion) wavelength. The data are derived from fibre parameters presented in [7.88], [7.90], [7.92], [7.93], and [7.94].

7.4 HIGH NUMERICAL APERTURE FIBRES

The capillary-stacking techniques that are generally used to make LMA-PCFs and HNL-PCFs can be readily adapted to allow the incorporation of high numerical aperture (NA) air-clad inner claddings within jacketed all-glass structures [7.69]. This technique was first described by Digiovanni *et al.* [7.70], in a US-patent from May 1999. The advantage in this connection is the large index step between air and silica, which make microstructured fibres with extremely high numerical aperture values (compared to standard fibres) possible. Fibres with NAs as high as 0.9 have been demonstrated. High numerical aperture fibres (which typically are strongly multimode) collect light very efficiently from a very broad space angle and distribute light in a broad angle at the output end. These fibres could find use for pig-tailing broad-area-emitting lasers, for lighting applications such as windmill warning signals, endoscopy, optical sensors, and several other applications.

Examples of high-NA photonic crystal fibre structures are shown in Figure 7-13. The picture to the left illustrates the fibre cross section of a fibre with a 25 micron multimode core diameter, whereas the picture to the right shows a fibre with a larger core (a 150 micron multimode core diameter is formed).

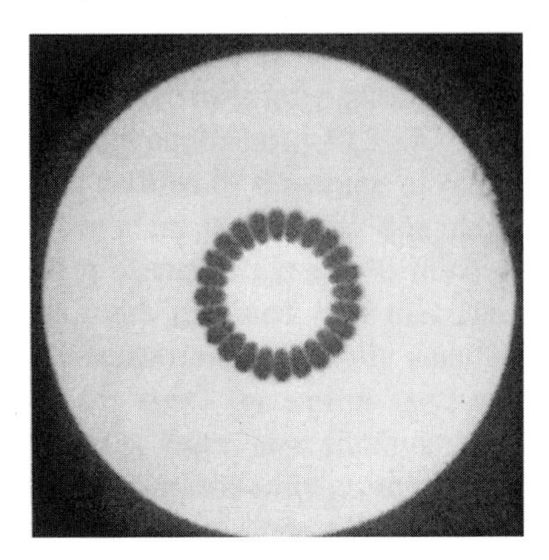
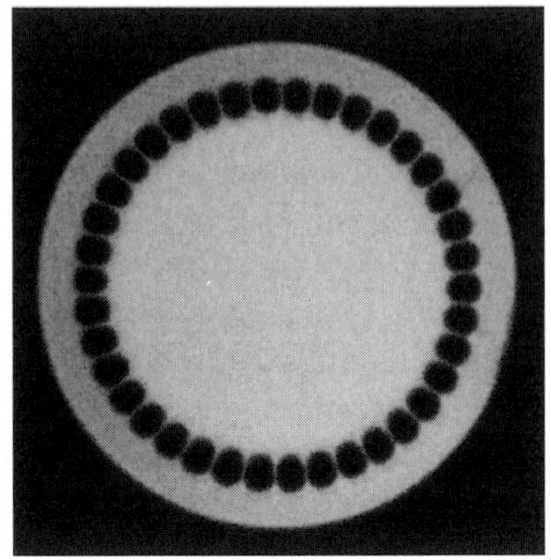

Figure 7-13. The microscope pictures shows cross-sections of pure silica fibres with high numerical aperture. The fibre to the left has a 25 micron multimode core diameter, whereas the fibre to the right has a 150 micron multimode core diameter. The numerical aperture is higher than 0.55. The photographs are kindly provided by Crystal Fibre A/S.

Another interesting issue concerning the potential possibilities of the use of air-ring isolation of the inner part of optical fibres was reported by Espindola *et al.* [7.72]. In this case, the focus was on the application of long-period gratings (LPGs) in optical fibres, which are used to couple power from the fundamental fibre mode into forward-propagating cladding modes. This allows for the construction of spectral filters and optical fibre sensors. However, a potential problem is that the LPG resonant wavelength and coupling strength may change dependent on the effective index and field distribution of the cladding modes, and this may be strongly influenced by thermal dependency of outer cladding materials. For this reason, Espindola *et al.* [7.72] suggested that insensitive LPGs could be fabricated using optical isolation of the inner parts of an optical fibre by the inclusion of a ring of air holes corresponding to the structures shown in Figure 7-13. A clear advantage of this approach is that the LPG can be recoated (after writing the grating structure), and it is insensitive to environmental or aging effects on the index of the coating.

7.5 PHOTONIC CRYSTAL FIBRE AMPLIFIERS

The potential for high-NA fibres is of predominant interest in the context of high power, rare-earth doped (e.g., Yb^{3+}, Nd^{3+}), LMA devices. The use of such rare-earth dopants in LMA PCFs is, however, challenging, because the presence of dopants and associated co-dopants (such as Germanium, Aluminium and Boron) that are required to incorporate the rare-earth ions at reasonable concentrations and to maintain laser efficiency modifies the refractive index of the host glass. This affects the NA of the fibre, and can lead to the loss of some of the most attractive LMA-PCF features, such as broadband single-mode guidance unless care is taken in the fibre designs.

PCF preforms are typically (at least for index-guiding fibres) fabricated by stacking capillaries around a solid rod, which ultimately forms the core. To produce active PCFs, it is thus necessary to produce a doped core-rod, which for example can be achieved by extracting the core region from a conventional doped fibre. We may for example look at the case described by Furusawa *et al.* [7.71], where the starting point was an alumina-silicate ytterbium-doped rod with an NA of 0.05 produced using conventional MCVD techniques. In addition, it is straightforward to adapt PCF fabrication technology to achieve an all-glass double-clad structure, which is advantageous for the efficient use of low brightness pump sources. In [7.71], a low index outer-cladding region is formed simply by inserting thin walled capillaries into the preform stack around the relatively thicker walled capillaries used to define the inner microstructured cladding region.

A schematic illustration of a fibre produced in such a manner is shown in Figure 7-14.

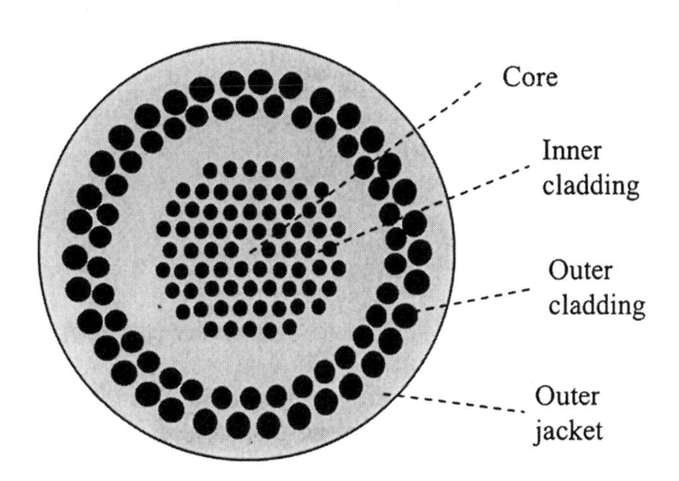

Figure 7-14. Schematic illustration of the cross-section of a double-clad microstructured optical fibre.

This kind of double-clad active fibre is very attractive for a number of reasons. First of all, the hole distribution and air-filling fraction in the inner cladding region may allow for a very significant flexibility concerning the spectral dependency of the effective refractive index. In this specific context, this may be used to match the refractive index of the core, which may be influenced by the inclusion of rare-earth materials and co dopants. This may allow for the design of structures that are single-mode at the signal wavelength. At the same time, high numerical-aperture values may be obtained through the use of large air-filling fractions in the outer cladding region. This permits an efficient coupling of pump power into the inner cladding region without having a structure in which the outer cladding is formed by a temperature sensitive coating material, such as it is the case in many double-clad fibres fabricated using standard technology. Among additional examples of flexibility is that an offset core may be used to break the symmetry of the cladding, and so to enhance the pump absorption

Both core and cladding-pumped lasers based on air-clad ytterbium-doped PCF were demonstrated by Furusawa *et al.* [7.71], and these results are briefly as follows:

A core-pumped LMA-PCF laser was demonstrated using a single-transverse-mode Ti:sapphire laser operating at 976 nm as a pump source. A conventional Fabry-Perot cavity configuration was formed by the 4% Fresnel reflection from the launch end of the fibre and a lens-coupled high reflector placed at the other end of the fibre. Slope efficiencies as high as 82% were reported, and these numbers were comparable to the best conventional ytterbium fibre lasers known at the time of publication [7.71]. Furthermore, the output beam was observed to be robustly single-mode.

A cladding-pumped LMA-PCF laser was formed from the same fibre. In this case, the pump was a low-brightness fibre-coupled laser diode at 915 nm. Using again a simple Fabry-Perot cavity with 4% Fresnel feedback, average powers in excess of 1 W were achieved using a 7.5 m long fibre with a measured slope efficiency of 70% [7.71]. It should be noted that as well as optically isolating the inner structure from the external environment, the air cladding also thermally isolates the laser. Although it might be imagined that this could lead to thermal problems under high-power operation, no such problems were reported in the experiments even at multi-Watt pump levels. Q-switching and mode-locking were also reported using this cladding-pumped LMA-PCF laser [7.71]. In the Q-switching experiment, a ~50 J stable pulse at repetition rates of a few kHz was obtained. The output pulse duration was ~10 ns, and the corresponding peak power was 5 kW. In the mode-locking experiments, fundamental mode-locking was obtained over a broad wavelength tuning range in excess of 60 nm. The pulse duration was estimated to be of order ~100 ps. An output power of more than 500 mW was achieved for a pump power of 1.33 W. Ultimately, it should be possible to develop compact ~ multi 10 nJ femtosecond pulse sources operating at 1 μm using LMA-PCFs in conjunction with the well established Kerr-nonlinearity-based stretched-pulse mode-locking technique.

The primary advantages of these forms of fibre relative to conventional polymer coated dual clad-fibres are that they will, ultimately, allow for all-glass structures, with larger inner cladding NAs (at least > 0.5) and good pump-mode mixing. In addition, they offer single-mode guidance in cores that are at least as large (but most likely larger) as those than can be made in conventional fibres. In device terms, these features will translate to advantages including, the possibility of higher coupled diode powers (for a given cladding dimension), shorter device lengths, and extended tuning ranges. The main drawback is likely to be thermal management, if truly-high power operation is required. These limits are, however, subjects for present and future investigations, and to our present knowledge, the thermal limits have not yet been reached.

An example of an advanced double-clad PCF structure is shown in Figure 7-15, in which a large-mode-area rare-earth-doped (RED) core is placed in a microstructured inner cladding. The effective refractive index is controlled by small air-holes in the inner cladding, and the outer cladding is formed by a ring of very large air holes providing a very well defined isolation of the inner cladding from the outside fibre sections.

Figure 7-15. A microscope picture of a double-clad rare-earth-doped photonic crystal fibre. The photograph was kindly provided by Jens Limpert, Friedrich Schiller University Jena, and similar results have been presented in [7.118].

7.6 TUNEABLE PHOTONIC CRYSTAL FIBRE COMPONENTS

We have already seen numerous examples on the fact that photonic crystal fibres provide large controllable index variations and, consequently, allow a wide range of guidance mechanisms. Most of the interest has been focussed on the guiding properties of the fundamental mode of the fibres, but we have also seen applications, where many cladding modes are used, e.g., such as the high NA-PCFs or the cladding-pumped fibre amplifiers. However, yet other fibre devices make use of properties of specific cladding modes, and among the most interesting are the fibres including longitudinal grating structures.

Some of the most fundamental work with respect to grating based PCFs has been done by Eggleton *et al.* [7.78 – 7.82], and it is relevant to note that in several publication by these authors, the term air-silica microstructured (ASM) optical fibres are used. In these applications, the air-regions are not necessarily involved in core mode guidance, but the micro structuring is designed to manipulate higher-order modes (HOM) that propagate predominantly in the cladding region of the fibre. Such modes are normally referred to as cladding modes. In the mentioned fibres, the fundamental mode guidance is achieved by changing the composition of the core, e.g., doping with germanium, which at the same time allows for the fabrication of photosensitive gratings. Eggleton *et al.*, has also outlined the possibility of introducing active materials into the air-hole regions yielding novel hybrid waveguide devices [7.80-7.82]. Among the most interesting of the presently known devices of this type are hybrid polymer-silica waveguides, including widely tuneable long-period grating (LPG) filters [7.81, 7.82] with on-fibre thin-film heaters. Eggleton *et al.* has demonstrated LPG resonances that are insensitive to external index values [7.79, 7.72], as well as a novel method for suppressing cladding mode loss in the fibre Bragg gratings (FBGs) [7.81, 7.82]. Some of these experiments, concerning grating resonances in ASM fibres [7.83], was done in the fibres having a Ge-doped photosensitive core and a close-packed (or triangular) distribution of air holes such as the fibre structures shown in Figure 7-2. Even more interesting (tuneable) devices has been developed by Eggleton *et al.* [7.84] on the so-called grapefruit fibre structure as schematically shown in Figure 7-16 (a) This grapefruit fibre also has a germanium-doped photosensitive core and due to the large air-holes in the cladding, the fibre cross section has an appearance close to that of a grapefruit. In Figure 7-16 (b), an index profile of a fibre of the grapefruit type with polymer-filled holes is shown. The interesting property is that, since the polymer refractive index is much more sensitive to temperature variations than that of silica, the effective index of the cladding modes will be modified as the temperature is tuned. The spectral location (and strength) of the coupling coefficient of the LPG written in the core of the fibre is hereby controlled, and Eggleton *et al.* [7.84] has shown that the transmission dip created by the grating may be tuned from 1425 nm to 1590 nm by changing the temperature of the fibre device from 25 ^{0}C to 120 ^{0}C.

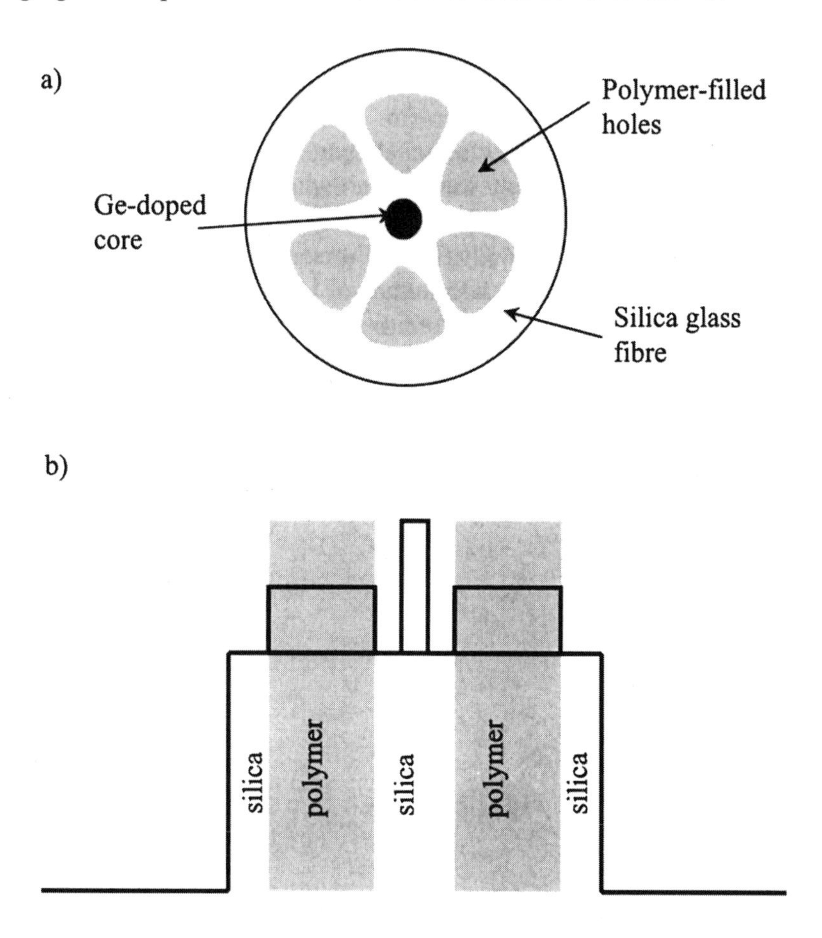

Figure 7-16. a) Schematic cross section of a grapefruit fibre as described by Eggleton *et al.* [7.84]. b) Index profile of grapefruit fibre having the air holes filled with temperature tuneable polymers.

7.7 HIGHLY BIREFRINGENT FIBRES

In the previous chapters (and examples presented in this chapter), it has been shown that the rod-and-tube stacking fabrication process allows for a large tailorability of the core geometries, but also highly asymmetric cores are obtainable by replacing one or more tubes by rods. The replacement of silica tubes with solid rods results in fibres having a high-index core region, which allow for light confinement through index guiding. One possible use of asymmetric-core fibres is as polarization-maintaining (PM) fibres. In standard fibre transmission systems, imperfections in the core-cladding interface introduce random birefringence that leads to light being randomly polarised (this is often referred to as polarization-mode dispersion (PMD)). These problems with random birefringence are in PM fibres overcome by deliberately introducing a larger uniform birefringence throughout the fibre. Current PM fibres realised by standard fibre technology include PANDA or bow-tie fibres [7.95], which achieve this goal by applying stress to the core region of a standard fibre. Thereby, a modal birefringence of up to $\Delta n \approx 5 \cdot 10^{-4}$ may be readily obtained [7.96]. However, as described by Hansen *et al.* [7.100], the intrinsically large index contrast in photonic crystal fibres in combination with asymmetric core designs may be used to create modal birefringence of at least one order of magnitude larger than for conventional PM fibres. It should be noted that Ortigosa-Blanch *et al.* [7.97] in 2000 reported a birefringence as high as $3.7 \cdot 10^{-3}$ in PM-PCFs.

Another interesting example of a PM-PCF fibre design was reported in 2001 by Suzuki *et al.* [7.101] leading to a fibre loss and modal birefringence at the wavelength of 1550 nm at 1.3 dB/km and $1.4 \cdot 10^{-3}$, respectively.

In one of the simplest design cases employing an asymmetric core PCF, the core consist of two neighbouring rods in a close-packed cladding structure of equally-sized air holes. This design was studied by Hansen *et al.* [7.100], and its cross section is shown in Figure 7-17. Note that the purpose of creating highly birefringent fibre is to reduce the coupling between the orthogonal states of the fundamental mode. This is mostly relevant, if no higher-order modes are supported. Triangular structured PCFs with a symmetric core consisting of one rod are endlessly single-mode for normalized hole sizes up to a value as large as $d/\Lambda = 0.45$ [7.98]. This is not the case for the fibre shown in Figure 7-17 due to the larger core area compared to the single-rod design.

Figure 7-17. Scanning electron micrograph of inner part of an asymmetric core photonic crystal fibre. The fibre cladding consists of a highly regular, triangular lattice of air holes with a pitch of 4.5 μm. The core is formed by the omission of two adjacent holes. The fibre has been reported by Hansen *et al.* [7.100], and the photograph is kindly provided by Crystal Fibre A/S.

The fibre structure shown in Figure 7-17 was simulated using the plane-wave method (see Chapter 3.4). For a relative hole size of $d/\Lambda = 0.40$, the design was found to support a second-order mode with a cutoff for a normalized frequency $\Lambda/\lambda = 1.67$. The calculated mode indices are shown as function of normalized frequency in Figure 7-18. The lifted degeneracy of the two fundamental modes may be noticed on the figure.

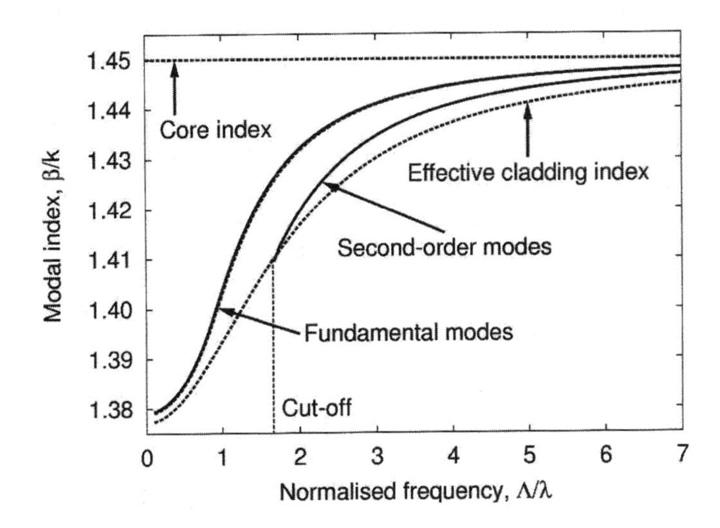

Figure 7-18. Modal index illustration of the operation of asymmetric core PCF. The fibre has an air-hole diameter of 0.4Λ, and it is seen to become multi-mode at normalized frequencies, $\Lambda/\lambda > 1.67$. The illustration is drawn from [7.100].

Hansen *et al.* [7.100] described, how the fibre structure shown in Figure 7-17, in the low-frequency limit, presented a normalized propagation constant tending to zero causing the field to extend far beyond the core region. The normalized propagation constant is given as:

$$B_{eff} = \frac{n_{eff} - n_{cl,eff}}{n_{co} - n_{cl,eff}} \tag{7.13}$$

where n_{co} is the refractive index of the core, equal to the index of fused silica, $n_{cl,eff}$ is equal to the index of the fundamental space-filling mode of the cladding , and n_{eff} denotes the effective index of a guided mode.

For the widely extended field, the asymmetric core shape has a vanishing influence on the polarization splitting, and the birefringence becomes negligible. It should, however, be noted that in the case of non-circular shaped cladding holes, it does become possible to achieve significant polarization effects due to an additional splitting of the degeneracy of the fundamental cladding mode [7.102]

As it was the case for the low-frequency limit, Hansen *et al.* [7.100] found that the birefringence becomes vanishing in the high-frequency limit. This means that an optimum frequency window exists for the design of high-birefringent PCFs. The theoretical predicted birefringence for a series of fibres with different hole sizes is illustrated in Figure 7-19. As expected, the birefringence is seen to strongly increase with increasing hole size. Also marked, on the figure is the second-order cutoff of the fibres. Here, it should also be noted that only fibres with a normalized hole size of $d/\Lambda = 0.30$ or smaller may be classified as endlessly single-mode, making cutoff analysis of high-birefringence fibres vital. For all studied hole sizes, however, the birefringence is seen to reach its maximum value, while the fibre is single mode. For a fibre with air holes of relative size $d/\Lambda = 0.70$, a maximum value of $\Delta n = 7.7 \cdot 10^{-3}$ is reached at a normalized frequency of size $\Lambda/\lambda = 0.66$. As this fibre reaches cutoff at a normalized frequency of $\Lambda/\lambda = 0.72$, strict requirements are impossed on fabrication tolerances (for a fibre to be operated at a wavelength around 1550nm, the pitch should be controlled within 100 nm.

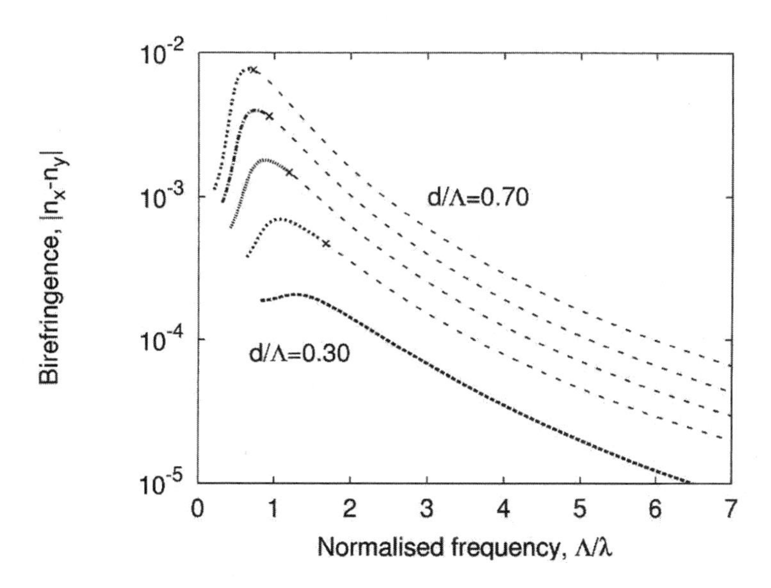

Figure 7-19. Illustration of the dependence of modal birefringence on relative size of air holes. The shown hole sizes are ranging from $d/\Lambda = 0.30$ to 0.70 in steps of 0.10. In each case, the second-order cutoff is shown, except for a hole size of 0.30, as this fibre is endlessly single mode (dashed curves indicate multi-mode operation). The illustration is drawn from [7.100].

7.8 DISPERSION MANAGED PHOTONIC CRYSTAL FIBRES

For transmission of optical signals over long distances, two key properties are of concern. One is the fibre attenuation, which we will return to in Section 7.10, and the other is the fibre dispersion. With the lifting of the loss limitation by the appearance of optical amplifiers in the early 1990'ies, dispersion became the primary bottleneck, and a huge effort was invested in methods for dispersion compensation. One of the most successful fibre components during the late 1990'ies was the dispersion compensating fibre (DCF) as described by Grüner-Nielsen *et al.* [7.103]. The fundamental appraoach in the design and development of DCFs is to provide a fibre having a large negative dispersion (typically in the order of -100 ps/(km·nm)) to counterbalance the typical dispersion (17 ps/(km·nm)) of standard optical fibres at the wavelength of 1550 nm. As described in [7.103], it is also a central objective to produce a DCF with a relative dispersion slope (RDS) comparable to that of standard fibres. Note that the RDS is defined as the ratio of dispersion slope to dispersion, which in a non-dispersion-shifted fibre has the typical value of 0.0034 nm^{-1}. Fibre dispersion does also become important in relation with short-length applications, e.g., in connection with optical pulse compression.

Being an important parameter in optical fibre development, the dispersion is naturally also of relevance for PCF technology. Moreover, because we already have seen that the PCFs often provide properties, which may not be obtained using standard optical fibres, we will look a little closer at the general dispersion properties of these fibres. To document the variation of dispersion that may be obtained using PCFs, Figure 7-20 illustrates the waveguide dispersion for a number of silica-air PCFs with varying hole sizes. The figure shows waveguide dispersion as function of normalized wavelength defined as λ/Λ. Hence, for a PCF with given pitch dimension, Λ, the total dispersion at a wavelength, λ, may be found (as a first approximation) by adding the material dispersion of silica to the waveguide dispersion of Figure 7-20.

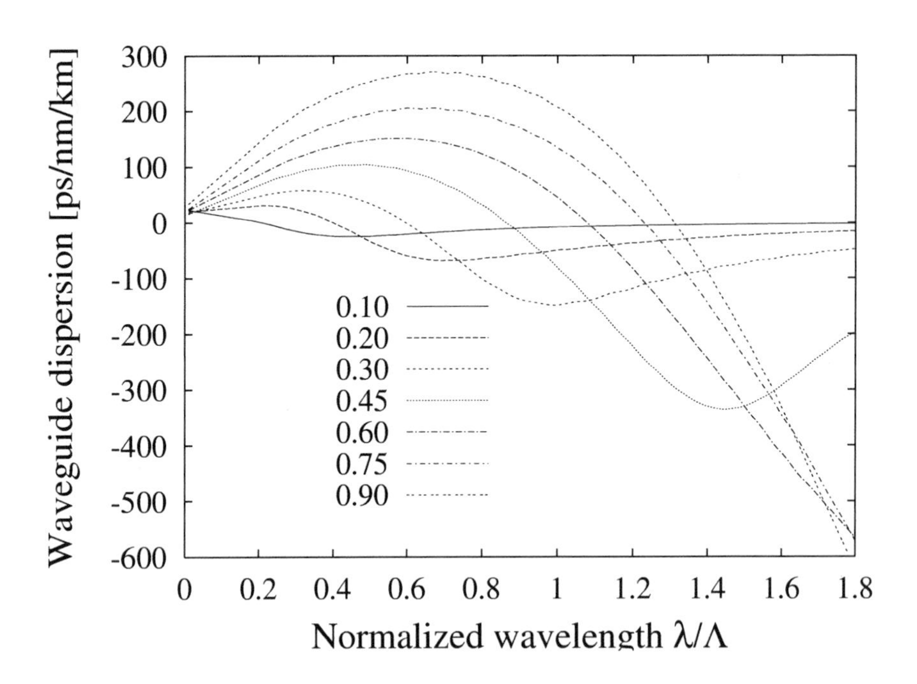

Figure 7-20. Waveguide dispersion as a function of normalized wavelength for a PCF with a close-packed cladding structure. Curves are shown for different ratios d/Λ.

Figure 7-20 indicates that for very small air-filling fractions, the waveguide dispersion is very limited (and a real fibre will have total dispersion close to the dispersion of the fibre material). As the diameter of the air holes is increased, the waveguide dispersion becomes increasingly strong and both normal and anomalous waveguide dispersion properties may be found. Hence, by appropriate design, PCFs may to a large degree be tailored with specifically desired dispersion properties. Note that, generally, for increased structure dimensions (small λ/Λ-values), the waveguide dispersion is found to be reduced. Intuitively, the reason for this being that large structure dimensions compared to the optical wavelength causes light to be guided with very little overlap with the holes. Hence, apart from material dispersion, a limited structure-introduced dispersive effect will be experienced. Strong dispersive effects are, therefore, generally found for PCFs of small dimensions, such as the non-linear fibres, whereas LMA PCFs exhibit dispersion properties dominated by the fibre material.

Having outlined general dispersion properties, we will focus on more specific aspects of dispersion in PCFs and shortly review some of the results presented in international literature over the past few years. One of the early contributions concerning dispersion compensation using photonic crystal fibres was made by Birks *et al.* [7.104], who took the approximate – but illuminating – approach of emulating a PCF with large air holes by a simple circular silica core in air. The simplified calculations lead to the prediction of waveguides having a dispersion as low as -2000 ps/(km·nm). The silica-strand equivalent of the PCF was further used by Knight *et al.* [7.105], and compared (with good agreement) with experimental observations of strictly single-mode PCFs having zero dispersion at a wavelength of 700 nm. These properties, which are unique for PCFs (and optical fibres tapered down to diameters of a few microns), allows for generation of solitons and supercontinua of new wavelengths in optical fibres.

It is, naturally, of very high relevance that the PCF technology provides not only large negative dispersion values, but also new wavelengths for obtaining anomalous dispersion. However, a further aspect of dispersion control in PCFs relates to the possibility of simultaneous control of dispersion and dispersion slope.

The possibility of making a very flat dispersion curve over a wide wavelength range was described by Bjarklev *et al.* [7.106], who applied a simplified scalar model to indicate the properties of PCFs having a silica core and a triangular distribution of air holes in the cladding. Such results were also reported by Broeng *et al.* [7.107]. The treatment of PCF designs having extremely flat dispersion curves was taken a significant step further by Ferrando *et al.* [7.108] in 2001, who numerically demonstrated very flat dispersion over a wavelength range of 200 nm or more, and in addition a tuneability of the level of dispersion from +45 ps/(km·nm) to -45 ps/(km·nm) as shown in Figure 7-21. Note that the allowed variation of the flattened dispersion profiles is set to be 2 ps/(km·nm), and their corresponding flattened dispersion bandwidths are (a) 270 nm, (b) 294 nm, (c) 259 nm, and

(d) 195 nm, respectively. Also note that the fibre parameters have to be carefully controlled in order to obtain the desired results.

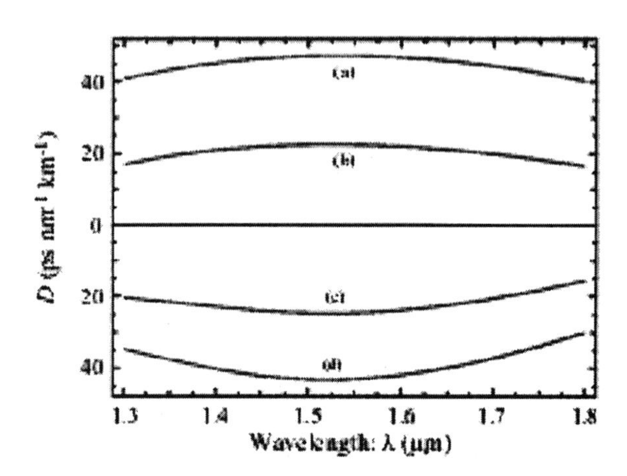

Figure 7-21. Four flattened dispersion curves shown as a function of wavelength. The curves correspond to different values of the dispersion centred near 1550 nm, and the simulated fibre parameters are as follows: (a) d = 0.98 μm and Λ = 2.32 μm, (b) d = 0.80 μm and Λ = 2.71 μm, (c) d = 0.56 μm and Λ = 2.16 μm, and (d) d = 0.54 μm and Λ = 1.99 μm. The curve was kindly provided by Professor Miguel Andrés Bou, and similar results have been presented in [7.108].

Other authors such as for example Sinha *et al.* [7.111] have made further theoretical predictions of detailed dispersion control in PCFs, but using different numerical methods such as the scalar model (applied in [7.111]) and more rigorous approaches.

Experimental results on dispersion-flattened (DF) photonic crystal fibres were presented in 2002 by Reeves *et al.* [7.109]. In this work, several PCFs was reported, and a first design containing 186 holes placed in a close-packed pattern of a target with a hole diameter of d = 0.63 μm and a pitch Λ = 2.64 μm was fabricated. The resulting dispersion was somewhat off the numerically predicted values, which Reeves *et al.* [7.109] concluded was due to an insufficient field confinement. Further modelling by Reeves *et al.* [7.109] indicated that for the fibre dispersion to converge towards that of a fibre with infinite cladding, at least 8 periods of air holes would be required. A second fibre with 455 holes (corresponding to 11 complete periods between the core and the external jacket) was then reported in [7.109] to get closer to the target, and a dispersion of D = 0 ± 0.6 ps/(km·nm) over a range from 1.24 μm to 1.44 μm was measured on a fibre length of 235 mm. For this fibre, the reported loss was measured to be in the order of 2 dB/m at a wavelength around 1500 nm.

Before we leave the subject of dispersion-managed PCFs, it is relevant also to look at the dispersion properties of photonic bandgap fibres. The number of publications on PBG fibres are today still limited to a few, and the specific issue of dispersion has almost not been treated yet in literature. Also for this reason, it was an interesting step forward, when Jasapara *et al.* [7.110] in the beginning of 2002 reported chromatic dispersion measurements in a photonic bandgap fibre. The specific fibre was realised by a PCF having a triangular lattice of holes in silica, and the effective index of the cladding structure was then raised above that of the silica core by sucking a high-index liquid (with a refractive index of 1.8) into the holes. The resulting high-index rods (with diameters of 2.4 μm and pitch of 4.3 μm) ensure that PBG is the only possible guiding mechanism. Using a spectral interferometric setup, the PBG fibre was characterised with respect to group delay, and near the bandgap edges, dispersion values ranging from around –1000 ps/(km·nm) at the lower wavelength edge to +7000 ps/(km·nm) at the upper wavelength edge was measured. The very strong waveguide dispersion in the studied PBG fibres makes them unique. However, as pointed out by Jasapara *et al.* [7.110], it is possible that by designing the crystal structure, fibres with tailored dispersion properties can be made for various telecommunication applications.

7.9 COUPLING AND SPLICING

One of the major concerns with respect to the practical application of PCFs in optical fibre systems was until quite recently the question of whether or not the PCFs could be spliced to standard optical fibres with reasonable loss levels. One of the first reports on this subject was published in 1999 by Bennett *et al.* [7.73], who spliced PCFs to conventional dispersion-shifted (DS) fibre by the use of a modified routine on a commercial splicer. The splicing loss measured by Bennett *et al.* [7.73] was around 1.5 dB at a wavelength of 1550 nm, and it was less than the loss observed, when the two fibres were butt coupled. The loss of 1.5 dB was in the specific case comparable to that predicted from the mode mismatch alone.

Splice loss was further studied from a numerical point of view by Lizier *et al.* [7.74] in 2001, and recent advances have further brought the splice loss down. An example of a high-quality splice is shown in Figure 7-22, where a standard fibre is spliced to a photonic crystal fibre. The photo on Figure 7-22 shows at the left side a PCF, and we note that the central part of the fibre appears to be more dark than the outer cladding, which is due to the scattering of the light by the many air-holes running along the length of the fibre. The fibre shown to the right of Figure 7-22 is a standard SMF128 single-mode fibre, and in the specific case, the splice loss is measured to 0.5 dB – corresponding to the mode mismatch between the specific PCF and the standard optical fibre.

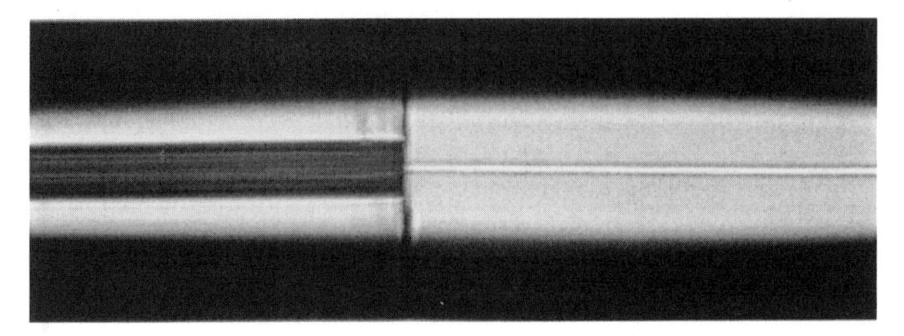

Figure 7-22. A photo of a splice between a photonic crystal fibre (left) and a standard optical fibre (right). The photograph is kindly provided by Crystal Fibre A/S.

7.10 LOSS PROPERTIES OF PHOTONIC CRYSTAL FIBRES

Since the photonic crystal fibres form a relatively new class of optical waveguides, a major effort over the past few years has been concentrated on structural control in the fabrication process rather than on loss reduction. However, as the PCF technology is becoming more and more mature, the reduction of the loss become of increasingly high importance. In this respect, it is very relevant to note that some of the first attempts of combining low loss and microstructuring was done by Hasegawa *et al.* [7.75] in connection with their hole-assisted lightguide fibre (HALF). The HALF waveguides still use co-doping of the silica glass, and the optical modes are not solely located by the air-hole structure in the fibre. However, it is very relevant to note that Hasegawa *et al.* [7.75] use a loss model based on the following formula:

$$\alpha(\lambda) = A\lambda^{-4} + B + \alpha_{OH}(\lambda) \tag{7.12}$$

where the SiOH absorption, $\alpha_{OH}(\lambda)$, according to Walker [7.76] may be expressed as:

$$\alpha_{OH}(\lambda) = \Delta\alpha_{OH} \cdot \sum_{n=1}^{6} a_n \cdot \exp\left[-\frac{1}{2}\left(\frac{\lambda - \lambda_n}{\sigma_n}\right)^2\right] \tag{7.13}$$

and the coefficients A and B are supposed to correspond to Rayleigh scattering loss and waveguide imperfection loss. In Eq. (7.13), the SiOH loss is described by amplitude values a_n, which are normalized so that $\Delta\alpha_{OH}$ gives the peak value of $\Delta\alpha_{OH}(\lambda)$, and the spectral location and width of the individual parts of the SiOH spectrum is given by λ_n and σ_n, respectively. According to [7.76], the relevant parameters to model the OH-loss peak are given as shown in Table 7-3.

n		1	2	3	4	5	6
a_n		0.040	0.604	0.377	0.184	0.035	0.017
λ_n	μm	1.347	1.379	1.390	1.402	1.243	1.256
σ_n	μm	0.0139	0.0070	0.0117	0.0277	0.0085	0.0136

Table 7-3. Parameters used in Eq. (7.13) for modelling of OH-loss peak. Values are according to [7.76].

Note that for the HALF fibres presented by Hasegawa *et al.* [7.75], the parameters A have values in the range from 0.93 to 1.1 dB/(km $\cdot\,\lambda^{-4}$) and B from 0.24 to 0.65 dB/km. The SiOH peak is described by peak values of $\Delta\alpha_{OH}$ from 4.0 to 7.8 dB/km.

Besides the Rayleigh loss, the SiOH loss, and the loss due to waveguide imperfections, more macroscopic properties also become relevant. One of these factors is the handling of the optical fibre with regard to macrobending.

As previously mentioned, the macrobending loss of photonic crystal fibres are quite unique compared to those of standard optical fibres. To get a closer understanding of these properties, we may look at the results presented in Figure 7-23, which illustrates the bending radius dependency of the operational windows for a specific LMA PCF with air-hole diameter d = 2.4 μm and hole-to-hole spacing Λ = 7.8 μm. As can be seen in Figure 7-23, a short-wavelength loss edge is evident for the LMA PCFs. This is in contrast to the standard fibre case, where only a long-wavelength loss edge is found. For so-called endlessly single mode PCFs – single mode fibres that can be designed with a very large mode area - it is, therefore, important to notice that macrobending losses in practice will limit the operational wavelength.

Figure 7-23 further includes the predictions of the effective index model, and as can be seen, the model is in this case capable of predicting accurately the spectral location of the (unusual) short-wavelength bend-loss edge [7.1]. The long wavelength bend loss edge of this specific fibre is positioned at mid-infrared wavelengths for all bending radii. Although this figure presents results for one particular LMA PCF, a number of generalisations can be made.

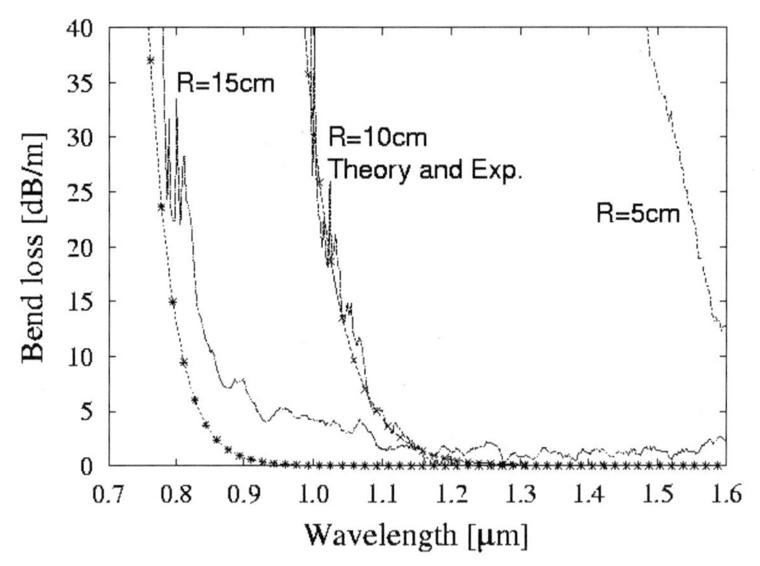

Figure 7-23. Calculated and measured values of spectral bending loss of LMA PCF. For operation around 1.5 μm, the critical bending radius is ~ 6 cm. The illustration corresponds to the one presented by Sørensen *et al.* [7.5].

One way of modelling the propagation of light in a fibre with a radius of curvature R_0 is to scale the refractive index using the transformation:

$$n(r) = n\sqrt{1 + \frac{2r\cos\phi}{R_0}} \tag{7.14}$$

where the coordinates according to Marcuse [7.77] are defined as follows: r, ϕ are the radial coordinates, R_0 is the radius of curvature, and n is the unperturbed refractive index profile. Using this transformation, the modes of the bent microstructured fibre can then be calculated using one of the full-structural techniques described in [7.112]. Note that the slant introduced by this transformation cannot be described in all of the models. Using the numerically determined modes, the transition loss can be calculated as the overlap between the straight and bent modes, and for pure bend loss it is assumed that the fraction of energy in the guided mode at $r > r_c$ (where r_c is defined as the radius, where the evanescent field turns into a radiating field) is lost over some distance scale. This model allows trends relating to the angular orientation of the fibre to be identified.

A general view has been that the operation of LMA-PCFs is limited by macrobending loss, but pushing technology to still larger effective areas, limitations can be set by microbending deformations as well [7.113]. Microbending deformations may be caused by external perturbations as studied recently [7.114], but even when there are no external perturbations, there may be still be residual microdeformations caused by frozen-in mechanical stress in e.g., the coating material. Recently, Mortensen *et al.* [7.113] reported results for the attenuation of a PCF with a 15 μm core diameter and an outer diameter of D = 125 μm, and found that in the visible regime, the performance was clearly suffering from microdeformation induced attenuation. Methods for reducing these microdeformation losses include a limited increase in fibre diameter and special coating design.

We have now briefly discussed several important loss factors, and we most note that their mutual size and absolute contribution to the total fibre loss is both complicated and strongly dependent on the fabrication technology. As the manufacturing methods are further refined, we may expect that the very clear trend of loss reduction as indicated in Figure 7-24 is going to continue until the (presently unknown) lower loss limits are reached. Note also that Figure 7-24, which illustrates various record-low loss values over time, shows an order-of-magnitude loss reduction for index-guiding as well as air-guiding PCFs. Although the air-guiding PCFs today show losses around an order of magnitude higher than those for index guiding PCFs, we might see comparable loss values of the two fibre classes in a not too distant future.

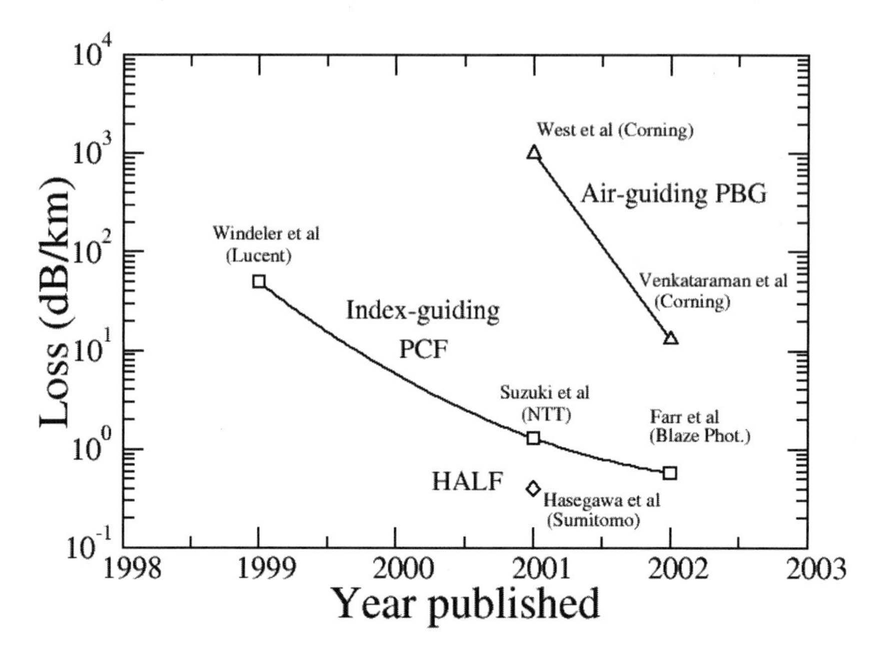

Figure 7-24. Reported fibre loss values shown versus the time of publication. The references are indicated by the names of the first authors on the papers. For index guiding fibres the relations are as follows: Windeler et al. – [7.117]. The results on HALF fibres were presented by Hasegawa et al. [7.75], and with respect to the air-guiding PBG fibres the relations are West et al. [7.115], and Venkataraman et al. [7.116]. The illustration was generated by Dr. J. Lægsgaard during 2002.

7.11 SUMMARY

The subjects and fibre structures described in this chapter has been selected to give a general understanding of the potential of photonic crystal fibres. We have in this connection discussed a number of applications, but it should also be noted that many other possibilities exist. Among some of these, we may mention the use of photonic crystal fibres in optical sensors [7.24] or even the application of the thin fibre channels for atom and particle guidance [7.85], using laser light to levitate and guide particles through the channels.

Considering the tremendous development within the area of photonic crystal fibres over the past few years, we are also convinced that numerous new applications will appear in the near future, and these findings will undoubtedly be the subject of subsequent papers and textbooks. However, we have chosen to let the examples of large-mode-area PCFs, highly-nonlinear PCFs, or PCFs with very high numerical aperture form the major part of this last chapter of this book. We have of course also briefly addressed issues such as PCF based amplifiers, tuneable PCF devices, and polarization maintaining PCFs, together with more basic properties such as dispersion, fibre attenuation and fibre coupling loss, but in order to limit the extend of the present text, we have chosen the more overall approach on these matters. Anyhow, it is our hope that this chapter has provided a good flavour of the present-day status and potential of photonic crystal fibre technology.

REFERENCES

[7.1] J.C Knight, T.A.Birks, R.F.Cregan, P.St.J.Russell, J.-P.De Sandro, "Large mode area photonic crystal fibre", *IEE Electronics Letters*, vol.34, pp.1347-1348, 1998.

[7.2] W. A. Gambling H. Matsumura, C. M. Ragdale, R.A. Sammut, "Measurement of radiation loss in curved single-mode fibres", *Microwaves Optics and Acoustics,* vol. 2, pp.134-140, 1978.

[7.3] J.-I. Sakai, T.Kimura, , "Bending loss of propagation modes in arbitrary-index profile fibers", *Applied Optics*, vol.17, pp. 1499-1506, 1978.

[7.4] E. Knudsen, A. Bjarklev, J. Broeng, S. E. Barkou, "Macro-bending loss estimation for air-guiding photonic crystal fibres", *14th International Conference on Optical Fiber Sensors,* OFS 2000, pp. 904-907, 2000.

[7.5] T. Sørensen, J. Broeng, A. Bjarklev, E. Knudsen, S.E. Barkou Libori, H.R. Simonsen, J. Riis Jensen, "Macrobending loss properties of photonic crystal fibres with different air filling fractions", *European Conference on Optical Communication*, ECOC'2001, Amsterdam, The Netherlands, Vol. 3, pp. 380-381, 30.Sep- 4. Oct. 2001

[7.6] J.C.Baggett, T.M.Monro, K.Furusawa, D.J.Richardson, "Comparative study of large mode holey and conventional fibers", *Optics Letters*, vol. 26, pp.1045-1047, 2001.

[7.7] K.S. Chiang and V. Rastogi, "Ultra-large-core single-mode fiber for optical communications: the segmented cladding fiber", *Conference on Optical Fiber Communication OFC'2002*, pp. 620-621, 2002.

[7.8] K. J. Ranka, S. R. Windeler, and A. J. Stentz
"Visible continuum generation in air-silica microstructure optical fibers with anomalous dispersion at 800 nm",
Optics Letters, vol.25, pp. 25-27, 2000.

[7.9] S. Coen, A. H. L. Chau, R. Leonhardt, and J. D. Harvey
"Supercontinuum generation via stimulated Raman scattering and parametric four-wave mixing in photonic crystal fibers",
The Journal of the Optical Society of America B, vol.26, pp.753, 2002.

[7.10] K. P. Hansen, J. Juul Larsen, J. Riis Jensen, S. Keiding, J. Broeng, H. R. Simonsen, A. Bjarklev.
"Super continuum generation at 800 nm in highly nonlinear photonic crystal fibers with normal dispersion",
LEOS Annual meeting, Vol.2, pp.703-704, 2001.

[7.11] I. Hartl, X. D. Li, C. Chudoba, R. K. Ghanta, T. H. Ko, and J. G. Fujimoto
"Ultrahigh-resolution optical coherence tomography using continuum generation in a air-silica microstructure optical fiber",
Optics Letters, vol. 26, pp.608-610, 2001.

[7.12] R. E. Drullinger, S. A. Diddams, K. R. Vogel, C. W. Oates, E. A. Curtis, W. D. Lee, W. M. Itano, L. Hollberg, and J. C. Bergquist,
"All-optical atomic clocks",
International Frequency Control Symposium and PDA Exhibition, pp. 69-75 (2001).

[7.13] P. A. Champert, S. V. Popov, and J. R. Taylor,
"Generation of multiwatt, broadband continua in holey fibers",
Optics Letters, vol.27, pp.122-124, 2002.

[7.14] A. V. Husakou and J. Herrmann,
"Supercontinuum generation of higher-order solitons by fission in photonic crystal fibers",
Physical Review Letters. vol. 87, 2001.

[7.15] J. Herrmann, U. Griebner, N. Zhavoronkov, A. Husakou, D. Nickel, J. C. Knight, W. J. Wadsworth, P. St. J. Russel, and G. Korn,
"Experimental evidence for supercontinuum generation by fission of higher-order solitons in photonic fibers",
Physical Review Letters, vol. 88, 2002.

[7.16] K. P. Hansen, J. Riis Jensen, D. Birkedal, J. M. Hvam, and A. Bjarklev,
"Pumping wavelength dependence of super continuum generation in photonic crystal fibers ",
Conference on Optical Fiber Communication, OFC, 2002.

[7.17] P. Petropoulos, T. M. Monro, W. Berlardi, K. Furusawa, J. H. Lee, and D. J. Richardson
"2R-regenerative all-optical switch based on a highly nonlinear holey fiber",
Optics Letters, Vol.26, pp.1233-1235, 2001.

[7.18] F. Futami, S. Watanabe, and T. Chikama,
"Simultaneous recovery of 20x20 GHz WDM optical clock using supercontinuum in a nonlinear fiber",
European Conference on Optical Communication, 2000.

[7.19] J. Hansryd and P. A. Andrekson
"Broad-band continuous-wave-pumped fiber optical parametric amplifier with 49-dB gain and wavelength-conversion efficiency",
IEEE Photonics Technology Letters, Vol. 13, pp.194-196, 2001.

[7.20] F. Druon, N. Sanner, G. Lucas-Leclin, P. Georges, R. Gaumé, B. Viana, K. P. Hansen, and A. Petersson
"Self-compression of 1-um femtosecond pulses in a photonic crystal fiber",
Conference on Lasers and Electro-Optics CLEO, pp.764-765, 2002.

[7.21] J. H. Lee, Z. Yusoff, W. Belardi, M. Ibsen, T. M. Monro, B. Thomsen, and D. J.Richardson,
"A holey fiber based WDM wavelength converter incorporating an apodized fiber Bragg grating filter",
CLEO, Paper CPDB5-1, 19-24, May 2002.

[7.22] J. E. Sharping, M. Fiorentino, P. Kumar, and R. S. Windeler
"All optical switching based on cross-phase modulation in microstructure fiber",
IEEE Photonics Technology Letters, Vol. 14, pp. 73-77, 2002.

[7.23] H. Takara, T. Ohara, K. Mori, K. Sato, E. Yamada, M. Abe, Y. Inoue, T. Shibata, T. Morioka, and K.-I. Sato.
"More than 1000 channel optical frequency chain generation from single supercontinuum source with 12.5 GHz channel spacing",
IEE Electronics Letters, Vol. 36, No. 25, pp. 2089-2090, 2000.

[7.24] K. P. Hansen, J. Riis Jensen, C. Jacobsen, H. R. Simonsen, J. Broeng, P. M. W. Skovgaard, A. Petersson, and A. Bjarklev,
"Highly Nonlinear Photonic Crystal Fiber with Zero-Dispersion at 1.55 μm",
OFC'02 post deadline paper FA9, 2002.

[7.25] K. S. Berg, L. K. Oxenløwe, A. Siahlo, A. Tersigni, A. T. Clausen, C. Peucheret, P. Jeppesen, K. P. Hansen, and J. R. Jensen,
"80 Gb/s transmission over 80 km and demultiplexing using a highly nonlinear photonic crystal fibre",
ECOC'2002, Paper 2.15, 2002.

[7.26] J. C. Knight, T. A. Birks, R. F. Cregan, P. St. J. Russell, and J.-P. de Sandro,
"Large mode area photonic crystal fibre",
IEE Electronics Letters, Vol.34, No.13, 1998, pp.1347-1348.

[7.27] L.K.Oxenløwe, A.Siahlo, K.S.Berg, A.Tersigni, C.Peucheret, A.T.Clausen, K.P.Hansen, and J.R.Jensen,
"160 to 10 Gb/s demultiplexing using a photonic crystal fibre",
CLEO, 2002.

[7.28] T.M.Monro, K.M.Kiang, J.H.Lee, K.Frampton, Z.Yusoff, R.Moore, J.Tucknott, D.W.Hewak, H.N.Rutt, and D.J.Richardson,
"High nonlinearity extruded single-mode holey optical fibers",
OFC, vol.1-3, 2002.

[7.29] G.P. Agrawal,
"Nonlinear Fiber Optics"
Academic, New York, 1989.

[7.30] T. Okuno, M. Onishi, T. Kashiwada, S. Ishikawa, and M. Nishimura,
"Silica-Based Functional Fibers with Enhanced Nonlinearity and Their Applications",
IEEE J. Sel. Top. Quant. vol. 5, pp. 1385—1391, 1999.

[7.31] P. Petropoulos, T.M. Monro, W. Belardi, K. Furusawa, J.H. Lee, and D.J. Richardson,
"2R-regenerative all--optical switch based on a highly nonlinear holey fiber",
Optics Lett. vol.26, pp. 1233-1235, 2001.

[7.32] V.Finazzi, T. M. Monro, and D. J. Richardson,
"Confinement loss in highly nonlinear holey optical fibers"
Proc. OFC, OSA Technical Digest, pp. 524-525, Anaheim, California, 2002.

[7.33] N. A. Mortensen, J. Riis Jensen, P.M.W. Skovgaard, and J. Broeng,
"Propagation and waveguiding - Numerical Aperture of Single-mode Photonic Crystal Fibers",
IEEE Photonics Technology Letters, Vol. 14. No. 8 , pp.1094-1096, 2002.

[7.34] A.K. Ghatak, and Thyagarajan,
"Introduction to fiber optics",
Cambridge University Press, Cambridge, 1998.

[7.35] N. A. Mortensen,
"Effective area of photonic crystal fibers",
Optics Express, Vol.10, No.7, April 2002, pp.341-348.

[7.36] N. A. Mortensen, and J. Riis Folkenberg,
"Near-field to far-field transition of photonic crystal fibers: symmetries and interference phenomena",
Optics Express, Vol.10, No.11, June 2002, pp.475-481.

[7.37] T. Niemi, H. Ludvigsen, F. Scholder, M. Legré, M. Wegmuller, N. Gisin, J. R. Jensen, A. Petersson, and P. W. M. Skovgaard,
"Polarization properties of single-moded, large-mode area photonic crystal fibers",
ECOC'2002, Paper S.1.9.

[7.38] J. Canning, E. Buckley, and K. Lyytikainen,
"Propagation in air by field superposition of scattered light within a Fresnel fiber",
Optics Letters, Vol.28, No.4, Febr.2003, pp.230-232.

[7.39] N. A. Mortensen, M. D. Nielsen, J. R. Folkenberg, A. Petersson, and H. R. Simonsen,
"Improved large-mode-area endlessly single-mode photonic crystal fibers",
Optics Letters, Vol.28, No.6, March.2003, pp.393-395.

[7.40] M. D. Nielsen, N. A. Mortensen, J. R. Folkenberg, A. Petersson, and A. Bjarklev,
"Improved all-silica single-mode photonic crystal fiber",
Conference on Optical Fiber Communication OFC'2003, Paper FI7, Vol.2, pp.701-702, 2003.

[7.41] T. M. Monro, D. J. Richardson, N. G. R. Broderick, P. J. Bennett,
"Holey optical fibers: an efficient modal model",
IEEE Journal of Lightwave Technology, vol. 17, pp. 1093-1101, 1999.

[7.42] J.C.Knight, J.Arriaga, T.A.Birks, A.Ortigosa-Blanch, J.W.Wadsworth, and P.St.J.Russell,
"Anomalous Dispersion in Photonic Crystal Fiber",
IEEE Photonics Technology Letters, Vol. 12, pp. 807-809,2000.

[7.43] J.H.V.Price, K.Furusawa, T.M.Monro, L.Lefort, D.J.Richardson,
"Tunable femtosecond pulse source operating in the range 1.06-1.33 micron based on an Yb3+-doped holey fiber amplifier",
J. Optical Soc. Am. B, Vol. 19, pp.1286-1294, 2002.

[7.44] J. K. Ranka, R. S. Windeler, A. J. Stentz,
"Optical properties of high-delta air-silica microstructure optical fibers",
Optics Letters, Vol. 25, no. 11, pp.796-798, 2000.

[7.45] T.A.Birks, D.Mogilevtsev, J.C.Knight and P.St.J. Russell,
"Dispersion compensation using single material fibers",
IEEE Photonics Technology Letters, Vol. 11, pp.674-676, 1999.

[7.46] M. Nakazawa, H. Kubota, and K. Tamura,
"Random evolution and coherence degradation of a high-order optical soliton train in the presence of noise"',
Optics Letters, vol.24, pp. 318-320, 1999.

[7.47] A. Carlsen, J. B. Jensen, L. H. Pedersen, P. E. Hoiby, L. B. Nielsen, A. Bjarklev, and T. P. Hansen,
"Evanescent-wave sensing of biomolecules in aquaous solution using a photonic crystal fiber",
NORTHERN OPTICS 2003, Espoo, Finland, p.16, 16-18 June 2003.

[7.48] K. Tajima, K. Nakajima, K. Kurokawa, N. Yoshizawa, and M. Ohashi,
"Low-loss photonic crystal fibers",
Proc. OFC, OSA Technical Digest, 523-524 (Anaheim, California, 2002).

[7.49] T. P. White, R. C. McPhedran, C. M. de Sterke, L. C. Botten, and M. J. Steel, "Confinement losses in microstructured optical fibers", Optics Letters, Vol.26, pp.1660-1662, 2001.

[7.50] T. Monro, A. Bjarklev, and J. Laegsgaard, "Microstructured optical fibres", *IEE Handbook of Optoelectronics*, 2003

[7.51] N. Nikolov, O. Bang, and A. Bjarklev, "Improving efficiency of supercontinuum generation in photonic crystal fibers by direct degenerate four-wave mixing", ECOC′2002, Paper P.1.36.

[7.52] R. R. Alfano and S. L. Shapiro, "Emission in the region 4000 to 7000Å via four-photon coupling in glass", *Phys. Rev. Lett.* vol.24, pp.584-587, 1970.

[7.53] R. R. Alfano and S. L. Shapiro, "Observation of self phase modulation and small-scale filaments in crystals and glasses", *Phys. Rev. Lett.* vol. 24, pp.592-594, 1970.

[7.54] R. R. Alfano, "The Supercontinuum Laser Source", ed. *Springer-Verlag*, New-York, 1989.

[7.55] C. Lin and R. H. Stolen, "New nanosecond continuum for excited-state spectroscopy", *Appl. Phys. Lett.* vol.28, pp.216-218, 1976.

[7.56] P. L. Baldeck and R. R. Alfano, "Intensity effects on the stimulated four photon spectra generated by picosecond pulses in optical fibers", *IEEE J.Lightwave Technol.* vol.5,pp.1712-1715, 1987.

[7. 57] S. Coen, A. H. L. Chao, R. Leonardt, J. D. Harvey, J. C. Knight, W. J. Wadsworth, and P. S. J. Russell, "White-light supercontinuum generation with 60-ps pulses in a photonic crystal fiber", *Opt. Lett.* vol.26, p.1356 ,2001

[7.58] K. Mori, H. Takara, S. Kawanishi and T. Morioka, "Flatly broadened supercontinuum spectrum generated in a dispersion decreasing fiber with convex dispersion profile", *IEE Electron. Lett.* vol.33, pp.1806-1808, 1997.

[7.59] K. Mori, H. Takara, and S. Kawanishi, "Analysis and design of supercontinuum pulse generation in a single-mode optical fiber", *J. Opt. Soc. Am.* B, vol. 18, 1780 , 2001.

[7.60] K. Tamura, H. Kubota, and M. Nakazawa,
"Fundamentals of stable continuum generation at high repetition rates",
IEEE J. Quantum Electron. vol. 36, pp. 773-779, 2000.

[7.61] A. Ferrando, E. Silvestre, J. J. Miret, and P. Andres,
"Nearly zero ultraflattened dispersion in photonic crystal fibers",
Opt. Lett. vol. 25, pp. 790, 2000.

[7.62] A. Ferrando, E. Silvestre, P. Andres, J. J. Miret, and M. V. Andres,
"Designing the properties of dispersion- flattened photonic crystal fibers",
Opt. Express, vol. 9, pp.687-697, 2001.

[7.63] T. A. Birks, W. J. Wadsworth, and P. St. J. Russell,
"Supercontinuum generation in tapered fibers",
Opt. Lett. vol.25, no. 19, pp.1415-1417, 2000.

[7.64] A. V. Gusakov, V. P. Kalosha, and J. Herrmann,
"Ultrawide spectral broadening and pulse compression in tapered and photonic fibers", *QELS*, pp. 29 (2001).

[7.65] J. M. Dudley, L. Provino, N. Grossard, H. Maillotte, R. S. Windler, B. J. Eggeleton and S. Coen,
"Supercontinuum generation in air-silica microstructured fibers with nanosecond and femtosecond pulse pumping",
J. Opt. Soc. Am. B, vol. 19, pp. 765, 2002.

[7.66] A. Ferrando, M. Zacarés, P. Fernández de Córdoba, and D. Binosi,
"Self-Trapped Localized Modes in Photonic Crystal Fibers",
Nonlinear Optics (NLO), Maui, Hawaii, July 29-Aug.2, 2002.

[7.67] Z. Yusoff, J. H. Lee, W. Belardi, T. M. Monro, P. C. Teh, D. J. Richardson,
"Raman effects in a highly nonlinear holey fiber: amplification and modulation",
Optics Lett., vol. 27, pp. 424-426, 2002.

[7.68] J. Nilsson, R. Selvas, W. Belardi, J. H. Lee, Z. Yusoff, T. M. Monro, D. J. Richardson, K. D. Park, P. H. Kim, N. Park,
"Continuous-wave pumped holey fiber Raman laser",
Proc. OFC, OSA Technical Digest, (Anaheim, California), (2002).

[7.69] J. K. Sahu, C. C. Renaud, K. Furusawa, R. Selvas, J. A. Alvarez-Chavez, D. J. Richardson, J. Nilsson,
"Jacketed air-clad cladding pumped ytterbium-doped fibre laser with wide tuning range",
IEE Electronics Letters, vol. 37, pp. 1116-1117, 2001.

[7.70] D. J. Digiovanni, and R. S. Windeler
"Article comprising an air-clad optical fiber",
United States Patent 5,907,652, May 25, 1999

[7.71] K. Furusawa, A. N. Malinowski, J. H. V. Price, T. M. Monro, J. K. Sahu, J. Nilsson, D. J. Richardson,
"A cladding pumped Ytterbium-doped fiber laser with holey inner and outer cladding", Optics Express, Vol. 9, pp.714-720, 2001.

[7.72] R. P. Espindola, R. S. Windeler, A. A. Abramov, B. J. Eggleton, T. A. Strasser, and D. J. Digiovanni,
"External refractive index insensitive air-clad long period fibre grating",
IEE Electron. Lett. vol.35, pp.327-328, 1999.

[7.73] P. J. Bennett, T. M. Monro, and D. J. Richardson,
"Toward practical holey fiber technology: fabrication, splicing, modeling, and characterisation",
Optics Letters, Vol.24, No. 17, pp.1203-1205, 1999.

[7.74] J. T. Lizier, and G. E. Town,
"Splice loss in holey optical fibres",
OECC/IOOC'2001, TUG, pp. 158-159, July 2001.

[7.75] T. Hasegawa, E. Sasaoka, M. Onishi, M. Nishimura, Y. Tsuji, and M. Koshiba,
"Hole-assisted lightguide fiber for large anonalous dispersion and low optical loss",
Optics Express, Vol.9, No.13, Dec. 2001, pp.681-686.

[7.76] S. S. Walker,
"Rapid modeling and estimation of total spectral loss in optical fibres",
IEEE Journal of Lightwave Technology, Vol. 8, 1990, pp.1536-1540.

[7.77] D. Marcuse,
"Influence of curvature on the losses of doubly clad fibres",
Appl. Opt., Vol. 21, 1982, pp.4208-4213.

[7.78] B. J. Eggleton, P.S. Westbrook, R.S. Windeler, S. Spälter, T.A. Strasser, and G.L.Burdge,
" Grating spectra in air-silica microstructured optical fibers",
Proc.Optic.Fiber Commun. Conf., Baltimore, MD, 2000, Paper ThI2

[7.79] B. J. Eggleton, P.S. Westbrook, R.S. Windeler, S. Spälter r, and T.A. Strasser,
"Grating resonances in air-silica microstructures",
Opt. Lett., vol.24, 1999.

[7.80] A.A. Abramov, B. J. Eggleton, J. A. Rogers, R .P. Spndola, A. Hale, R.S. Windeler, S. Spälter, and T.A. Strasser,
"Electrically tunable efficient broad-band fiber filter",
IEEE Photon. Technol. Lett., vol. 11, pp. 495-497, 2000.

[7.81] P. S. Westbrook, B. J. Eggleton, R.S. Windeler, A. Hale, T.A. Strasser, and G.L.Burdge,
"Control of waveguide properties in hybrid polymer-silica microstructured optical fiber gatings",
Poc. optic. Fiber Commun., Baltimore, MD, 2000. Paper I3.

[7.82] P. S. Westbrook, B. J. Eggleton, R. S. Windeler, A. Hale, T. A. Strasser, and G. L. Burdge,
"Cladding mode resonances in hybrid polymer-silica microstructured optical fiber gratings",
IEEE Photonics Technology Letters, vol. 12, pp 495-496, 2000.

[7.83] B. J. Eggleton, , P.S. Westbrook, R.S. Windeler, S.Spälter, T.A. Strasser,
"Grating resonances in air-silica microstructured optical fibers"
Opt. Lett., vol. 24, No.21. pp. 1460-14662, November 1999

[7.84] B. J. Eggleton, P. S. Westbrook, C. A. White, C. Kerbage, R. S. Windeler, and G. L. Burdge.
"Cladding-Mode-Resonances in Air-Silica Microstructure Optical Fibers"
IEEE Journal of Lightwave Technology, Vol. 18, No.8, pp. 1084-1100, August 2000

[7.85] Philip Russell,
"Photonic Crystal Fibers"
Science, vol. 229, pp.358-362, January 2003

[7.86] W. J. Wadsworth, J. C. Knigth, A. Ortigosa-Blanch, J. Arriaga, E. Silvestre and P. St. J. Russell,
"Soliton effects in photonic crystal fibres at 850nm"
IEE Electronics Letters, vol. 36, pp. 53-55, 2000.

[7.87] W. J. Wadsworth, A. Ortigosa-Blanch, J. C. Knight, T. A. Birks, T-.P. Martin, and P. St. J. Russell,
"Nonlinear optics of photonic crystals – photonic crystal fibers – supercontinuum generation in photonic crystal fibers and optical fiber tapers: A novel light source",
J. OSA. B., Vol. 19, No. 9, pp. 2148-2155, 2002.

[7.88] N. G. R. Broderick, T. M. Monro, P. J. Bennett, and D. J. Richardson
"Nonlinearity in holey optical fibres: measurement and future opportunities",
IEE Electronics Letters, vol. 24, no. 20, pp. 1395-1397, October 2000.

[7.89] N. G. R. Broderick, P. J. Bennett, D. Hewak, T. M. Monro, D. J. Richardson and, Y. D. West,
"Nonlinearity in holey optical fibres",
LEOS'2000, pp.591-592.

[7.90] S. Coen, A. H. L. Chau, R. Leonhardt, and J. D. Harvey,
"Single-mode whith-ligth supercontinuum with 60 ps pump pulses in a photonic crystal fiber"
Nonlinear Guided Waves and their Applications, pp. 405-407, March 25–28., 2001, Clearwater, Florida.

[7.91] J. H. Price, W. Belardi, L. Lefort, T. M. Monro, D.J. Richardson
"Nonlinear pulse compression, dispersion compensation, and solition propagation in holey fiber at 1μm"
Nonlinear Guided Waves and their Applications, pp. 430-432, March 25-28., 2001, Clearwater, Florida.

[7.92] K. P. Hansen, J. R. Folkenberg, C. Peucheret, and A. Bjarklev,
"Fully dispersion controlled triangular-core nonlinear photonic crystal fiber",
OFC'03, March 27, Atlanta, GA, USA, Post deadline paper PD2, 2003.

[7.93] P. Petropoulos, T. M. Monro, H. Ebendorff-Heidepriem, K. Frampton, R. C. Moore, H. N. Rutt, and D. J. Richardson,
"Soliton-self-frequency-shift effects and pulse compression in an anomalous dispersive high nonlinearity lead silicate holey fibre",
OFC'03, March 27, Atlanta, GA, USA, Post deadline paper PD3, 2003.

[7.94] J. H. Lee, Z. Yusoff, W. Belardi, T. M. Monro, P. C. The, and D. J. Richardson,
"A holey fibre Raman amplifier and all-optical modulator",
European Conference on Optical Communication ECOC'2001, Amsterdam, The Netherlands, Sept.30-Oct.4, 2001, Post Deadline paper PD.A.1.1.

[7.95] K.-H. Tsai, K.-S. Kim, and T. F. Morse,
"General solution for stress-induced polarization in optical fibres",
Journal of Lightwave Technology, vol. 9, no. 1, pp. 7-17, January 1991.

[7.96] G. P. Agrawal,
"Fiber-Optic Communications Systems",
John Wiley & Sons, Inc., 1997

[7.97] A. Ortigosa-Blanch, J. C. Knigth, W. Wadsworth, J. Arriaga, B. J. Mangan, T. A. Birks and P. St. J. Russel,
"Highly birefringerent photonic crystal fibres",
Optics Letters, vol. 25, no. 18, pp. 1325-1327, September 2000

[7.98] T. A. Birks, J. C. Knight, and P. St. J. Russell,
"Endlessly single-mode photonic crystal fiber",
Optics Letters, vol.22, pp. 961-963, 1997

[7.99] J. Hansryd, F. Dross, M. Westlund, P. A. Andrekson and S. N. Knudsen,
"Increase of the SBS threshold in a short highly nonlinear fiber by applying a temperature distribution",
IEEE Journal of Lightwave Technology, vol.19, no. 11, pp.1691-1697, November 2001

[7.100] T P. Hansen, J. Broeng, S.E. B, Libori, E.Knudsen, A.Bjarklev, J. R. Jensen, and H. Simonsen,
"Highly birefringent index-guiding photonic crystal fibres",
IEEE Photonics Technology Letters, vol. 13, no. 6, pp. 588-90, June 2001.

[7.101] K. Suzuki, H.Kubota, S. Kawanishi,
"Optical properties of a low-loss polarization-maintaining photonic crystal fiber",
Optics Express, vol. 9, no. 13, pp. 676-680, December 2001.

[7.102] M. J. Steel and R. M. Osgood, Jr.,
"Elliptical-hole photonic crystal fibers",
Optics Letters, vol. 26, no.4, 2001.

[7.103] L. Grüner-Nielsen, S. Nissen Knudsen, B. Edvold, T. Veng, D. Magnussen,
C. C. Larsen, and H. Damsgaard,
"Dispersion compensating fibers",
Optical Fiber Technology, Vol. 6, pp. 164-180, 2000.

[7.104] T. A. Birks, D. Mogilevtsev, J. C. knight, and P. St. J. Russell,
"Dispersion compensation using single-material fibers",
IEEE Photonics Technology Letters, Vol. 11, No. 6, pp. 674-676, June 1999.

[7.105] J. C. Knight, J. Arriaga, T. A. Birks, A. Ortigosa-Blanch, W. J. Wadsworth, and
P. St. J. Russell,
"Anomalous dispersion in photonic crystal fiber",
IEEE Photonics Technology Letters, Vol. 12, No. 7, pp. 807-809, July 2000.

[7.106] A. Bjarklev, J. Broeng, S. E. Barkou, and K. Dridi,
"Dispersion properties of photonic crystal fibers",
ECOC'98, Madrid, Spain, Sept. 1998.

[7.107] J. Broeng, D. Mogilevtsev, S. E. Barkou, and A. Bjarklev,
"Photonic crystal fibers: A new class of optical waveguides",
Optical Fiber Technology, Vol. 5, pp. 305-330, 1999.

[7.108] A. Ferrando, E. Silvestre, P. Andrés, J. J. Miret, and Miguel V. Andrés,
"Designing the properties of dispersion-flattened photonic crystal fibers",
Optics Express, Vol. 9, No. 13, pp. 687-697, Dec. 2001.

[7.109] W. H. Reeves, J. C. Knight, P. St. J. Russell, and P. J. Roberts,
"Demonstration of ultra-flattened dispersion in photonic crystal fibers",
Optics Express, Vol. 10, No. 14, pp. 609-613, July 2002.

[7.110] J. Jasapara, R. Bise, and R. Windeler,
"Chromatic dispersion measurements in a photonic bandgap fiber",
OFC'2002, Anaheim, California, USA, 2002.

[7.111] R. K.Sinha, and S. K. Varshney,
"Dispersion properties of photonic crystal fibers",
Microwave and Optical Technology Letters, Vol. 37, No. 2, pp. 129-132, April 2003.

[7.112] J. C. Baggett, T. M. Monro, K. Furusawa, and D. J. Richardson,
"Distinguishing transition and pure bend losses in holey fibers",
CLEO'2002, Long Beach, California, USA, Paper CMJ6, 2002.

[7.113] N. A. Mortensen, and J. R. Folkenberg,
"Low-loss criterion and effective area considerations for photonic crystal fibers",
J. Opt. A: Pure and Appl. Opt., Vol.5, 2003.

[7.114] M. D. Nielsen, G. Vienne, J. R. Folkenberg, and A. Bjarklev,
"Investigation of micro deformation induced attenuation spectra in a photonic crystal fiber",
Optics Letters, Vol. 28, p. 236, 2003.

[7.115] J. A. West, N.Venkataraman, C. M. Smith, M. T. Gallagher,
"Photonic Crystal Fibres".
ECOC'2001, Amsterdam, Netherlands, Paper Th.A.2,pp. 582-585, 2001.

[7.116] N. Venkataraman, M. T. Gallagher, C. M. Smith, D. Müller, J. A. West, K. W. Koch, J. C. Fajardo,
"Low loss (13 dB/km) air core photonic band-gap fibre",
ECOC'2002, Copenhagen, Denmark, Post deadline paper PD1.1, 2002.

[7.117] R. S. Windeler, J. L. Wagener, D. J. DiGiovanni,
"Silica-air microstructured fibers: Properties and applications",
Conference on Optical Fiber Communication OFC'99, 1999.

[7.118] J. Limpert, T. Schreiber, S. Nolte, H. Zellmer, A. Tunnermann, R. Iliew, F. Lederer, J. Broeng, G. Vienne, A. Petersson, and C. Jakobsen,
"High-power air-clad large-mode-area photonic crystal fiber laser",
Optics Express, Vol. 11, p. 818, 2003.

ACRONYMS

ABC	Adjustable-Boundary-Condition
ASM	Air-Silica Microstructured
BBM	Biorthonormal-Basis Method
BPM	Beam-Propagation Method
CP	Close Packed
CW	Continuous Wave
DC	Dispersion-Compensating
DCF	Dispersion-Compensating Fibre
DF	Dispersion Flattened
DWDM	Dense Wavelength Division Multiplex
EAIM	Equivalent Average Index Method
EIM	Effective-Index Method
ESC	Electric Short Circuit
FBG	Fibre Bragg Grating
FD	Finite-Difference
FDM	Fourier Decomposition Method
FDFD	Finite-Difference Frequency-Domain
FDTD	Finite-Difference Time-Domain
FEM	Finte-Element Method
FFT	Fast Fourier Transformation
FOM	Figure Of Merit
FSM (fsm)	Fundamental Space-filling Mode
FWM	Four-Wave Mixing
GaAs	Gallium Arsenide
GLS	Gallium Lanthanum Sulphide (glass)
GVD	Group Velocity Dispersion

HALF	Hole-Assisted Lightguide Fibre
HNL	Highly NonLinear
HOM	Higher-Order Modes
InP	Indium Phosphide
LBF	Localized Basis Function
LFM	Localized Function Method
LMA	Large Mode Area
LPG	Long Period Grating
MCVD	Modified Chemical Vapour Deposition
MFD	Mode Field Diameter
MPM	Multipole Method
MPOF	Microstructured Polymer Optical Fibres
MSC	Magnetic Short Circuit
MTIR	Modified Total Internal Reflection
NA	Numerical Aperture
OCDMA	Optical Code Division Multiple Accsess
OPA	Optical Parametric Amplifier
OVD	Outside Vapour Deposition
PBG	Photonic Band Gap
PCF	Photonic Crystal Fibre
PM	Polarization Maintaining
PMD	Polarization-Mode Dispersion
PML	Perfectly Matched Layer
PMMA	Polymethyl Methacrylate
PWM	Plane-Wave Method
RDS	Relative Dispersion Slope
RED	Rare-Earth Doped
SC	Super Continuum
SEM	Scaning Electron Microscope
SIF	Step-Index Fibre
SPM	Self-Phase Modulation
SRS	Stimulated Raman Scattering
TE	Transverse Electric
TIR	Total Internal Reflection
TM	Transverse Magnetic
VAD	Vapour Axial Deposition
WDM	Wavelength Division Multiplexed
1D	One Dimensional
2D	Two Dimensional
3D	Three Dimensional

SYMBOLS

Symbol	Significance	Unit
A_e	Amplitude coefficient of the cladding electric field	-
A_i	Sellmeier coefficient	-
A_{mn}	Expansion coefficients of mode field decomposition	-
a_m	Higher-order Fourer terms in EAIM	-
$a_m^{(l)}$	Coefficient of Bessel function solution related to cylinder l	Vm^{-1}
B_i	Sellmeier coefficient	m^2
b	Centre hole radius in supersquare PBG structure	m
$b_m^{(l)}$	Coefficient of Bessel function solution related to cylinder l	Vm^{-1}
c	Velocity of light in vacuum	ms^{-1}
c_{ab}	Coefficient in index decomposition	-
$c_m^{(l)}$	Coefficient of Bessel function solution related to cylinder l	Vm^{-1}
D	Central area of fibre	-
D	Core diameter for step index fibre	m
D	Group velocity dispersion (GVD)	ps/(km·nm)
D_{eff}	Effective core diameter for PBG fibre	m
d	Diameter of air hole	m
d_{cl}	Diameter of air hole (in cladding)	m
d_{co}	Diameter of air hole (in core)	m
d_p	Diameter of air hole (in polarisation maintaining fibre)	m
E_x	Transversal electric field component	Vm^{-1}
E_y	Transversal electric field component	Vm^{-1}
E_z	Longitudinal electric field component	Vm^{-1}
f	Air-filling fraction	-
f_{int}	air-filling fraction of interstitial holes	-

f_R	Reflection frequency	s^{-1}
f_T	Transmission frequency	s^{-1}
H_m	Hermite polynomial of the order *m*	-
$H_m^{(1)}$	Hankel function of first kind and order *m*	-
H_x	Transversal magnetic field component	$A\ m^{-1}$
H_y	Transversal magnetic field component	$A\ m^{-1}$
H_z	Longitudinal magnetic field component	Am^{-1}
h_x	Transversal magnetic field component on symbolic form	$A\ m^{-1}$
h_y	Transversal magnetic field component on symbolic form	$A\ m^{-1}$
J_m	m-order Bessel function	-
J_1	1-order Bessel function	-
K	Symmetry point of the reciprocal lattice	-
K_m	Modified Bessel function of the order *m*	-
k, k_0	Free-space wave number	m^{-1}
k_x	Wave vector component in *x* direction	m^{-1}
k_y	Wave vector component in *y* direction	m^{-1}
k_z	Wave vector component in *z* direction	m^{-1}
$k_\perp^e$	Transverse wave number outside cylinder region	m-1
$k_\perp^i$	Transverse wave number inside cylinder region	m-1
L	Nonself adjoint operator	-
$L_{k,l}^{m,n}$	Matrix coefficients in Hermite-Gaussian basis	m^{-2}
$L^\dagger$	Operator in biorthonormal-basis method	-
l	Indicator of cylinder	-
l_i	Eigenvalue in biorthonormal-basis method	-
l_i^*	Complex conjungate eigenvalue in biorthonormal-basis method	-
M	Symmetry point of the reciprocal lattice	-
m	Order of Hermite polynomial	-
M	Number of Bessel function terms	-
N	Number of guided modes	-
N	Number of lattice vectors	-
N_{conv}	Number of index-guided modes for step-index fibre	-
N_{PBG}	Number of PBG-guided modes	-
N_{PCF}	Number of index-guided modes	-
N_{PW}	Number of expansion terms	-
n	Refractive index (or mode index of PBG)	-
n_{co}	Core refractive index	-
n_{cl}	Cladding refractive index	-
$n_{cl,eff}$	Effective cladding refractive index	-
n_e	Silica background refractive index	-
n_{eff}	Effective refractive index of cladding	-
$n_{eff,co}$	Effective refractive index of core region	-
n_h	Refractive index of homogeneous medium	-
n_J	Complex refractive index of jacket material	-

n_T	Modal index of PBG in triangular structure	-
n_{TI}	Modal index of PBG in triangular structure with interstitial holes	-
n_1	Refractive index of material one	-
n_2	Refractive index of material two	-
n_3	Refractive index of material three	-
n_{s1}	Refractive index of doped silica type one	-
n_{s2}	Refractive index of doped silica type two	-
n_{s3}	Refractive index of doped silica type three	-
n_x	x-projection of unit-vector normal to the dielectric interface	-
n_y	y-projection of unit-vector normal to the dielectric interface	-
P	Number of terms representing holes (in LFM)	-
P	Propagating power carried by fundamental mode	W
$P()$	Longitudinal (mode-field) profile function in BPM	-
p_{ab}	Coefficient in index decomposition	-
R	Radius of curvature of fibre	m
R_J	Outer radius of fibre jacket material	m
r_l	Local polar coordinate	m
s	Coordinate normal to cladding cell edge	m
t	Time	s
U	Normalised propagation constant (in core)	-
U_{eff}	Effective normalised propagation constant	-
V	Constant depending on effective mode index	-
V	Normalised frequency	-
V_{eff}	Effective normalised frequency	-
W	Normalised decay constant (in cladding)	-
W	Constant depending on effective mode index	-
x	Coordinate defining direction in transversal plane	m
y	Coordinate defining direction in transversal plane	m
Z_0	Free-space impedance	Ω
z	Coordinate defining position along the fibre	m
α	Macrobending loss value	-
α	Parameter used in cutoff boundary expression (Eq.5.6)	-
α_i	Higher-order Fourer terms in EAIM	-
α_m	Parameter in Fourier-decomposition method	-
α_{mn}	Parameter in Fourier-decomposition method	-
α_{rad}	Confinement loss per unit length	m^{-1}
α_2	Out-of-plane angle	radians
α_3	Out-of-plane angle	radians
β	Propagation constant	m^{-1}
β_{fsm}	Propagation constant of fundamental space-filling mode	m^{-1}
β_i	Propagation constant of mode *i*	m^{-1}

β_{mn}	Parameter in Fourier-decomposition method	-
β_H	Upper propagation constant boundary of a PBG (at fixed frequency)	m^{-1}
β_L	Lower propagation constant boundary of a PBG (at fixed frequency)	m^{-1}
β_{PBG}	Propagation constant of PBG-guided mode	m^{-1}
β_{TIR}	Propagation constant of TIR-guided mode	m^{-1}
Δ	Relative index difference	-
Δt	Time increment in FDTD method	s
Δx	Length increment in FDTD method	m
Δy	Length increment in FDTD method	m
δ	Imaginary part of refractive index in fibre jacket material	-
δ_{ij}	Kronecker delta	-
ε_r	Relative dielectric constant	-
ε_t	Eff. dielectric constant of triangular periodic hole arrangement	-
ε_1	Effective dielectric constant (of material in holes)	-
ε_2	Effective dielectric constant (of background material)	-
$\varepsilon_\perp$	Wiener limit	-
$\varepsilon_\parallel$	Wiener limits	-
ϕ_{mn}	Hermite-Gaussian basis function	-
ϕ_{mn}	Fourier-decomposed normalised mode field	-
ϕ_l	Local polar coordinate (azimuthal)	-
γ	Parameter used in cutoff boundary expression (Eq.5.6)	-
Γ	Gamma point	-
ℓ	Transverse extent of the structure of fibre cross section	m^2
Λ	Pitch (hole distance or lattice spacing)	m
$\Lambda_{H,node}$	Separations between nodes in honeycomb structure	m
$\Lambda_{T,node}$	Separations between nodes in triangular structure	m
$\Lambda_{TI,node}$	Separations between nodes in triangular structure with interstitial holes	m
λ	Free-space wavelength	m
λ_{max}	Wavelength corresponding to centre frequency of bandgap	m
$\lambda_{H,max}$	Wavelength corresponding to bandgap centre frequency in honeycomb fibre	m
$\lambda_{T,max}$	Wavelength corresponding to centre frequency of primary bandgap in triangular structure.	m
$\lambda_{TI,max}$	Wavelength corresponding to centre frequency of primary bandgap in triangular structure with interstitial holes	m
μ_r	Relative permeability	-
μ_0	Free-space permeability	Hm-1
ω	Angular frequency (of optical field)	s^{-1}
ω_d	Characteristic width	m

ω_{fsm}	Lowest possible angular frequency of allowed mode	s^{-1}
ρ	Effective core radius	m
σ	Conductivity of material	Sm^{-1}
ψ	Cladding mode field	-
ψ_a^d	Element of orthonormal set of Hermite-Gaussian	-
ψ_{mn}	Fourier-decomposed normalised mode field	-
$\overline{B}$	Magnetic flux density	T
$\overline{D}$	Electric flux density	$C\ m^{-2}$
$\overline{E}$	The electric field vector	$V\ m^{-1}$
$\overline{e}$	The electric field in the transverse coordinate	Vm^{-1}
$\hat{e}_1$	Unit vector	-
$\hat{e}_2$	Unit vector	-
$\overline{e}_t$	Transversal component of electric field	Vm^{-1}
$\overline{G}$	Reciprocal Lattice vector	m^{-1}
$\overline{H}$	The magnetic field vector	$A\ m^{-1}$
$\overline{h}$	The field in the transverse coordinate	$A\ m^{-1}$
$\overline{H}_{\overline{k}}$	Eigenvector	$A\ m^{-1}$
$\overline{h}_{\perp}$	Transversal vector component of magnetic field	$A\ m^{-1}$
$\overline{h}_t$	Transversal components of magnetic field	$A\ m^{-1}$
$\overline{h}_{\perp}^{mn}$	Transversal vector component of magnetic field in the Hermite-Gaussian basis	$A\ m^{-1}$
$\overline{I}$	Square identity matrix	-
$\overline{k}$	Wave vector of the solution	m^{-1}
$\overline{k}_{\parallel}$	Wave vector component in periodic plane	m^{-1}
$\overline{n}$	Unit-vector normal to the dielectric interface	-
$\overline{R}_1$	Real-space primitive lattice vector	m
$\overline{R}_2$	Real-space primitive lattice vector	m
$\overline{r}$	Vector representing a point in space	m

$\overline{r}_{\parallel}$	Vector representing a point in space	
	parallel with the periodic plane	m
$\overline{U}_k$	Periodic function	-
$\overline{U}_x$	Square matrix depending on boundary conditions in FDFD	-
$\overline{U}_y$	Square matrix depending on boundary conditions in FDFD	-
$\overline{V}_k$	Periodic function	-
$\overline{V}_x$	Square matrix depending on boundary conditions in FDFD	-
$\overline{V}_y$	Square matrix depending on boundary conditions in FDFD	-
$\overline{X}_i$	Eigenvectors	-
$\overline{x}_t$	Transversal position vector	m
$\hat{x}$	Coordinate defining direction in transversal plane	-
$\hat{y}$	Coordinate defining direction in transversal plane	-
$\overline{\theta}_i$	Eigenmode in biorthonormal-basis method	-
$\overline{\theta}_j$	Eigenmode in biorthonormal-basis method	
∇_t	The transversal gradient operator	m^{-1}
∇_t^2	The Laplacian gradient operator	m^{-2}
$\nabla_{\perp}$	Gradient in the periodic transversal plane	m^{-1}

INDEX

A

absorption losses 212
acoustic waves 25
additional air gaps 117
adjacent mesh grids 94
air-filling fraction 31, 40, 63, 138, 155, 167
air-guided modes 162
  number 208
air-guiding 10
  fibres 12, 206
  PBG-fibres 182
air-hole 5
  lattice 65
air-line 166, 167, 182, 206
air-ring isolation 246
air-silica 6
air-suspended
  central core region 87
  rod 237
  waveguides 161
air suspension 86
algebraic problem 80
all-optical switching 232
  nonlinear optical loop mirror 232
amplifiers 247
amplitude strength 100
analogy 143
angle of incidence 23
angular orientations 224
anisotropic 126
annealing 126
anomalous
  dispersion 150, 191, 192, 234
  waveguide dispersion 150-151, 192
antenna applications 25
anti-Stokes lines 239
applications 219
artificial
  initial field distribution 93
  material 151
  super-periodicity 39
asymmetric core region 201
attenuation 205, 242
  level 115
  parameter 83
auxiliary system 79
averaged

electric field amplitude 102
refractive index 96

B

band
diagram 40
folding 40
structure 90
bandgap edges 163
bandgap-guiding fibres 9
background material 74, 175
basis function expansion technique 87
beam-propagation method 12, 55, 94, 99-100, 105
bend related loss 223
bending losses 191
Bessel function 227
coefficients 146
first order 74
modified 88
regular 82
biorthogonality 78
relation 79
biorthonormal-basis method 12, 54, 75-81, 105
birefringence 66, 153, 161, 197, 200, 202, 204, 252-255
properties 41
bit-error rates 203
Bloch´s
state 29, 30
theorem 29, 69
bound mode 25, 61, 90
boundary condition 56, 76, 82, 83, 226
adjustable 89
outward radiating 87
periodic 76
Bragg
fibres 11
gratings 20
stacks 4, 20
Bravais Lattice 24
bridges of silica 6
Brillouin
effect 241
first 72, 73
irreducible 34, 72, 73
zone 30, 40, 41
bulk glass losses 127
butterfly wings 3, 4

C

capillary tubes 116
Cartesian grid 66
central
enlargement 133
hole 11
index defect 64
centre frequency 163
chalcogenides 12,
fibres 124
characteristic width 64
chromatic dispersion 75
circular
hole assumption 82
holes/rods 75
symmetric mode solution 56
preforms 116
cladding 5, 11
capillaries124
filling fraction 184
index 57
mode field 56
requirements 208
structure 56
closed-packed structure 118, 177
close-to-Gaussian field 227
close-packing 206
circular-symetric preforms 116
coating procedure 115
coherence degradation 235
complete
basis set 65

in-plane 2D PBG 31
PBGs 162
2D PBG 32
complex
effective mode index 87
propagation constant 82, 142
refractive index 83
compound glasess 124
computation time 102
computational domain 88
computer
memory 91
processing time 36, 91
conductivity 91
confined modes 150
confinement 86
loss 55, 81, 237
connecting regions 173
constitutive equations 68-69
continuity conditions 95
continuous light generation 193
conventional periodic lattice 152
convergence 34
core 6
design 183
index 57
modes 181
radius 57
requirements 208
correlation function 100
coupling 12, 259
coefficients 211
loss 242
of optical power 123
splicing loss 220
strength 246
critical bend radius 146, 224
cross section 120
curl operaton 38
cutoff
properties 12, 132, 138, 142
wavelength 221
cylinders 23
cylindrical harmonic function 82

D

decomposition of the wave vector 146
defect 11, 39, 61
diameter 64
mode 41, 179, 181
region 26, 40
depressed index ring 154
design-route 174
diagonalisation 80
diagonal matrices 96
dielectric
constant 31, 94
discontinuity 95
interface 39
material 68
scatterers 3
differential equations 80
diffractive effects 28
discrete translational symmetry 23
discretization
points 91
scheme 94
discretized equations 96
dispersion 12
compensating fibre 244
compensation 99
flattened 149
characteristics 194
managed PCFs 149, 255-259
manipulating fibres 194
properties 132, 148, 149, 161, 191, 232
relations 21, 164
distinct homogeneous subspaces 97
dopant
levels 175
materials 133
doped
fibre material 134

glass 60
doping techniques 115
double-clad fibres 220, 248
doubly-degenerate
defect mode 179
mode 40, 197
drawing 119
process 178
tower 119
dual-core PCF structure 123

E

effective
area 127, 227
criterion 142
core radius 138, 142
index 28, 36, 54, 56, 85
approach 11
considerations 164
contrast 222
interpretation 154
method 105
mode area 220, 237
mode index 227
nonlinearity 127, 233
normalized core parameter 141
refractive index 11
contrasts 9
v-value 140, 141
eigenstate 30
eigenvalue 38, 75, 80
problem 62, 70
eigenvector 30, 38, 78-80
electric
field 68
flux density 68
electron-beam lithography 26
electronic bandgap 2
electromagnetic
field 75
scattering problem 54
wave propagation 21
elementary
matrices 97
subspaces 97
end-reflection 226
endlessly single-mode 141, 142, 230
guidance regime 61
equivalent
average-index method 12, 55, 101-103, 105
core radius 60
core region 183
fibre 60
equivalent-index method 56
etching techniques 25
evolution time 93
excited mode 93
excess bending loss 146
exciton 2
expansion
functions 89
methods 82
terms 37, 189
number 38
experimental demonstration 177
extrudes non-silica glass fibres 126
extrusion
method 86,
technic 125, 127

F

fabrication 12, 25
false birefringence 66
fast Fourier transformation 38
fibre
-based nonlinear devices 233, 234
drawing 12, 115
fabrication conditions 153
preform extruded 232
field
component 91

distribution
harmonic 77
linear 77
figure-of-merit 244
filling fraction 73
finite cladding region 237
finite-difference
frequency-domain method 94-96, 105
method 12, 55, 90-96, 105
time-domain method 90-93, 105
finite-element method 12, 55, 97-99, 105
finite-size PCFs 142
first Brillouin zone 33
Floquet mode 29
fluoride glass fibres 126, 212
forbidden
gaps 2
frequency intervals 22
region 36
form birefingence 66
Fourier
coefficient 66, 70, 74
decomposition 89
method 12, 55, 85-90, 105
series 66, 74, 101
expansion 37, 70
transform 93
transformation 37, 70
four-wave-mixing 232, 239
free-space
impedance 95
permeability 69
permittivity 69
propagation constant 56
wavelength 21, 150
wave number 62, 77, 82
frequency
bands 163
domain 90, 93
analysis 68
intervals 21, 35
Fresnel fibre 228
full confinement 182
full-periodic structures 39
full-vectorial
numerical method 7, 209
plane-wave expansion method 54, 67-75
fundamental
cladding mode 58
guided mode 197
mode 183
space-filling mode 56, 59, 165

G

gallium lanthanum sulphide 124
Gaussian field 226
geometrical birefringence 236
Gibbs phenomenon 37
glass base material 6
gradient 62
Graf's addition theorem 83
graphite lattice structure 168
grating structure 100
gravity 121
grazing angle 167
grid points 39, 92, 187
group
representation theory 153
theory 66, 153
group-velocity dispersion 75, 191, 193
guided mode expression 101

H

halides glasess 124
hand-stacking technique 116
Hermite
-Gaussian function 62, 63
polynomial 62, 65
Hermitian
eigenvalue problem 63
matrix 71
heavy metal oxide glasess 124

hexagonal
- lattice 23
- photonic crystals 167
- structures 33, 161, 167
- symmetry 24, 71, 168

high-birefringent fibres 197
high-frequency regime 146
high-index
- contrast system 36
- core fibres 9, 12, 131
- core PCFs 154
- core region 152
- core triangular PCF 54
- defect 230
- material 11

high NA PCFs 12, 245-246
high-frequency limit 165
high-index regions 173, 175
higher frequency PBG 32
higher order eigensolutions 38
- modes 138
- PBG-mode 187

high-numerical-aperture fibres 9
highly birefringent PCFs 12, 220
highly non-linear fibres 9
- PCFs 220, 232

high-frequency
- limit 140
- regime 62

high-power fibre lasers 12
high-reflection mirrors 20
Hilbert space 78
hole 23
- -assisted lightguide 115
  - fibres 124, 132, 154-155
- diameter 141

hole/rod-axis 31
hole spacing 168, 222, 238
holey fibres 9
hollow-core 206
- fibres 11

homogeneous
- algebraic equation 83
- medium 164, 207

honeycomb
- cells 40
- lattice structure 168
- PBG-fibre 39, 198
- structure 118, 167, 168
- structured fibres 177

hybrid 154
- approach 65
- fibre types 115
- implementation 66

I

imperfection 199
index
- averaging technique 94
- contrast 23, 26, 33, 220
- decomposition 65
- guidance 7, 28
- -guided modes 155
- guiding 1,7, 131, 139, 147, 148
- fibres 9, 53, 142
- PCFs 151, 154

index-reducing voids 156
index-raised doped glass 115
injection moulding 125
in-plane 34
- photonic crystals 31
- wave propagation 24, 168

integrated
- optics 24, 55
- photonics 5

interaction length 238
interfacial cells 96
intermediate
- drawing process 119
- step 178

interpolation of the refractive index 39
intersection points 133

interstitial holes 117, 170
intra-band guidance 204
intrinsic nonlinearity 233
invariant direction 27, 163, 181
inversion symmetry 73
isolated modes 173
iterative scheme 38

J

jacketing material 83

K

kagomé strcture 118

L

Laplacian 77
large-bandgap photonic crystals 173, 175
large-mode-area
 fibres 9
 PCF 12, 219, 221
lattice 3
 constant 29
 spacing 64
 structure 23
layer-by-layer structure 26
leakage-free bound modes 25
leakage
 free operational window 209
 free spectral ranges 210
 loss 54
leaky mode 83, 86, 88, 100, 209
light intensities 233
linear isotropic material 91, 98
lithography 25
local polar coordinates 82
localized basis function 61
 method 11, 105
localization of light 2
longitudinal electrical field component 82
 modulations 230
long-period grating 246
loss 133, 260-264
 mechanisms 206
 properties 213
lower bend edges 146
lowest-frequency photonic bandgap 22, 32
lowest-order mode 36
low-frequency limit 165
low-index
 contrast PBG fibres 12
 core fibres 11
 core region 186
 region 179
Low-loss
 PBG-guidance 181
 waveguides 162
low-melting point glasses 12, 115
low-transmission intervals 21

M

macrobending loss 132, 146
macroporous silicon 26
magnetic
 field 29, 68, 83, 205, 227
 flux density 68
manufacturing 25
material
 combinations 123
 dispersion 149, 150, 192
matrix equation 37, 71
Maxwell's equations 21, 54, 61, 68-69, 77
 first-curl 92
 time-dependent 91
melting temperature 128
mesh 91
metrology 232
microbending resistant fibre structure 223
micro cavities 27
microscopic bend loss 223
microstructured

fibres 9, 53, 152
optical fibres 9, 100
polymer optical fibres 125
silica fibres 6
microstructures 3
microstructuring 115, 132
microwave
frequencies 25
regime 25
milling 117
minimum
bend loss 224
effective mode ares 237
minimization functional 37
modal-index 166
modal-representation method 76
modal
index 174
intra-band dispersion 205
spectrum 100
mode 21
area 230
confinement 9, 154, 220
degeneracy 153
evolution 100
field 81, 97
confinement 12
diameter 227, 230
distribution 208
localization 61
frequencies 30
index 41
curves 191
number 142
profile 182
propagation properties 55
properties 65
modelling 53
modified
chemical vapour deposition 116, 221
honeycomb structure 172, 175
photonic crystals 170
total internal reflection 5, 11, 131
morphology 178
multi core photonic crystal fibres 123
multilayer stack 20
multi-moded 141
fibres 220
multiple
clock recovery 232
scattering 21
multipole method 12, 54, 75, 81-85, 105, 142

N

nature 3
near field mode profile 222
picture 222
new materials 12, 123
nodes 173
separation 173
node-veins concept 173
non-circular
hole shapes 153
symmetry 116
non-degenerated orthogonal fibre modes 220
nonlinear
coefficient 220, 235
effects 191
properties 233
threshold 222
non-magnetic dielectric medium 76
non-perfect structures 151
non-periodic
circular cladding structures 151-153
cladding structures 12
non-self-adjoint operators 79
non-uniformities 198
normal
dispersion 235
incidence 21
normalized
attenuation constant 59
centre frequency 163

cut-of frequency 138, 169
decay parameter 146
frequency 22, 31, 146, 173
hole size 147
propagation constant 40
transverse propagation constant 59
wavelength 184
novel dispersion characteristics 155
number
of discretization points 93
of guided modes 140
of lattice vectors 34
of (air-hole) rings 82, 84
numerical
analysis 187
aperture 221, 225, 227, 233
methods 11
modelling 53
precision 80
representation 153
simulations 189

O

octave-spanning spectrum broadening 240
octaves 232
off-lattice air hole 204
offset core 248
one-dimensional photonic crystals 20
opal-based 3D photonic crystals 27
operational principle 155
wavelength 242
windows 224
optical
atom waveguide 3
code division multiple access system 241
coherence tomography 232
components 4
comucatios 6
gratings 4
power 11
signal processing 220
waveguides 78
wavelengths 25, 174, 237
optimum
basis cladding-structure 162
effective core radius 145
operatinal frequency 173
orientationally dependent behaviour 224
orthogonal 38
field component 62
orthonormal Hermite-Gaussian functions 66
outside vapour deposition 116
outer cladding tube 133
out-of-plane 34
angles 24
PBGs 24, 163
propagation 24, 36
properties 168
wave propagation 162
wave vector 23
component 28, 165
outward propagating field 88
overlap integral 62, 65, 66

P

PANDA-fibres 200
parametric amplifiers 232
peak powers 238
perfectly matched layer 93
period 21
of the lattice 141
periodic
array 6
cladding 5
structures 134
function 63, 69
plane 23, 28
structure 4, 20, 54
periodical
boundary conditions 80
materials 2
periodically spaced holes 177

periodicity 11, 68, 80, 151
- -breaking air hole 118
- core region 119, 178

permeability 91
permittivity 29, 91
photonic
- band diagram 21, 35
- bandgap 1, 4, 21, 31, 161
  - fibre 54, 185
    - length 186
  - boundaries 179
  - effect 2, 7, 9, 11
  - edges 35
  - guided modes 181
  - structures 2, 161
- bandgap-guiding fibre 7, 191
- crystal 69
  - cladding 6
  - fibre (PCF) 5, 6, 7, 10
    - analogy 145
  - optical waveguides 11
  - technology 29
- crystals 2, 19,26

planar
- applications 25
- optical structures 25
- photonic crystal waveguides 5
- technology 25, 26

plane of periodicity 24, 28
plane-wave
- expansion 65
- method 11, 36, 105

point defects 27
polarization 98
- dispersion 203
- -mode coupling 231
  - dispersion 231
- properties 197
- states 31, 199
  - splitting 201

polymer
- based PBG fibres 213
- casting 125
- chains 126
- PCFs 125
- preforms 125
- processing 126

polymerization 125
polymers 12, 115
post-processing 116
power loss coefficient 146
preform 12, 115
- canes 119, 178
- fabrication 115

preconditioned conjugate gradient method 38
primary PBG 172, 185
primitive lattice vector 29, 30, 72
process control 152
propagation 77
- constant 28, 56, 59, 62, 75, 77, 82, 94, 102
  - boundaries 183

pseudo-ID periodic structure 33
pulse compression 232, 234
pure macrobending loss 223

Q

qualitative guiding mode description 57
quasi-linear polarization states 208

R

radius curvature 146
Raman effect 241
random cladding-hole distributions 151
rare-earth ions 124, 247
Rayleigh scattering 241
reactive ion etching 26
real-space 38
- dielectric distribution 74
- primitive lattice vector 72

reciprocal

lattices 30, 74
lattice vectors 30, 70, 73
space 33, 38, 41, 69
reflection amplitude 226
refractive index 7, 56, 63
contrast 23
distribution 38, 81
profile 77
regeneration 241
regular triangular lattice 174
re-iteration 90
relative
dielectric constant 69
dispersion slope (RDS) 155
hole size 143
permeability 69
size of a PBG 32
residual interstitial air holes 187
rod 23
-plate technique 133
rotational symmetry 89, 153

S

sampling points 39
sapphire fibres 212
secondary PBG 172
second-order mode 138, 140
self
-aligned 26
-phase modulation 232, 238
-trapped localized modes 240
segmented-cladding 227
scalability 25
scalar
approach 101
effective-index method 56
model 54
wave equation 56, 87, 102
scattering losses 5
Schrödinger equation 54
sharply-bent waveguides 5, 27
silicate fibre 242
six-fold symmetry 153
six-lobed mode pattern 190
self-frequency shift 234
Sellmeier
coefficients 58
equation 192
formula 57
semiconductors 2, 36
sensor applications 232
short circuit
electric 98
magnetic 98
silica 6
air honeycomb photonic crystal 40
air photonic crystals 31
-based PCF technology 11
subclasses 9
waveguide 6
silicon-on-isulator 26
simulated Raman scattering 232
single
-material 6
fibres 6, 132, 133
mode-dual mode boundary 143
mode fibres 192
operation 221
polarization 162
state 202
fibres 200
single-rod PCF 230
slab
-coupled waveguides 133
modes 133
slope efficiency 241
small-core
fibre 234
regime 237
softening temperature 127
solid core region 131
solid-state crystalline materials 36
soliton

fission 232
generation 234
propagation 191
spatial defects 26, 28
spatial mode quality 221
spatially localized modes 54
spectral
filters 246
properties 151
width 224
window 146, 167
spectroscopy 232,
speed of light 69
splicing 12, 259
spontaneous emission 27
spot-size 182, 220, 227
value 226
squared index distribution 64
square identity matrix 96
staircase approximation 94
standard
optical fibres 154
step-index fibre 147
state-space argument 184
stationary values 140
step-index fibre (SIF) 144, 145
analogy 144
parameters 60
stimulated Raman scattering 239
Stokes lines 239
stop bands 22
storage requirements 36
strand of silica 61
stress-induced refractive index changes 116
structural
control 152
fluctuations 26
non-uniformities 197, 198
sub-micrometer wavelengths 26
sulphides glasess 124
super-cell approximation 36
supercontinuum generating media 220
generation 232, 238
super-square photonic crystal fibres 204
support filaments 86
supporting slab 133
suppression of the secondary band gap 171
surface
integral 74
plasmons 28
tension 119
forces 121
symmetry 66, 153
points 74
synthetic opal-based crystals 26

T

tapered fibres 239
theoretical methods 11
thermal limit 249
three-dimensional
periodicity 25
photonic bandgap 3, 25
crystal waveguides 28
periodic structures 25
threshold frequency 204
threshold power 27
time
dependence 82
domain 90, 93
increment 92
total
dispersion 150
internal reflection (TIR) 1, 9, 25, 31, 167
based fibres 179
guided modes number 185
translation symmetry 29, 73
transmission
amplitude 226
loss 155
spectrum 22
transition loss 223
transversal gradient 77

transverse
- electric 31
  - field 94
  - polarization 31
- extent of structure 64
- magnetic 31
  - field component 62
  - polarization 32
- modal profile 100
- optical field 9
- 2D structure 75
- wave number 82

trial vector 38
triangular
- arrangement of air holes 6
- core 230
- high-index core 148
- lattice 31, 33, 35
- photonic crystal 7, 163, 207
- structure 72, 81, 118, 161, 162

triply periodic materials 23
truly bandgap-guiding photonic crystal fibres 7
truly single-mode 202
tuneable fibres devices 220, 250
two dimensional
- photonic crystals 24, 28, 71, 72
  - periodicity 23

two-fold degenerate pairs 153

U

ultra-refractivity 28
un-doped silica-air fibres 86
uniformity 167, 213
unique wavelength dependency 151
unit
- cell 72, 91, 170
- vector 70
- upper bend edge 146

UV-writing 25

V

vacuum 207
variational method 38
vapour axial deposition 116
vapour deposition techniques 116
vectorial method 12
veins 173
velocity of light 21, 62
vertical
- confinement 25
- losses 28
- scattering losses 5

viscosity 121
visible light 4
voids 11
V-parameter 139

W

waveguidance in air 206
waveguiding 2,9
- principle 179
- properties 12

wave
- component 24
- equation 69
- interference 21
- number 94
- propagation 22, 164
- vector 21,30, 69

waveguide dispersion 150
wavelength 4
- conversion 232, 241
- dependent cladding index 60
- dependent core index 60
- -division multiplexed 27
- independent loss 155
- -scale lattice dimension 23
- selectivity 4

weakly guiding field assumption 57
weighting function 88

width parameter 66
Wiener limits 39

Y

Yee-cell technique 91
Yee's mesh 94
ytterbium-doped PCF 234

Z

zero-dispersion wavelengths 149, 150, 192, 235